T0298018

Introductory MEMS

Thomas M. Adams · Richard A. Layton

Introductory MEMS

Fabrication and Applications

Thomas M. Adams
Department of Mechanical Engineering
Rose-Hulman Institute of Technology
5500 Wabash Ave.
Terre Haute IN 47803
USA
thomas.m.adams@rose-hulman.edu

Richard A. Layton
Department of Mechanical Engineering
Rose-Hulman Institute of Technology
5500 Wabash Ave.
Terre Haute IN 47803
USA
layton@rose-hulman.edu

ISBN 978-0-387-09510-3 ISBN 978-0-387-09511-0 (eBook)
DOI 10.1007/978-0-387-09511-0
Springer New York Dordrecht Heidelberg London

Library of Congress Control Number: 2009939153

Printed on acid-free paper

Springer is part of Springer Science+Business Media (www.springer.com)

Dedication

For Diedre, whose love and companionship is no small thing.

–Thom

To Gary, my first guitar teacher. Thanks, Dad.

–Richard

Contents

Preface xiii

Part I—Fabrication

Chapter 1: Introduction 3

1.1 What are MEMS? 3
1.2 Why MEMS? 4
1.2.1. Low cost, redundancy and disposability 4
1.2.2. Favorable scalings 5
1.3 How are MEMS made? 8
1.4 Roadmap and perspective 12

Essay: The Role of Surface to Volume Atoms as Magnetic Devices Miniaturize 12

Chapter 2: The substrate and adding material to it 17

2.1 Introduction 17
2.2 The silicon substrate 17
2.2.1 Silicon growth 17
2.2.2 It's a crystal 19
2.2.3 Miller indices 20
2.2.4 It's a semiconductor 24
2.3 Additive technique: Oxidation 35
2.3.1 Growing an oxide layer 35
2.3.2 Oxidation kinetics 37
2.4 Additive technique: Physical vapor deposition 40
2.4.1 Vacuum fundamentals 41
2.4.2 Thermal evaporation 46
2.4.3 Sputtering 51
2.5 Other additive techniques 57

2.5.1 Chemical vapor deposition 57
2.5.2 Electrodeposition 58
2.5.3 Spin casting 58
2.5.4 Wafer bonding 58

Essay: Silicon Ingot Manufacturing 59

Chapter 3: Creating and transferring patterns—Photolithography 65

3.1 Introduction 65
3.2 Keeping it clean 66
3.3 Photoresist 69
3.3.1 Positive resist 69
3.3.2 Negative resist 70
3.4 Working with resist 71
3.4.1 Applying photoresist 71
3.4.2 Exposure and pattern transfer 72
3.4.3 Development and post-treatment 77
3.5 Masks 79
3.6 Resolution 81
3.6.1 Resolution in contact and proximity printing 81
3.6.2 Resolution in projection printing 82
3.6.3 Sensitivity and resist profiles 84
3.6.4 Modeling of resist profiles 86
3.6.5 Photolithography resolution enhancement technology 87
3.6.6 Mask alignment 88
3.7 Permanent resists 89

Essay: Photolithography—Past, Present and Future 90

Chapter 4: Creating structures—Micromachining 95

4.1 Introduction 95
4.2 Bulk micromachining processes 96
4.2.1 Wet chemical etching 96
4.2.2 Dry etching 106
4.3 Surface micromachining 108
4.3.1 Surface micromachining processes 109
4.3.2 Problems with surface micromachining 111
4.3.3 Lift-off 112
4.4 Process integration 113
4.4.1 A surface micromachining example 115

4.4.2 Designing a good MEMS process flow 119
4.4.3 Last thoughts 124

Essay: Introduction to MEMS Packaging 126

Chapter 5: Solid mechanics 131

5.1 Introduction 131
5.2 Fundamentals of solid mechanics 131
5.2.1 Stress 132
5.2.2 Strain 133
5.2.3 Elasticity 135
5.2.4 Special cases 138
5.2.5 Non-isotropic materials 139
5.2.6 Thermal strain 141
5.3 Properties of thin films 142
5.3.1 Adhesion 142
5.3.2 Stress in thin films 142
5.3.3 Peel forces 149

Part II—Applications

Chapter 6: Thinking about modeling 157

6.1 What is modeling? 157
6.2 Units 158
6.3 The input-output concept 159
6.4 Physical variables and notation 162
6.5 Preface to the modeling chapters 163

Chapter 7: MEMS transducers—An overview of how they work 167

7.1 What is a transducer? 167
7.2 Distinguishing between sensors and actuators 168
7.3 Response characteristics of transducers 171
7.3.1 Static response characteristics 172
7.3.2 Dynamic performance characteristics 173
7.4 MEMS sensors: principles of operation 178

7.4.1 Resistive sensing 178
7.4.2 Capacitive sensing 181
7.4.3 Piezoelectric sensing 182
7.4.4 Resonant sensing 184
7.4.5 Thermoelectric sensing 186
7.4.6 Magnetic sensing 189
7.5 MEMS actuators: principles of operation 193
7.5.1 Capacitive actuation 193
7.5.2 Piezoelectric actuation 194
7.5.3 Thermo-mechanical actuation 196
7.5.4 Thermo-electric cooling 201
7.5.5 Magnetic actuation 202
7.6 Signal conditioning 204
7.7 A quick look at two applications 206
7.7.1 RF applications 207
7.7.2 Optical applications 207

Chapter 8: Piezoresistive transducers 211

8.1 Introduction 211
8.2 Modeling piezoresistive transducers 212
8.2.1 Bridge analysis 213
8.2.2 Relating electrical resistance to mechanical strain 215
8.3 Device case study: Piezoresistive pressure sensor 221

Chapter 9: Capacitive transducers 231

9.1 Introduction 231
9.2 Capacitor fundamentals 232
9.2.1. Fixed-capacitance capacitor 232
9.2.2. Variable-capacitance capacitor 234
9.2.3. An overview of capacitive sensors and actuators 236
9.3 Modeling a capacitive sensor 239
9.3.1. Capacitive half-bridge 239
9.3.2. Conditioning the signal from the half-bridge 243
9.3.3. Mechanical subsystem 246
9.4 Device case study: Capacitive accelerometer 250

Chapter 10: Piezoelectric transducers 255

10.1 Introduction 255
10.2 Modeling piezoelectric materials 256
10.3 Mechanical modeling of beams and plates 261
10.3.1 Distributed parameter modeling 261
10.3.2 Statics 262
10.3.3 Bending in beams 268
10.3.4 Bending in plates 274
10.4 Case study: Cantilever piezoelectric actuator 276

Chapter 11: Thermal transducers 283

11.1 Introduction 283
11.2 Basic heat transfer 284
11.2.1 Conduction 286
11.2.2 Convection 288
11.2.3 Radiation 289
11.3 Case study: Hot-arm actuator 294
11.3.1 Lumped element model 295
11.3.2 Distributed parameter model 300
11.3.3 FEA model 306

Essay: Effect of Scale on Thermal Properties 310

Chapter 12: Introduction to microfluidics 317

12.1 Introduction 317
12.2 Basics of fluid mechanics 319
12.2.1 Viscosity and flow regimes 320
12.2.2 Entrance lengths 324
12.3 Basic equations of fluid mechanics 325
12.3.1 Conservation of mass 325
12.3.2 Conservation of linear momentum 326
12.3.3 Conservation equations at a point: Continuity and Navier-Stokes equations 329
12.4 Some solutions to the Navier-Stokes equations 337
12.4.1 Couette flow 337
12.4.2 Poiseuille flow 339
12.5 Electro-osmotic flow 339
12.5.1 Electrostatics 340

12.5.2 Ionic double layers 346
12.5.3 Navier-Stokes with a constant electric field 355
12.6 Electrophoretic separation 357

Essay: Detection Schemes Employed in Microfluidic Devices for Chemical Analysis 362

Part III—Microfabrication laboratories

Chapter 13: Microfabrication laboratories 371

13.1 Hot-arm actuator as a hands-on case study 371
13.2 Overview of fabrication of hot-arm actuators 372
13.3 Cleanroom safety and etiquette 375
13.4 Experiments 377
Experiment 1: Wet oxidation of a silicon wafer 377
Experiment 2: Photolithography of sacrificial layer 384
Experiment 3: Depositing metal contacts with evaporation 388
Experiment 4: Wet chemical etching of aluminum 392
Experiment 5: Plasma ash release 395
Experiment 6: Characterization of hot-arm actuators 397

Appendix A: Notation 405

Appendix B: Periodic table of the elements 411

Appendix C: The complimentary error function 413

Appendix D: Color chart for thermally grown silicon dioxide 415

Glossary 417

Subject Index 439

Preface

We originally wanted to call this book *Dr. Thom's Big Book about Little Things*, but, apart from being perhaps too playful a title, we didn't like the word "big" in it. That's because the book was intended to be an introduction to the world of science and engineering at the microscale, not a comprehensive treatment of the field at large. Other authors have already written books like that, and they have done a wonderful job. But we wanted something different. We wanted an introductory MEMS text accessible to *any* undergraduate technical major, students whose common background consists of freshman level physics, chemistry, calculus and differential equations. And while a "little" book about little things might not suffice to that end, we at least desired a somewhat compact book about things micro.

When we taught an introductory MEMS course for the first time in the spring of 2002 to just such an audience, it was, at least to our knowledge, a unique endeavor. We attempted to cover way too much material though, and we threw one of those big comprehensive books at the students. It nearly knocked them out. In subsequent installments, we cut back on the material and started using instructor notes in lieu of a text. Those notes, outlines, bulleted lists, and fill-in-the-blank handouts became the skeleton around which this text was formed.

As creating microstructures requires such a different set of tools than those encountered in the macro-world, much of learning about MEMS rests squarely in learning the details of how to make them. Part I of this text therefore deals mainly with introducing the reader to the world of MEMS and their fabrication. Actuation and sensing are also treated from a generic standpoint, with MEMS devices used as examples throughout. In Chapter 7 of Part II, an overview of some of the most common MEMS transducers is given from a mainly qualitative, non-mathematical standpoint. Hence, the first seven chapters should suffice for the majority of introductory courses.

Following Chapter 7 are specific treatments and modeling strategies for a handful of selected MEMS. The mathematical modeling is more detailed than in previous chapters, covering a number energy domains. Though the models given can be a bit involved, the necessary tools are developed for the reader. Part II is therefore better suited for a follow-up course, or perhaps a standalone course for students with the appropriate background. Alternatively, an introductory course covering Chapters 1-7 could culminate with one modeling chapter from Part II. The last chapter on microfluidics is in some sense a standalone treatment of the field.

Just as no one is able to design a functioning power plant after having taken an introductory course in thermodynamics, no one will be able to design, build and test a successful MEMS device after only reading this text. However, the reader will have acquired the new skill of considering microtechnology-based solutions to problems, as well as the ability to speak intelligently about MEMS and how they are modeled. The text can therefore serve as both a springboard for further study or as an end in itself.

One of the challenges in writing such a text is that it is a bit like writing a book entitled "Introduction to Science and Engineering," as this is what MEMS really is – science and engineering – at the microscale, that is. It can therefore be quite difficult to keep it truly general. By making the intended audience an undergraduate technical major in any field, and therefore not assuming any other specialized background, we have avoided slanting the text in some preferred direction. That is to say, we have done our best to keep the text a true introduction to MEMS as a whole rather than an introduction to, say, dynamic systems modeling of MEMS devices, or materials engineering aspects of MEMS.

The opposite danger, of course, is not including enough material to really understand MEMS. To address this, where needed we have included introductions to fields that are generally not part of the common experience of all technical majors. The introductions are kept brief, as they are intended to give the reader just enough background to understand the field in context of the MEMS device(s) at hand. Naturally these sections can be omitted when tailoring a course for specific majors.

In reading the text, most readers will find themselves outside of their comfort zone at some point. At other times the reader may find themselves reading things that seem obvious. Which things are which will be different for different readers, depending on their backgrounds. What's more, readers may occasionally find themselves a trifle disoriented even within a field in which they have heretofore considered themselves well-versed. A prime example is Chapter 12 on microfluidics, in which electro-osmotic flow is treated. In electro-osmotic flow the traditional fields of fluid mechanics and electrostatics, topics usually thought of as having little to do with each other, are coupled and of equal importance. Throw in a smattering of mass transfer and chemistry and you have a topic in which very few of us can hit the ground running.

Such is the world of MEMS. Scientists and engineers of all fields are necessarily drawn to one another in order to make things work. The cliché that the world is getting smaller has found new metaphorical meaning with the miniaturization of technology. It is for this very reason that we feel so strongly that this text and the types of courses it is designed to accompany are vitally necessary in the education of scientists and engineers. Gone are the days when technical professionals could pigeonhole themselves into not venturing outside of narrow areas of expertise. Multidisciplinary endeavors are all the buzz anymore, and rightfully so. For MEMS they are its very lifeblood.

ACKNOWLEDGEMENTS

Such a book is never truly the work of only it authors. Countless people have come together to make this happen. Many thanks go to the members of the Rose-Hulman MiNDS Group for their numerous selflessly given hours to make MEMS more assessable to a general audience: Jameel Ahmed, Daniel Coronell, Pat Ferro, Tina Hudson, Elaine Kirkpatrick, Scott Kirkpatrick, Michael McInerney, Daniel Morris, Jerome Wagner and Ed Wheeler. Thanks also to the students over the years with whom we've worked and learned. We wish we had delivered you this text earlier. Special thanks to Azad Siahmakoun, director of the Micro-Nano Devices and Systems Facility at Rose-Hulman Institute of Technology, without whom undergraduate MEMS education would not exist at all at our institution, let alone a textbook. Thom extends extra special thanks to Diedre Adams for all the corrections and wonderful suggestions, and also for simply putting up with him. That's for both the little things and the big things.

THOMAS M. ADAMS
RICHARD A. LAYTON

TERRE HAUTE, INDIANA

Part I—Fabrication

Chapter 1. Introduction

1.1 What are MEMS?

Strictly speaking, MEMS is an acronym for *micro-electro-mechanical systems*. The first M (micro) indicates the small size of MEMS devices. The micrometer, or micron - one one-millionth of a meter - is the base unit of measure in MEMS. (To gage this length, consider that a human hair, on average, is about 50-70 μm in diameter.) The E (electro) refers to electricity, often in the form of electrostatic forces. The second M (mechanical) refers to the fact that these tiny devices have moving parts. Lastly, S (systems) indicates that "electro" and "mechanical" go together, that the electricity and moving parts are integrated into a single system on a MEMS device. And so, when many people think of MEMS, they think of tiny machines - little gear trains, motors and the like - the largest dimension of which is no more than a millimeter. Figure 1.1 shows a MEMS device befitting this description. Also shown in Fig. 1.1 is a spidermite, a creature typically measuring 1 mm or less from end to end, which gives perspective to how small the device really is.

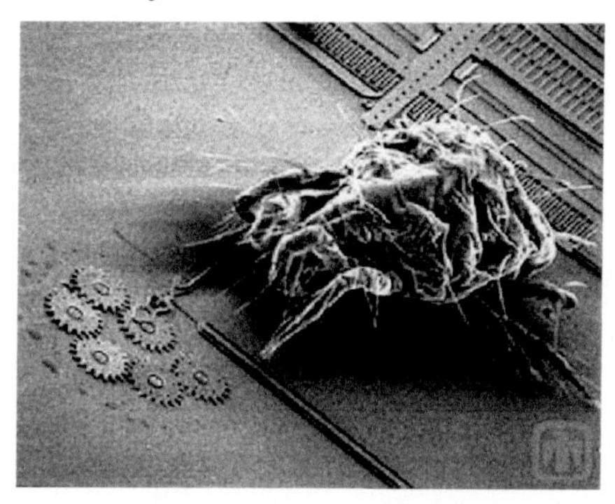

Fig. 1.1. MEMS gear train with spidermite (Courtesy of Sandia National Laboratories, SUMMiT™ Technologies, www.mems.sandia.gov)

T.M. Adams, R.A. Layton, *Introductory MEMS: Fabrication and Applications*,
DOI 10.1007/978-0-387-09511-0_1, © Springer Science+Business Media, LLC 2010

Many MEMS are exactly that, tiny little machines. But the letter-for-letter interpretation of the MEMS acronym can be misleading. Many MEMS devices have no moving parts whatsoever. Some MEMS have no electrical components. (Some microfluidic channels are good examples of MEMS that have neither!) The term MEMS has thus outgrown its etymology to mean the science and engineering of things smaller than a millimeter, but larger than a micron. Indeed, in much of Europe the field is known simply as microsystems or microsystems technology (MST).

We find MEMS devices in a huge variety of applications, including sensors for air-bag deployment in automobiles, microscale gyroscopes for aerospace applications, ink jet printer heads, radio frequency (RF) switching devices in cell phones and other wireless communication devices, optical switches for fiber-optic communication, chemical and biological detection, drug discovery, and a host of others too numerous to list. Everyone owns a few MEMS whether they know it or not!

Whether it is called MEMS, microsystems, or by any other name, the field represents one of the most exciting endeavors within science and technology today. At the microscale, many physical phenomena that are neglected at normal length scales become dominant features. These phenomena are often exploited in clever ways, resulting in MEMS devices with operating principles completely different than their macroscale counterparts. Furthermore, fabricating, say, a cantilever measuring 100 μm in length requires a much different set of skills and processes than does making a macroscale cantilever such as a diving board. Thus, a rather small number of off-the-shelf solutions exist within the field, and the MEMS technologist must draw upon knowledge of fundamental physical concepts from many disciplines in order to complete their task.

1.2 Why MEMS?

1.2.1 Low cost, redundancy and disposability

Of course one obvious question immediately arises: Why go micro? The answers are numerous. For starters, MEMS require very little material for fabrication, therefore depleting fewer resources and tending to make them less expensive than macroscale devices. Furthermore, most MEMS are fabricated using **batch processes**, processes in which large quantities of devices are produced as the result of a single operation, making MEMS cheaper still. For these reasons, some applications employ several MEMS devices where a single, larger device was once used. This increased redundancy results in increased system reliability.

Sub-millimeter devices are also minimally invasive, and are therefore often treated as disposable, which offers obvious advantages for chemical and biological applications involving hazardous agents.

1.2.2 Favorable scalings

As already mentioned, many physical phenomenona are favored at the microscale, allowing for new ways to achieve specific results, often with greater accuracy and dependability. Finding ways of exploiting the different relative importance of physical phenomena at the microscale, that is, finding *favorable scalings*, is fundamental to MEMS.

As an example of what we mean by scaling, consider two systems, one macroscale and one microscale. Let the macroscale system be a cubic building block with a side length of L = 30 mm, whereas the microscale system is a building block with side length of only L = 30 μm, or 30×10^{-6} m. For this situation we would say that the *characteristic length scale* of the microscale system is 10^{-3} that of the macroscale system. (See Fig. 1.2, which, ironically, is not to scale.)

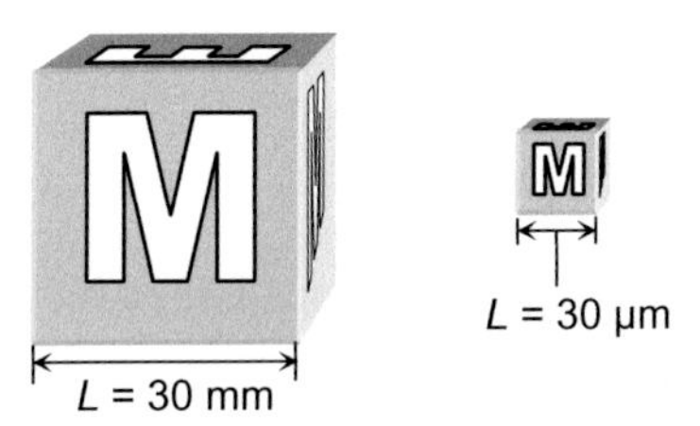

Fig. 1.2. Macroscale and microscale building blocks

If we consider the density of the blocks in Fig. 1.2 to be constant and the same for each, then we can say that the masses of the two blocks are proportional to their volumes. Since the volume of a cube is L^3, we can write that the mass of the micro-block compared to the macro-block is

$$\frac{m_{micro}}{m_{macro}} = \frac{L^3{}_{micro}}{L^3{}_{macro}} = \frac{(30\times10^{-6}\,\mathrm{m})^3}{(30\times10^{-3}\,\mathrm{m})^3} = 10^{-9}\,. \tag{1.1}$$

Even though the length of one side of the micro-block is 10^{-3} times that of the macro-block, its mass is only 10^{-9} that of the macro-block! A shorthand way of representing this idea is to write that the mass *scales with the characteristic length cubed*:

$$m \sim L^3 . \tag{1.2}$$

Let us now consider the surface area of the blocks. The actual surface area of a block is given by $6L^2$. For the purposes of scaling, the constant 6, or any constant on the order of one to ten, is not all that important to us. Thus we would write that the surface areas scales with L^2,

$$A_{surface} \sim L^2 . \tag{1.3}$$

Comparing the surface areas of the two blocks to each other,

$$\frac{A_{micro}}{A_{macro}} = \frac{L^2{}_{micro}}{L^2{}_{macro}} = \frac{(30 \times 10^{-6}\,\mathrm{m})^2}{(30 \times 10^{-3}\,\mathrm{m})^2} = 10^{-6} . \tag{1.4}$$

This means that the surface area of the micro-block is 10^{-6} times that of the macro-block, a much larger fraction than 10^{-9}. We see, then, that the micro-block has a lot more surface area compared to its mass than the macro-block. Again, using our short hand,

$$\frac{A_{surface}}{m} \sim \frac{L^2}{L^3} \sim \frac{1}{L} . \tag{1.5}$$

Equation (1.5) is a specific example of one of the most important generalizations when miniaturizing systems. That is, as the characteristic length scale decreases, physical parameters associated with surface area grow larger with respect to those associated with volume:

$$\frac{A_{surface}}{V} \sim \frac{L^2}{L^3} \sim \frac{1}{L} . \tag{1.6}$$

As another example, consider turning a glass of water upside down. It is no surprise that the water spills out under the action of gravity. However, there is in fact a physical phenomenon, surface tension, which opposes the gravity force. This surface tension can be thought of as residing in a thin membrane stretched between the surface of the water and the surrounding air. The gravity force is a volume parameter, whereas the surface tension is an area parameter. Hence, the importance of gravity to surface tension scales with $1/L$. For an ordinary glass of water, the gravity force is orders of magnitude larger than the surface tension, so much so that the surface tension is virtually negligible. If you were to turn over a micro-sized glass of water, however, you would have a very difficult time getting the water out at all even if you were to shake it vigorously. For a small L, the surface tension dominates. In fact, one usually does not consider gravity in micro-devices at all.

The idea of scale analysis that we have used here comes in handy when considering the physics of miniaturization. It does not provide enough detail to accurately model MEMS devices, but it is does a great job in helping us discern which physical parameters are important to consider with relatively little effort. The following example illustrates the power of this technique for a more complicated system.

Example 1.1

Time required to decelerate a micro-flywheel

A flywheel is a mechanical device used to store energy in the form of a rotating mass. For a flywheel with a radius of 90 mm spinning at an angular velocity ω it is observed that the angular velocity takes five minutes to decelerate to half its original value. If the system is to be miniaturized so that the radius is only 90 μm, estimate the time required to decelerate the flywheel.

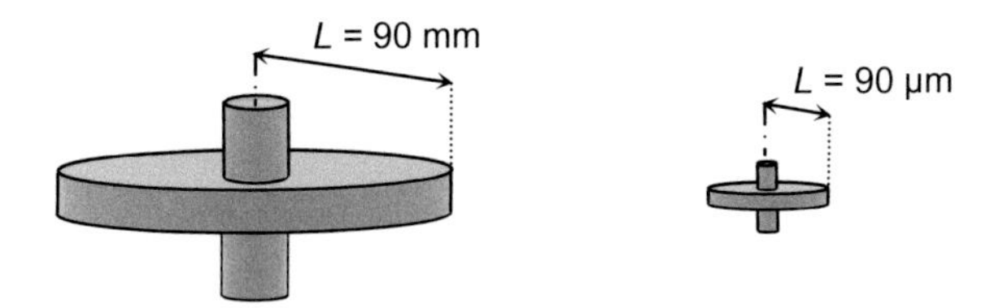

Solution

The time required to decelerate the flywheel is related to the law of conservation of angular momentum, which you would have encountered in a fundamental physics course. The magnitude of the angular momentum of the flywheel about its center of mass is given by

$$\mathscr{L} = I\omega = \omega \int_{m_{total}} r^2 dm \tag{1.7}$$

where $\mathscr{L}$ is the angular momentum, I is the second mass moment of inertia of the flywheel about its center of mass, and r is the distance from the center of mass to an arbitrary location on the wheel. The law of conservation of angular momentum relates the time rate of change of the wheel's angular momentum to an applied moment or torque:

$$\frac{d\mathscr{L}}{dt} = M , \tag{1.8}$$

where M is the magnitude of the applied moment.

Solving equations similar to Eqs. (1.7) and (1.8) strikes fear in the hearts of freshmen physics students. However, performing scale analysis on Eqs. (1.7) and (1.8) is actually quite easy. Rather than solve Eq. (1.7), we are simply after how $\mathscr{L}$ scales with a typical length of the system, here, the radius. Hence, we can essentially ignore the integral and consider I to be some length squared multiplied by the mass. Furthermore, the actual geometry of the system is not particularly important to us either. We can therefore take the characteristic length of the wheel to be the radius L and ignore the other details of the geometry. Since we already know that mass scales with L^3, the scaled version of Eq. (1.7) becomes

$$\mathscr{L} \sim \omega L^2 L^3 \sim \omega L^5 . \tag{1.9}$$

Next we separate the variables of Eq. (1.9) and perform scale analysis on the result:

$$dt = \frac{d\mathscr{L}}{M} \tag{1.10}$$

$$t \sim \frac{\omega L^5}{M} \tag{1.11}$$

Equation (1.11) gives us our result, namely that the deceleration time scales with L^5. If the macro-flywheel with a radius of 90 mm takes 5 minutes to slow to half its original velocity, then the micro-flywheel with radius 90 μm should take only $(90\times10^{-6}/90\times10^{-3})^5 \times 5$ minutes $= 10^{-15} \times 5$ minutes. This is a mere 83×10^{-18} seconds! The favorable scaling encountered in this example is a much faster *response time* for the microscale device.◄

1.3 How are MEMS made?

Just how does one go about fabricating a device no larger than an ant?

In recent years the tools found in a standard machine shop, tools such as lathes, milling machines, band saws and CNCs, have grown vastly in sophistication and precision. At exceedingly high speed (~60,000 rpm or greater) some of these machines are capable of producing features in the sub-millimeter range, and so-called "micromilling" is becoming a popular microfabrication technique.

Most MEMS, however, are fabricated with a very different set of tools than are found in a standard machine shop. The first MEMS, and indeed the bulk of MEMS today, are fabricated using techniques borrowed and adapted from integrated circuit fabrication and semiconductor processing. Such techniques create structures on thin, flat **substrates** (usually **silicon**) in a series of layered processes. This fabrication in layers is best illustrated by a couple of specific examples. Let us first consider creating a thin, flexible diaphragm that may ultimately be used as part of a MEMS pressure sensor. A typical process for doing this is illustrated in Fig. 1.3.

We start with a thin silicon substrate, called a **wafer**, typically measuring 200-400 μm thick (Fig. 1.3 (a)). A thin layer of silicon dioxide (SiO_2) is then "grown" on the wafer by placing it in a furnace at an elevated temperature (Fig. 1.3 (b)). Next, a thin layer of photosensitive material called **photoresist**, or simply resist, is deposited on the SiO_2 layer in a process called **spinning** (Fig. 1.3 (c)). A transparent plate with selective opaque regions called a **mask** is then brought in close proximity to the wafer (Fig. 1.3 (d)). Ultraviolet light is shown through the mask (Fig. 1.3 (e)). On the regions of the photoresist that make contact with the UV light, the resist undergoes a photochemical process in which it hardens and becomes less soluble. (This is true for a **negative resist**. If a **positive resist** were used, then the exposed regions would become more soluble.) The unexposed resist is removed by using a chemical called a **developer**, leaving a portion of the SiO_2 layer exposed (Fig. 1.3 (f)). The result is a window through the resist to the SiO_2. This exposed region is then chemically **etched** with buffered hydrofluoric (HF) acid, which is a fancy way of saying that the acid eats away the SiO_2 layer. The presence of the photoresist on certain regions, however, protects the SiO_2 beneath it from being etched (Fig. 1.3 (g)). And so, we have a chemical reaction taking place only on certain regions of the wafer where we want it to occur. This selective control of where things happen on the wafer forms the cornerstone of creating MEMS devices.

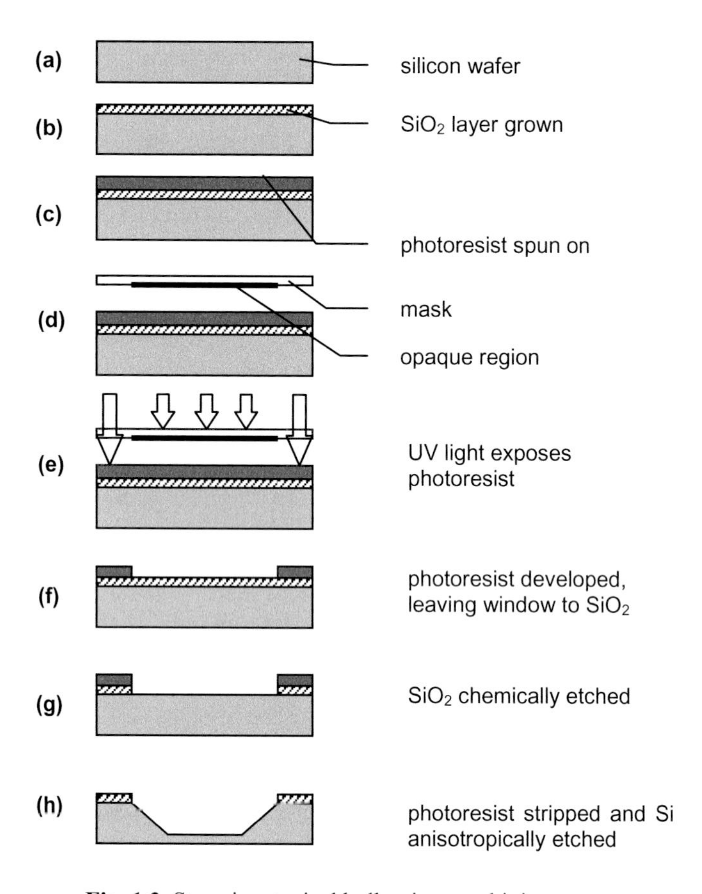

Fig. 1.3. Steps in a typical bulk micromachining process

In yet another chemical process, the remaining photoresist is stripped from the wafer leaving the patterned SiO_2. This time we have a window through the SiO_2 layer directly to the original silicon wafer. The substrate itself is now etched using a potassium hydroxide (KOH) solution (Fig. 1.3 (h)). The SiO_2 covered regions protect the silicon substrate from this etching, just like the resist covered regions protected the SiO_2 in the previous step. Unlike the previous etching process, however, the KOH-Si reaction proceeds at different rates in different spatial directions due to the crystalline structure of the silicon substrate and its orientation. Such a direction-dependent etching process is known as **anisotropic etching**. For the particular process illustrated here, this results in a pit with an inverted pyramid-like shape. The reaction is allowed to proceed long enough so that the bottom of the pit is fairly close to the bottom of the substrate. The resulting thin layer of substrate becomes the flexible membrane of our pressure sensor.

The previous description was simplified and should leave you with many questions. What should the thickness of SiO_2 be and how do you achieve it? How was the mask made? How do you spin photoresist? How do we know that HF eats away SiO_2 and not resist, and that KOH eats away Si but not SiO_2? Would a different crystal orientation result in a differently shaped pit? How do you get the KOH reaction to stop when you want it to? What about sensing the movement of the diaphragm and correlating it to pressure? A great portion of the remaining chapters of this book are geared toward answering questions like these.

The previous example illustrates what has become known as **bulk micromachining**. In bulk micromachining the substrate itself becomes part of the structure for the MEMS device. In another technique called **surface micromachining**, the structure of the MEMS consists of layers of material built on top of the substrate. The substrate in this case serves as a foundation on which to build the MEMS, much like the large thin green piece in a Lego® set, onto which interlocking blocks are placed to form a structure.

A good example of surface micromachining is the creation of an aluminum cantilever beam on a silicon substrate. Figure 1.4 illustrates a simplified version of this process.

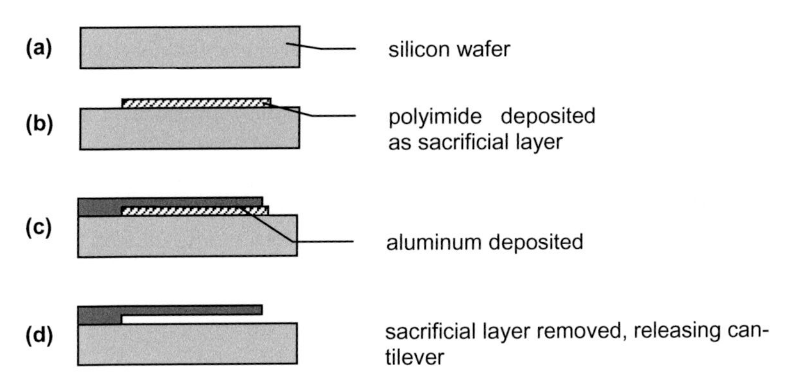

Fig. 1.4. Steps in a typical surface micromachining process

First, a polymer known as polyimide is selectively deposited on the silicon substrate. The polyimide itself will not form any structural part of the cantilever, but rather is a temporary layer used to build around, ultimately being removed from the substrate completely. Such a temporary layer is called a **sacrificial layer**. Next a thin film of aluminum is deposited via **physical vapor deposition** (PVD) on the sacrificial layer. The polyimide is then chemically removed in a process called **release**, leaving behind the aluminum cantilever.

Once more, you should be left with many questions. What is physical vapor deposition and how is it accomplished? How do you get the aluminum to go where you want? Will the cantilever really look like that, or will it curl up—or fall over? How do you get the cantilever to move once you have created it? Again, much of this text is devoted to these details.

1.4 Roadmap and perspective

This text is divided into two main sections: 1) Microfabrication and 2) Applications. In Part 1, we will focus on those unanswered questions that arose during our simplified fabrication examples, introducing the basic processes of microfabrication. Part 2 focuses on MEMS functionality and device physics, examining what MEMS devices do, how they move, and/or how they sense physical quantities. Simplified models are given for each major class of device, as well as more detailed models for some selected devices. Parts 1 and 2 together should give the reader an opportunity to gain a holistic, "big picture" perspective of MEMS.

If it has not already become apparent, MEMS is necessarily a multidisciplinary endeavor, drawing upon physics, chemistry, and all branches of engineering - all in healthy doses. And if one considers BIOMEMS, we can add life sciences to the mix as well! This text, however, assumes a background only in introductory level chemistry, physics, calculus and differential equations, and should therefore be accessible to undergraduate students in most technical majors. (That is not to say that there will not be *any* complicated concepts and/or sticky math along the way, but hopefully these will be introduced in a gentle way.) As a whole the text is intended to give the reader a good overview of MEMS and to act as a springboard for further study.

Essay: The Role of Surface to Volume Atoms as Magnetic Devices Miniaturize

Elaine Kirkpatrick
Department of Physics and Optical Engineering, Rose-Hulman Institute of Technology

As demand drives products from cell phones to hard drives ever smaller, technology must find ways to fabricate smaller components. Devices that were once only macroscopic are now microscopic and are used in a variety of applications including: accelerometers used to determine when to deploy airbags, gyroscopes to activate dynamic stability control, and biosensors to detect a variety of cells and compounds. As the quest for smaller devices continues the length scales are moving from micro- to nano- and beyond. However, as things go into the nanometer size range, the ratio of surface atoms to interior atoms increases drastically. Assuming the atoms are arranged so that each atom forms the corner of a cube—a *cubic lattice structure*—with a spacing of around 3 Å, the ratio of surface atoms to total atoms goes from less than 0.1% to 18% as the particle size goes from 1 μm to 10 nm! Although the exact percentage of surface atoms in a particle will depend on crystal structure and atom size, surface effects due to difference in *coordination number* (the integer number of nearest neighbors of an atom) in general become significant in the nanometer size regime, and become the dominate contributor for particles around 1 nm in size. The mechanical, optical, electrical, and magnetic properties of the material can therefore change when the particle size approaches the nanometer regime.

In magnetism, the largest property change as you move from bulk magnetism to fine particle magnetism (the particle size where "fine particle magnetism" starts depends upon the specific magnetic material, but is usually around 10-100 nm) is the *coercivity*, or the magnetic field necessary to bring the total magnetization of the material back to zero. The mechanism responsible for coercivity is different for bulk materials and fine particles. In bulk magnetic materials, like a large chunk of iron, there are several magnetic domains. The magnetic spins on all atoms in a single domain will be aligned, but the spins on atoms in neighboring domains will point in another direction (See figure.) The regions in between the domains are called *domain walls*, and when a magnetic field is applied the domains in a direction similar to the applied field will expand while the other domains shrink. For example, if a field directed upward was applied to the material in the figure the domain wall between the black and white arrows would move to the right, growing the "black domain" while shrinking the "white domain". Since the domain wall just moves, only a few spins have to switch at a time, and it can be fairly easy to for bulk magnets to switch their magnetization state. Defects in the crystal may cause the domain wall to pin, or not move, temporarily, and this domain wall pinning is the mechanism that gives bulk magnetic materials their coercivity.

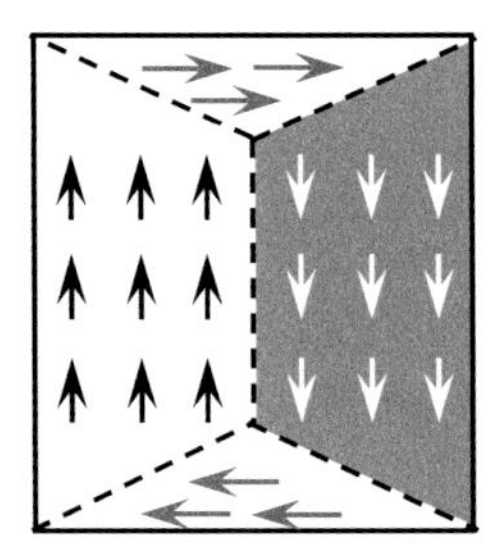

Schematic of a magnetic material with four domains. The arrows represent the spins on the individual atoms and the dashed lines represent the domain walls.

As magnetic particles become smaller it becomes energetically favorable for them to be single domain. The size at which this happens will depend on the shape and composition of the particles, but for the elemental magnets ranges from ~70 nm for cobalt to ~14 nm for iron. (Note some compounds with very high anisotropy can have large single domain particles sizes, for instance up to 2 μm in certain samarium-cobalt magnets.) The mechanism for reversal of the particle then changes from domain wall motion to Stoner-Wohlfarth rotation where all the spins reverse at the same time. Since it is harder to reverse the magnetization of several spins at once rather than reversing a few spins at a time (like in domain wall motion) the coercivity of a material increases as the particle size decreases until the single domain size is reached. The coercivity will then decrease with a further reduction in particles size since smaller particles have fewer spins to reverse and therefore it takes less energy to reverse the entire particle.

As applications require smaller magnetic components, whether that is a smaller bit size for data storage or just for making a miniature magnet, the coervicities of the materials will be affected. For applications such as data storage the coercivity of the particles must be low enough to "write" the data, but high enough so that the data isn't easily corrupted by stray fields. The research in this area therefore focuses on fabricating compounds that have the "correct" coercivity with the smallest possible particle size. No matter what the application: magnetic, electrical, optical or mechanical, as devices become smaller and smaller the increasing number of surface atoms will affect device performance and it is up to researcher and device fabricators to adjust the processes and materials accordingly.

References and suggested reading

"Microsystems Science, Technology and Components Dimensions." *Sandi National Laboratories*. 2005. http://www.mems.sandia.gov/

Madou MJ (2002) Fundamentals of Microfabrication, The Art and Science of Miniaturization., 2nd edn. New York, CRC Press

Schmidt-Nielsen K (1984) Scaling: Why Is Animal Size So Important? Cambridge, Cambridge University Press

Questions and problems

1.1 What does MEMS stand for?
1.2 Give two examples of commonly used MEMS devices.
1.3 Do all MEMS devices have electrical components? Do all MEMS have moving parts? Explain.
1.4 Give two advantages of MEMS devices.
1.5 Will a hotdog cook more quickly or more slowly if it is cut into small pieces? Why?
1.6 One of the most common structures in MEMS devices is a cantilever beam. When a force is applied to the tip of such a beam, the tip deflects. The tip deflection is proportional to the magnitude of the force, such that the cantilever can be modeled as a liner spring with

$$F = k_{eq} x ,$$

where F is the applied force, x is the tip displacement and k_{eq} is the equivalent spring constant given by

$$k_{eq} = \frac{Ewt^3}{4L^3} .$$

Here, E is Young's modulus of elasticity (an invariant material property), w is the beam width, t is the thickness and L is the length. How does the equivalent spring constant scale? Do you expect a micro-sized cantilever to be stiffer or less stiff than a macro-sized cantilever?
1.7 What is photoresist?
1.8 What is a mask?
1.9 What is meant by etching?
1.10 How does surface micromachining differ from bulk micromachining?
1.11 What is a sacrificial layer?
1.12 Find an example of a MEMS device via a journal publication and/or the Internet whose operating principle makes use of a favorable scal-

ing. Explain the operating principle of the device and how its micro-scale dimensions figure into it.

1.13 Find an example of a MEMS device via a journal publication and/or the Internet. Explain in general terms how the device is fabricated. Categorize the fabrication processes into those that you recognize from this chapter and ones that are new to you.

Chapter 2. The substrate and adding material to it

2.1 Introduction

One of the most important techniques employed in microfabrication is the addition of a thin layer of material to an underlying layer. Added layers may form part of the MEMS structure, serve as a mask for etching, or serve as a sacrificial layer.

Additive techniques include those occurring via chemical reaction with an existing layer, as is the case in the oxidation of a silicon **substrate** to form a silicon dioxide layer, as well as those techniques in which a layer is deposited directly on a surface, such as **physical vapor deposition** (PVD) and **chemical vapor deposition** (CVD). The addition of impurities to a material in order to alter its properties, a practice known as **doping**, is also counted among additive techniques, though it does not result in a new physically distinct layer.

Before exploring the various additive techniques themselves we will first examine the nature of the silicon substrate itself.

2.2 The silicon substrate

2.2.1 Silicon growth

In MEMS and microfabrication we start with a thin, flat piece of material onto which (or into which – or both!) we create structures. This thin, flat piece of material is known as the **substrate**, the most common of which in MEMS is crystalline silicon. Silicon's physical and chemical properties make it a versatile material in accomplishing structural, mechanical and electrical tasks in the fabrication of a MEMS.

Silicon in the form of silicon dioxide in sand is the most abundant material on earth. Sand does not make a good substrate, however. Rather, al-

T.M. Adams, R.A. Layton, *Introductory MEMS: Fabrication and Applications*,
DOI 10.1007/978-0-387-09511-0_2, © Springer Science+Business Media, LLC 2010

most all crystalline silicon substrates are formed using a process call the Czochralski method.

In the Czochralski method ultra pure elemental silicon is melted in a quartz crucible in an inert atmosphere to temperatures of 1200-1414°C. A small "seed" is introduced to the melt so that as is cools and solidifies, it does so as a crystal rather than amorphously or with a granular structure. This is accomplished by slowly drawing and simultaneously cooling the melt while rotating the seed and the crucible the silicon melt in opposite directions. The size of the resulting silicon ingot is determined by carefully controlling temperature as well as the rotational and vertical withdrawal speeds. Once the crystalline silicon is formed, it is cut into disks called wafers. The thicknesses of silicon wafers vary from 200 to 500 μm thick with diameters of 4 to 12 inches. These are typically polished to within 2 μm tolerance on thickness. Figure 2.1 shows single crystal silicon being formed by the Czochralski method.

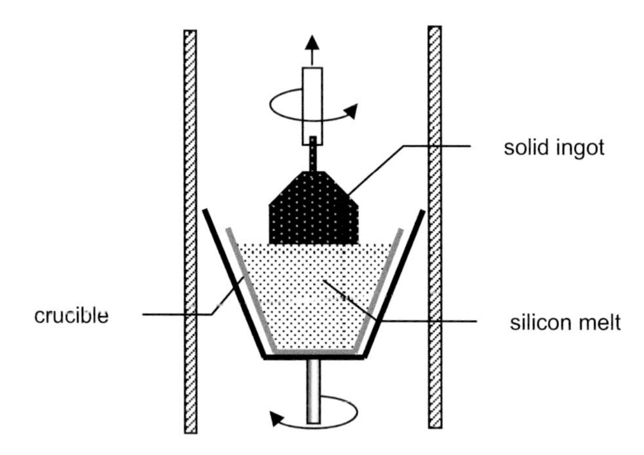

Fig. 2.1. Single crystal silicon formed by the Czochralski method

Sometimes different atmospheres (oxidizing or reducing) are utilized rather than an inert atmosphere to effect crystals with different properties. Furthermore, controlled amounts of impurities are sometimes added during crystal growth in a process known as doping. Typical dopant elements include boron, phosphorous, arsenic and antimony, and can bring about desirable electrical properties in the silicon. We will discuss doping more thoroughly soon.

2.2.2 It's a crystal

Atoms line up in well-ordered patterns in crystalline solids. In such solids, we can think of the atoms as tiny spheres and the different crystal structures as the different ways these spheres are aligned relative to the other spheres. As it turns out, there are only fourteen different possible relative alignments of atoms in crystalline solids. For the semiconductor materials in MEMS, the most important family of crystals are those forming cubic lattices. Figure 2.2 shows the three types of cubic **unit cells,** the building blocks representative of the structure of the entire crystal.

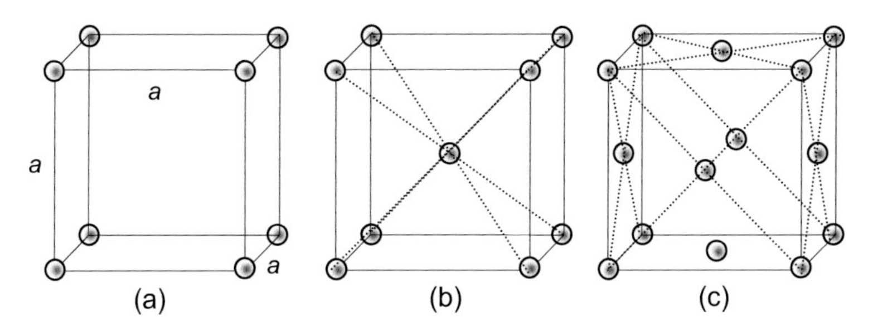

Fig. 2.2 Cubic lattice arrangements: (a) Cubic; (b) Body-centered cubic; (c) Face-centered cubic

Figure 2.2 (a) shows the arrangement of atoms in a simple cubic lattice. The unit cell is simply a cube with an atom positioned at each of the eight corners. Polonium exhibits this structure over a narrow range of temperatures. Note that though there is one atom at each of the eight corners, there is only one atom in this unit cell, as each corner contributes an eighth of its atom to the cell. Figure 2.2 (b) illustrates a body-centered cubic unit cell, which resembles a simple cubic arrangement with an additional atom in the center of the cube. This structure is exhibited by molybdenum, tantalum and tungsten. The cell contains two atoms. Finally, Fig. 2.2 (c) illustrates a face-centered cubic unit cell. This is the cell of most interest for silicon. There are eight corner atoms plus additional atoms centered in the six faces of the cube. This structure is exhibited by copper, gold, nickel, platinum and silver. There are four atoms in this cell. In each unit cell shown, the distance a characterizing the side length of the unit cell is called *the lattice constant.*[1]

[1] For non-cubic materials there can be more than one lattice constant, as the sides of the unit cells are not of equal lengths.

Silicon exhibits a special kind of face-centered cubic structure known as the diamond lattice. This lattice structure is a combination of two face-centered cubic unit cells in which one cell has been slid along the main diagonal of the cube one-forth of the distance along the diagonal. Figure 2.3 shows this structure. There are eight atoms in this structure, four from each cell. Each Si atom is surrounded by four nearest neighbors in a tetrahedral configuration with the original Si atom located at the center of the tetrahedron. Since Si has four valence electrons, it shares these electrons with its four nearest neighbors in covalent bonds. Modeling Si atoms as hard spheres, the Si radius is 1.18Å with a lattice constant of 5.43 Å. The distance between nearest neighbors is $d = (3)^{1/2}a/4 = 2.35$Å.

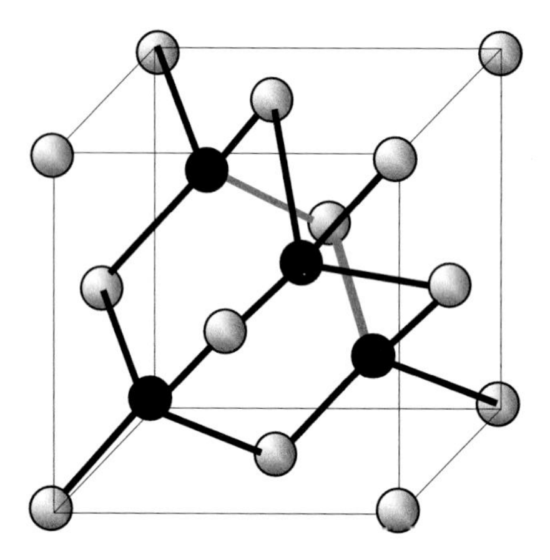

Fig. 2.3 The diamond lattice structure

2.2.3 Miller indices

There are varieties of **crystal planes** defined by the different atoms of a unit cell. The use of **Miller indices** helps us designate particular planes and also directions in crystals. The notation used with Miller indices are the symbols *h*, *k* and *l* with the use of various parentheses and brackets to indicate individual planes, families of planes and so forth. Specifically, the notation $(h\ k\ l)$ indicates a specific plane; $\{h\ k\ l\}$ indicates a family of equivalent planes; $[h\ k\ l]$ indicates a specific direction in the crystal; and $<h\ k\ l>$ indicates a family of equivalent directions.

There is a simple three-step method to find the Miller indices of a plane:

1. Identify where the plane of interest intersects the three axes forming the unit cell. Express this in terms of an integer multiple of the lattice constant for the appropriate axis.
2. Next, take the reciprocal of each quantity. This eliminates infinities.
3. Finally, multiply the set by the least common denominator. Enclose the set with the appropriate brackets. Negative quantities are usually indicated with an over-score above the number.

Example 2.1 illustrates this method.

Example 2.1

Finding the Miller indices of a plane

Find the Miller indices for the plane shown in the figure.

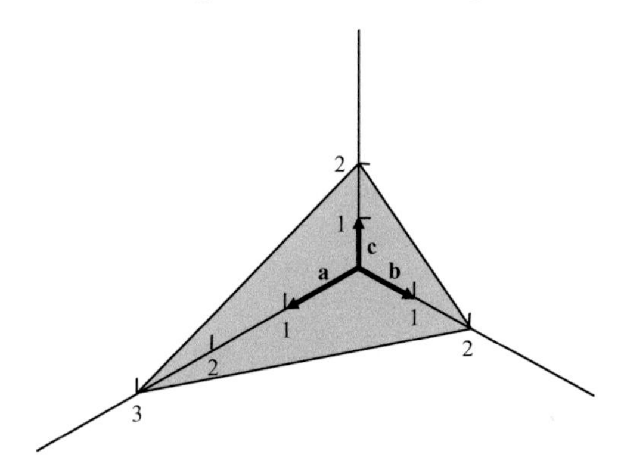

Solution

The plane intersects the axes at 3a, 2b and 2c.
The reciprocals of the theses numbers are 1/3, 1/2 and 1/2.
Multiplying by the least common denominator of 6 gives 2, 3, 3.

Hence the Miller indices of this plane are (2 3 3). ◄

For cubic crystals the Miller indices represent a direction vector perpendicular to a plane with integer components. That is, the Miller indices of a direction are also the Miller indices of the plane normal to it:

$$[h\,k\,l] \perp (h\,k\,l).$$

(This is not necessarily the case with non-cubic crystals.) Figure 2.4 shows the three most important planes for a cubic crystal and the corresponding Miller indices.

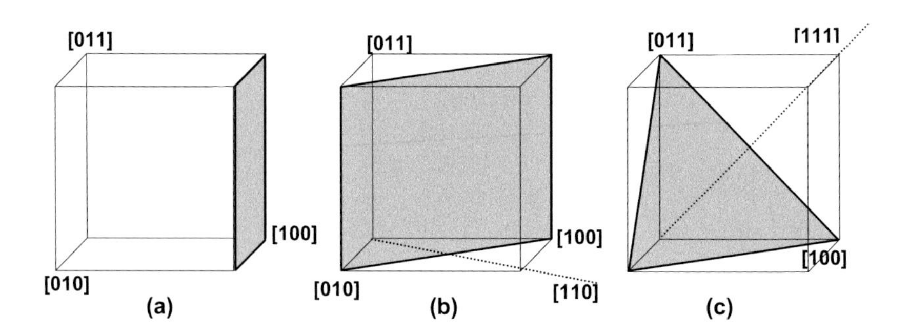

Fig. 2.4. Miller indices of important planes in a cubic crystal (a) (100); (b) (110); (c) (111)

Silicon wafers are classified in part by the orientation of their various crystal planes with relation to the surface plane of the wafer itself. In what is called a (100) wafer, for example, the plane of the wafer corresponds a {100} crystal plane. Likewise, the plane of a (111) wafer coincides with the {111} plane. The addition of straight edges, or flats to the otherwise circular wafers helps us identify the crystalline orientation of the wafers. These flats also tell us if and/or whether the wafers are p-type or n-type. (P-type and n-type refer to wafer's doping, a process affecting the wafer's semiconductor properties. We will explore doping soon.) A few examples are given in Fig. 2.5. Figure 2.6 shows the relative orientations of the three important cubic directions for a {100} wafer.

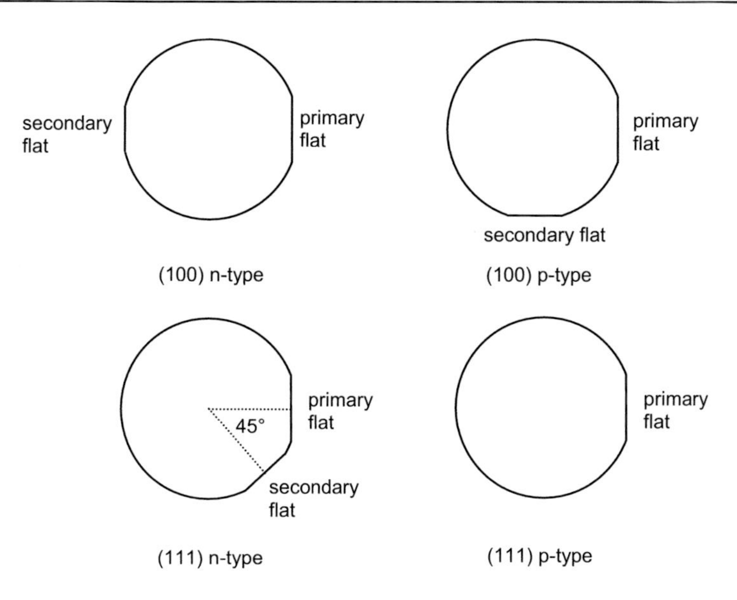

Fig. 2.5. Wafer flats are used to identify wafer crystalline orientation and doping.

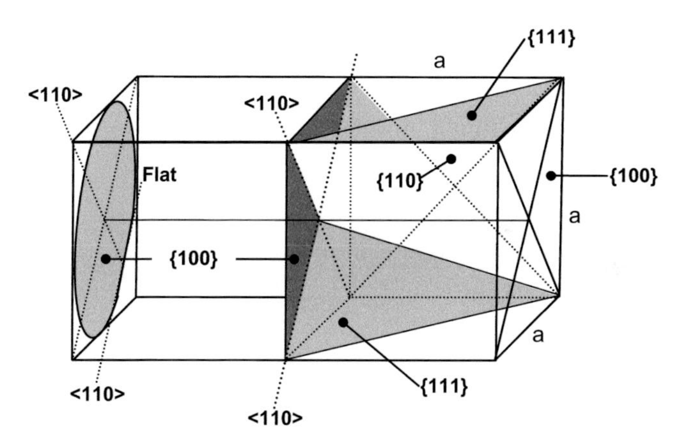

Fig. 2.6. Orientations of various crystal directions and planes in a (100) wafer (Adapted from Peeters, 1994)

2.2.4 It's a semiconductor

In metals there are large numbers of weakly bound electrons that move around freely when an electric field is applied. This migration of electrons is the mechanism by which electric current is conducted in metals, and metals are appropriately called electrical **conductors**. Other materials by contrast have valence electrons that are tightly bound to their atoms, and therefore don't move much when an electric field is applied. Such materials are known as **insulators**. The group IV elements of the periodic table have electric properties somewhere in between conductors and insulators, and are known as **semiconductors**. Silicon is the most widely used semiconductor material. Its various semiconductor properties come in handy in the MEMS fabrication process and also in the operating principles of many MEMS devices.

Figure 2.7 shows the relative electron energy bands in conductors, insulators and semiconductors. In conductors there are large numbers of electrons in the energy level called the conduction band. In insulators the conduction band is empty, and a large energy gap exists between the valence band and the conduction band energy levels, making it difficult for electrons in the valence band to become conduction electrons. Semiconductor materials have only a small energy gap in between the valence band and the otherwise empty conduction band so that when a voltage is applied some of the valence electrons can make the jump to the conduction band, becoming charge carriers. The nature of how this jump is made is affected by both temperature and light in semiconductors, which precipitates their use as sensors and optical switching devices in MEMS.

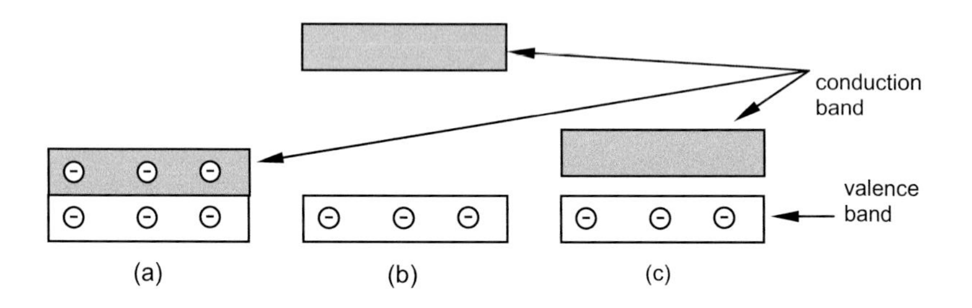

Fig. 2.7. Valence and conduction bands of various materials: (a) Conductor; (b) Insulator; (c) Semiconductor

The **electrical conductivity** or simply conductivity σ of a material is a measure of how easily it conducts electricity. The inverse of conductivity is **electrical resistivity** ρ. Typical units of conductivity and resistivity are $(\Omega\cdot m)^{-1}$ and $\Omega\cdot m$, respectively. As you might expect, semiconductors have resistivity values somewhere in between those of conductors and insulators, which can be seen in Table 2.1. Resistivity also exhibits a temperature dependence which is exploited in some MEMS temperature sensors.

Table 2.1. Resistivities of selected materials at 20°C

Material	Resistivity ($\Omega\cdot m$)
Silver	1.59×10^{-8}
Copper	1.72×10^{-8}
Germanium	4.6×10^{-1}
Silicon	6.40×10^{2}
Glass	10^{10} to 10^{14}
Quartz	7.5×10^{17}

Conductivity or resistivity can be used to calculate the electrical resistance for a given chuck of that material. For the geometry shown in Fig. 2.8 the electrical resistance in the direction of the current is given by

$$R = \frac{L}{\sigma A} = \rho \frac{L}{A}. \tag{2.1}$$

The trends of Eq. (2.1) should make intuitive sense. Materials with higher resistivities result in higher resistances, as do longer path lengths for electrical current and/or skinnier cross sections. The electrical resistance of a MEMS structure can therefore be tailored via choice of material and geometry.

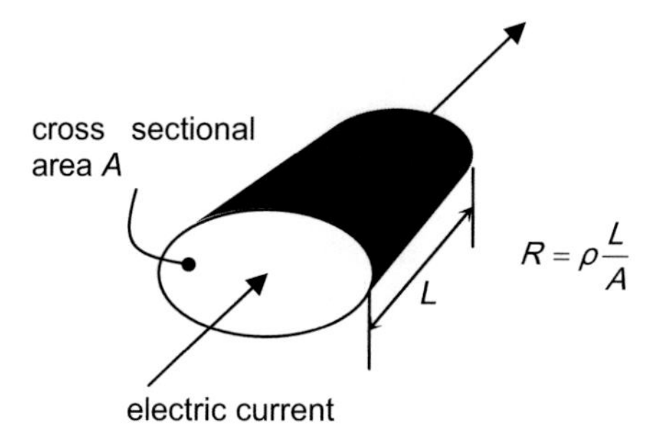

Fig. 2.8. Electrical resistance is determined from resistivity and geometry considerations.

Doping

The properties of semiconductors can be changed significantly by inserting small amounts of group III or group V elements of the periodic table into the crystal lattice. Such a process is called **doping**, and the introduced elements, **dopants**. Careful control of the doping process can also bring about highly localized differences in properties, which has numerous uses both in MEMS device functionality and the microfabrication process itself.

As an example of doping, consider silicon, which has four valence electrons in its valence band. Phosphorous, however, is a group V element and therefore has five valence electrons. If we introduce phosphorous as a dopant into a silicon lattice, one extra electron is floating around and is therefore available as a conduction electron. The dopant material in this case is called a **donor**, as it donates this extra electron. The doped silicon is now an **n-type** semiconductor, indicating that the donated charge carriers are negatively charged. Should the dopant be an element such as boron from group III of the periodic table, however, only three valence band electrons exist in the dopant material. Effectively, a **hole** has been introduced into the lattice, one that is intermittently filled by the more numerous valence electrons of the silicon. This type of dopant is called an **acceptor**, as it accepts electrons from the silicon into its valence band. The electric current in such a semiconductor is the effective movement of these holes from atom to atom. Hence, such semiconductors are called **p-type**, indicating positive charge carriers in the material. By controlling the concentration of the dopant material, the resistivity of silicon can be varied over a range of about 1×10^{-4} to 1×10^{8} $\Omega\cdot$m!

Doping can be achieved in several ways. One way is to build it right into the wafer itself by including the dopant material in the silicon crystal growth process. Such a process results in a uniform distribution of dopant material throughout the wafer, forming what is called the background concentration of the dopant material. Doping already existing wafers is usually achieved by one of two methods, implantation, or thermal diffusion. These methods result in a non-uniform distribution of dopant material in the wafer. Often implantation and/or diffusion are done using an n-type dopant if the wafer is already a p-type, and using a p-type dopant if the wafer is already an n-type. Where the background concentration of dopant in the wafer matches the newly implanted or diffused dopant concentration, a *p-n junction* is formed, the corresponding depth being called the junction depth. P-n junctions play a significant role in microfabrication, often serving as *etch stops*, mechanisms by which a chemical etching process can be halted.

Often implantation and diffusion are done through masks on the wafer surface in order to create p-n junctions at specific locations. Silicon dioxide thin films and photoresist are common masking materials used in this process. Figure 2.9 shows ion implantation occurring though a photoresist mask.

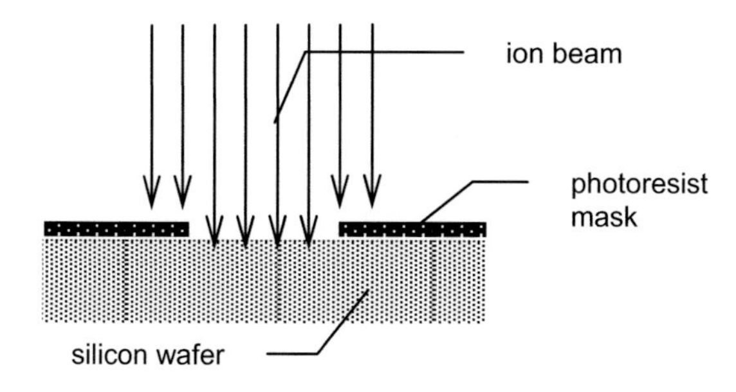

Fig. 2.9. Ion implantation through a photoresist mask

Doping by diffusion

When you pop a helium-filled balloon, the helium doesn't stay in its original location defined by where the balloon was. Rather, it disperses through the surrounding air, eventually filling the entire room at a low concentration. This movement of mass from areas of high concentration to low concentration is appropriately called **mass diffusion**, or simply diffusion. Doping can be achieved by diffusion as well.

When doping by diffusion, the dopant material migrates from regions of high concentration in the wafer to regions of low concentrations in a process called *mass diffusion*. The process is governed by *Fick's law of diffusion*, which in this case simply states that the mass flux of dopant is proportional to the concentration gradient of the dopant material in the wafer, and a material constant characterizing the movement of the dopant material in the wafer material. Flux refers to amount of material moving past a point per unit time and per unit area normal to the flow direction. Flux has dimensions of [amount of substance]/[time][area], typical units for which would be moles/s-m^2 or atoms/s-m^2. Fick's law of diffusion is given by

$$j = -D\frac{\partial C}{\partial x}, \tag{2.2}$$

where j is the mass flux, D is the **diffusion constant** of the dopant in the wafer material , C is the concentration of the dopant in the wafer material, and x is the coordinate direction of interest. In doping, x is usually the direction perpendicular to the surface of the wafer going into the wafer. The negative sign of Eq. (2.2) indicates that the flow of dopant material is in the direction of decreasing concentration.

The diffusion constant D is a temperature dependent constant that characterizes the diffusion of one material in another.[2] It has dimensions of length squared divided by time. For general purposes the constant is well calculated by

$$D = D_0 e^{-\frac{E_a}{k_b T}}, \tag{2.3}$$

where D_o is the **frequency factor**, a parameter related to vibrations of the atoms in a lattice, k_b is Boltzmann's constant (1.38×10^{-23} J/K) and E_a is the **activation energy**, the minimum energy a diffusing atom must overcome in order to migrate. Table 2.2 gives of D_0 and E_a for the diffusion of boron and phosphorus in silicon.

Table 2.2. Frequency factor and activation energy for diffusion of dopants in silicon

Material	D_0 [cm^2/s]	E_a [eV]
Boron	0.76	3.46
Phosphorus	3.85	3.66

[2] The *form* of Ficks's law of diffusion also applies to the transfer of heat and momentum within a material. If you consider only one-dimensional fluxes, heat flux and viscous stress in a flowing fluid are given by $q = -\kappa \cdot dT/dx$ and $\tau = \eta \cdot dV/dx$, respectively. The thermal conductivity κ and the viscosity η play the same role in the transfer of heat and momentum as does the diffusion constant in mass transfer. One difference, however, is that D not only depends on the diffusing species, but also the substance through which it diffuses.

We see, then, that mass travels in the direction of decreasing concentration, heat flows in the direction of decreasing temperature and momentum travels in the direction of decreasing velocity. The three areas are therefore sometimes collectively referred to as **transport phenomena**. (The interpretation of force as a momentum transport, however, is not as common. Thus, the usual sign convention for stress is opposite of that which would result in a negative sign in the stress equation.)

In doping by diffusion the dopant is first delivered to the surface of the wafer after which it diffuses into the wafer. The resulting distribution of dopant within the wafer is therefore a function of both time and depth from the wafer surface. In order to determine this distribution, one solves an equation representing the idea that mass is conserved as the dopant diffuses within the wafer. The version of conservation of mass that applies at a point within a substance is called the *continuity equation*, deriving its name from the assumption that the material can be well modeled as a continuum; that is, it makes sense to talk about properties having a value at an infinitesimally small point. This assumption is valid for most MEMS devices, though it often breaks down at the scales encountered in nanotechnology where the length scales are on the order of the size of molecules. In any case, the continuity equation in regards to mass diffusion reduces to

$$\frac{\partial j}{\partial x} = -D\frac{\partial^2 C}{\partial x^2} = \frac{\partial C(x,t)}{\partial t}. \tag{2.4}$$

As this partial differential equation is first order in time and a second order in depth, we require one initial condition and two boundary conditions in order to solve it. The initial condition is that the concentration of dopant at any depth is zero at $t = 0$, or $C(x, t = 0) = 0$. There are several possibilities for the boundary conditions, however.

One common set of boundary conditions is that the surface concentration goes to some constant value for $t > 0$ and remains at that constant value for all times after that. The second boundary condition comes from the wafer thickness being significantly larger than the depths to which the dopant diffuses. In equation form theses two boundary conditions are, respectively

$$C(x = 0, t > 0) = C_s \tag{2.5}$$

$$C(x \to \infty, t > 0) = 0 \tag{2.6}$$

The solution to Eq. (2.4) incorporating these boundary conditions yields

$$C(x,t) = C_s erfc\left(\frac{x}{2\sqrt{Dt}}\right), \tag{2.7}$$

where C_s is the surface concentration and $erfc(\lambda)$ is the *complementary error function* given by

$$erfc(\lambda) \equiv \frac{2}{\sqrt{\pi}}\int_x^{\infty} e^{-\lambda^2} d\lambda \tag{2.8}$$

The integral of the complementary error function does not have a closed form solution, making it a transcendental function. Appendix C gives values of the complementary error function. Many modern calculators include $erfc(\lambda)$ as a built-in feature as do many software packages.

Figure 2.10 gives the concentration of boron in silicon as a function of depth with time as a parameter due to thermal diffusion at 1050°C assuming a constant surface concentration boundary condition. Also shown in the figure is the characteristic length called the **diffusion length**, given by

$$x_{diff} \approx \sqrt{4Dt}\ . \tag{2.9}$$

The diffusion length gives a rough estimate of how far the dopant has diffused into the substrate material at a given time. As the concentration profiles are asymptotic, no single length gives a true cut-off at which point the dopant concentration is actually zero.

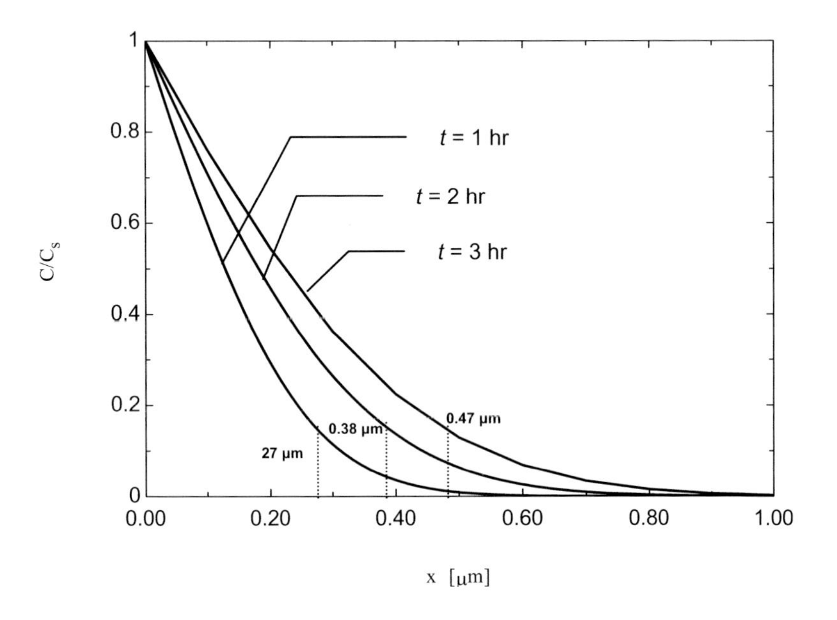

Fig. 2.10. Diffusion of boron in silicon at 1050°C for various times. Diffusion lengths are also shown.

Another quantity of interest in diffusion is the total amount of dopant that has diffused into the substrate per unit area. This can be calculated by integrating Eq. (2.7), the result being

$$Q(t) = \int_0^\infty C_s erfc\left(\frac{x}{2\sqrt{Dt}}\right)dx = \frac{2\sqrt{Dt}}{\sqrt{\pi}}C_s \tag{2.10}$$

where Q is the amount of dopant per area, sometimes called the **ion dose**. Often a more appropriate boundary condition for solving Eq. (2.4) is that the total amount of dopant supplied to the substrate is constant, or that Q is constant. Retaining the same initial condition and the second boundary condition, $C(\infty, t) = 0$, the solution to Eq. (2.4) yields a Gaussian distribution,

$$C(x,t) = \frac{Q}{\sqrt{\pi Dt}}\exp\left(\frac{-x^2}{4Dt}\right) = C_s \exp\left(\frac{-x^2}{4Dt}\right). \tag{2.11}$$

The Gaussian distribution results from the fact that the concentration of ions at the surface is depleted as the diffusion continues. This is reflected in Eq. (2.11) in that the surface concentration is given by $C_s = Q/(\sqrt{(\pi Dt)})$.

Doping by implantation

In implantation a dopant in the form of an ion beam is delivered to the surface of a wafer via a particle accelerator. One may liken the process to throwing rocks into the sand at the beach. The wafer is spun as the accelerator shoots the dopant directly at the surface to ensure more uniform implantation. (Fig. 2.11.) The ions penetrate the surface and are stopped after a short distance, usually within a micron, due to collisions. As a result, the peek concentration of the implanted ions is actually below the surface, with decreasing concentrations both above and below that peek following a normal (Gaussian) distribution. After impact implantation takes only femtoseconds for the ions to stop.

Theoretically this process can occur at room temperature. However, the process is usually followed by a high temperature step (~900°C) in which the ions are "activated," meaning to ensure that they find their way to the spaces in between the atoms in the crystal.[3] Furthermore, the ion bombardment often physically damages the substrate to a degree, and the high temperature anneals the wafer, repairing those defects.

[3] These spaces are called *interstitial sites*.

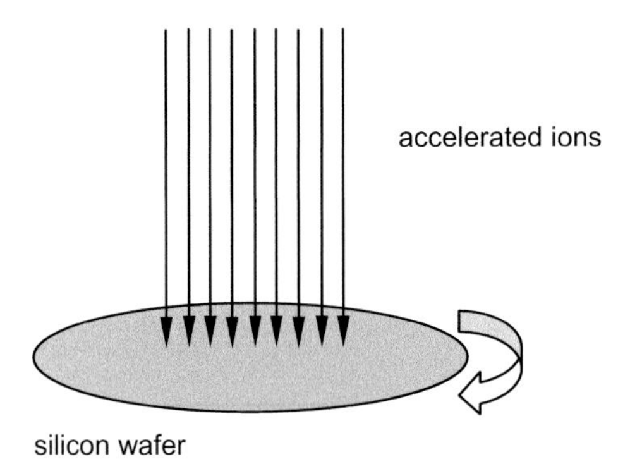

Fig. 2.11. Doping by ion implantation

In ion implantation, the bombardment of ions on the surface results in a Gaussian distribution of ions given by

$$C(x) = C_P \exp\left(-\frac{(x - R_P)^2}{2\Delta R_P^2} \right) \tag{2.12}$$

where C_P is the peak concentration of dopant, R_P is the **projected range** (the depth of peak concentration of dopant in wafer) and ΔR_P is the standard deviation of the distribution. The range is affected by the mass of the dopant, its acceleration energy, and the stopping power of the substrate material. The peak concentration can be found from the total implanted dose from

$$C_P = \frac{Q_i}{\sqrt{2\pi}\Delta R_P} \tag{2.13}$$

where Q_i is the total implanted ion dose. Figure 2.12 gives some typical profiles of the ion implantation of various dopant species.

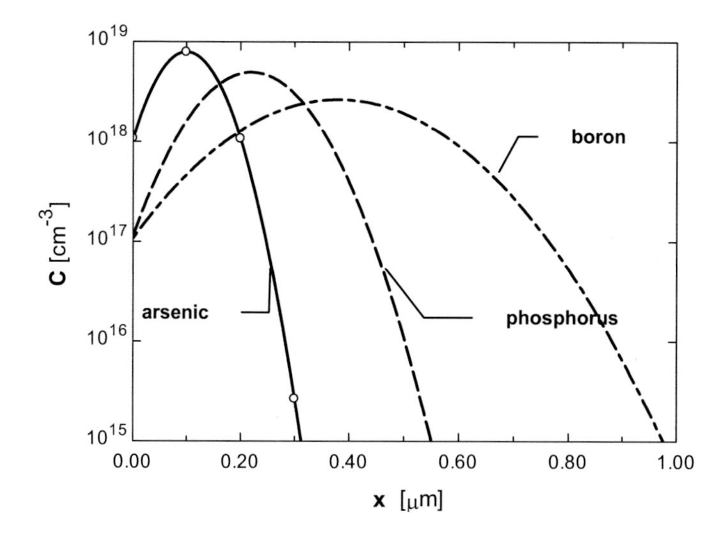

Fig. 2.12 Typical concentration profiles for ion implantation of various dopant species

After ion implantation the dopant material is often thermally diffused even further into the substrate. As such, implantation is often a two step process consisting of the original bombardment of the surface with ions, called *pre-deposition*, and the thermal diffusion step called *drive-in*. The high temperature drive-in step tends to be at a higher temperature than the annealing step that typically follows the implantation process. If the pre-deposition step results in a projected range that is fairly close to the surface, Eq. (2.12) can be used to predict the resulting distribution of dopant after drive-in using $Q = Q_i$.

P-n junctions

One of the major reasons for doping a wafer by thermal diffusion and/or ion implantation is to create a **p-n junction** at specific locations in the wafer. To create a p-n junction, the newly introduced dopant must be of the opposite carrier type than the wafer. That is, if the wafer is already a p-type, the implantation and/or diffusion are done using an n-type dopant, and an n-type wafer would be doped with a p-type dopant. A p-n junction refers to the location at which the diffused or implanted ion concentration matches the existing background concentration of dopant in the wafer. The corresponding depth of the p-n junction called the **junction depth**.

On one side of the junction the wafer behaves as a p-type semiconductor, and on the other it behaves as an n-type. This local variation of semiconductor properties is what gives the junction all its utility. In microelectronic applications p-n junctions are used to create diodes (something like a check valve for electric current) and transistors. In MEMS applications p-n are used to create things such as piezo-resistors, electrical resistors whose resistance changes with applied pressure, enabling them to be used as sensors. P-n junctions can also be used to stop chemical reactions that eat away the silicon substrate, in which case the junction serves as an etch stop. We will learn more about these chemical reactions in Chapter 4 where we discuss bulk micromachining.

For now the important thing to know is how to approximate the location of a p-n junction. Graphically the junction location is at the intersection of the implanted/diffused dopant concentration distribution and the background concentration curves. (Fig. 2.13.) Mathematically the location is found by substituting the background concentration value into the appropriate dopant distribution relation and solving for depth. In the case of a thermally diffused dopant, Eq. (2.11) yields the following for the junction depth:

$$x_j = \sqrt{4Dt \ln\left(\frac{Q}{C_{bg}\sqrt{\pi Dt}}\right)} \tag{2.14}$$

where x_j is the junction depth and C_{bg} is the background concentration of the wafer dopant.

Care must be taken when using the relations in this section, as they are all first order approximations. When second order effects are important, such as diffusion in the lateral direction in a wafer, these relations may not yield sufficiently accurate estimates for design. In such cases numerical modeling schemes using computer software packages are usually employed. Nonetheless, the relations given here are sufficient in many applications and will at least give a “ball park” estimate.

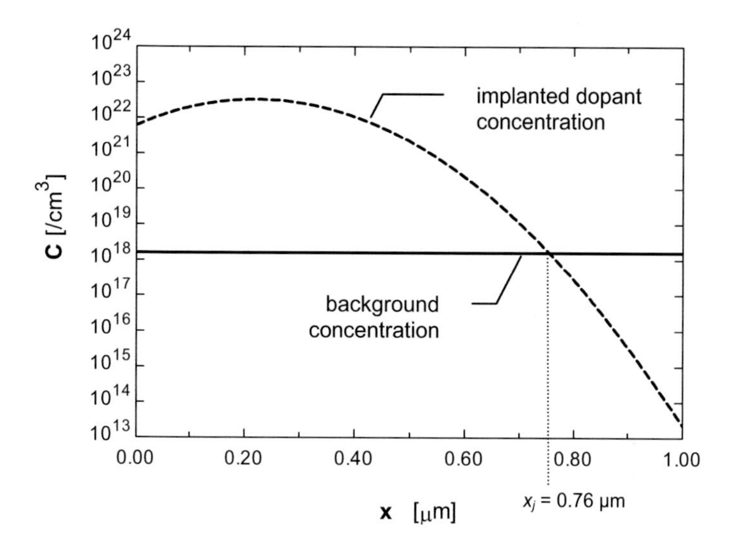

Fig. 2.13. Determination of junction depth

2.3 Additive technique: Oxidation

We have learned much about the nature of the silicon substrate itself and ways of doping it. We next turn our attention to methods of adding physically distinct layers of material on top of the substrate. Sometimes these added materials will serve as masks, at other times as structural layers, and sometimes as sacrificial layers.

2.3.1 Growing an oxide layer

One method of adding a layer of material to the silicon substrate is to "grow" a layer of silicon dioxide (SiO_2) on it. The process is naturally called oxidation. In oxidation, the silicon on the surface of the substrate reacts with oxygen in the environment to form the SiO_2 layer.

The resulting SiO_2 layer is often called an oxide layer for short, or simply oxide. It does a great job as a sacrificial layer or as a hard mask. Oxides can also provide a layer of electrical insulation on the substrate, providing necessary electrical isolation for certain electrical parts in a MEMS.

From simply being in contact with air the silicon substrate will form a thin layer of oxide without any stimulus from the MEMS designer. Such a layer is called a native oxide layer, typical thicknesses being on the order of 20-30 nm. For an oxide layer to be useful as a sacrificial layer or a mask, however, larger thicknesses are typically required. Thus active methods of encouraging the growth of oxide are used.

There are two basic methods used to grow oxide layers: dry oxidation and wet oxidation. Both methods make use of furnaces at elevated temperatures on the order of 800-1200°C. Both methods also allow for careful control of the oxygen flow within the furnace. In dry oxidation, however, only oxygen diluted with nitrogen makes contact with the wafer surface, whereas wet oxidation also includes the presence of water vapor. The chemical reactions for dry and wet oxidation, respectively, are given by

$$\text{(dry) } Si + O_2 \rightarrow SiO_2 \text{ and} \tag{2.15}$$

$$\text{(wet) } Si + 2H_2O \rightarrow SiO_2 + 2H_2 \,. \tag{2.16}$$

Dry oxidation has the advantage of growing a very high quality oxide. The presence of water vapor in wet oxidation has the advantage of greatly increasing the reaction rate, allowing thicker oxides to be grown more quickly. The resulting oxide layers of wet oxidation tend to be of lower quality than the oxides of dry oxidation. Hence, dry oxidation tends to be used when oxides of the highest quality are required, whereas wet oxidation is favored for thick oxide layers. An oxide layer is generally considered to be thick in the 100 nm – 1.5 μm range.

Oxidation is interesting as an additive technique in that the added layer consists both of added material (the oxygen) and material from the original substrate (the silicon). As a result, the thickness added to the substrate is only a fraction of the SiO_2 thickness. In reference to Fig. 2.14, the ratio of the added thickness beyond that of the original substrate (x_{add}) to the oxide thickness itself (x_{ox}) is

$$\frac{x_{add}}{x_{ox}} = 0.54 \,. \tag{2.17}$$

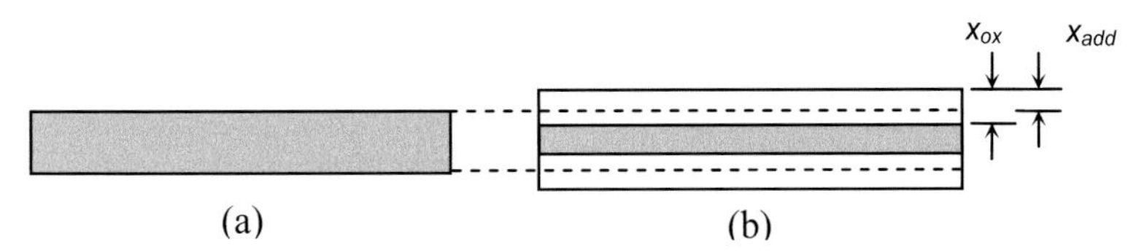

Fig. 2.14 A silicon wafer (a) before oxidation and (b) after oxidation

2.3.2 Oxidation kinetics

As the oxidation reaction continues, oxygen delivered to the surface must diffuse through thicker and thicker layers of silicon dioxide before it can react with the underlying silicon. As such, the time required to grow an additional thickness of oxide becomes longer as the layer thickens. The **Deal-Grove model** of oxidation kinetics is the most widely used model to relate oxide thickness to reaction time. The model comes from the solution to an ordinary differential equation that describes the mass diffusion of oxygen through the oxide layer. The solution is given by

$$x_{ox} = \frac{A}{2}\left\{-1 + \sqrt{\frac{4B}{A^2}(t+\tau)+1}\right\}, \quad (2.18)$$

where t is the oxidation time, A and B are temperature dependent constants, and τ is a parameter that depends on the initial oxide thickness, x_i. The value of τ is found by solving Eq. (2.18) with t=0:

$$\tau = \frac{x_i^2}{B} + \frac{A}{B}x_i . \quad (2.19)$$

Table 2.3 and Table 2.4 gives the values of A and B for both dry and wet oxidation using the Deal-Grove model for (100) and (111) silicon, respectively. The value of τ in the tables is the recommended value when starting with an otherwise bare wafer, and accounts for the presence of a native oxide layer, assumed to be 25 nm. The Deal-Grove model does not predict oxide growth accurately for thicknesses less than 25 nm, in which case oxide growth occurs much more quickly. Hence, the inclusion of τ can also be viewed as a correction term.

Table 2.3. Suggested Deal-Grove rate constants for oxidation of (100) silicon

Temperature (°C)	A (μm)		B (μm²/hr)		τ (hr)	
	Dry	Wet	Dry	Wet	Dry	Wet
800	0.859	3.662	0.00129	0.0839	17.1	1.10
900	0.423	1.136	0.00402	0.172	2.79	0.169
1000	0.232	0.424	0.0104	0.316	0.616	0.0355
1100	0.139	0.182	0.0236	0.530	0.174	0.0098
1200	0.090	0.088	0.0479	0.828	0.060	0.0034

Table 2.4. Suggested Deal-Grove rate constants for oxidation of (111) silicon

Temperature (°C)	A (μm)		B (μm²/hr)		τ (hr)	
	Dry	Wet	Dry	Wet	Dry	Wet
800	0.512	2.18	0.00129	0.0839	10.4	0.657
900	0.252	0.6761	0.00402	0.172	1.72	0.102
1000	0.138	0.252	0.0104	0.316	0.391	0.0220
1100	0.0830	0.1085	0.0236	0.530	0.114	0.0063
1200	0.0534	0.05236	0.0479	0.828	0.0401	0.0023

Equation (2.18) can be simplified for short time or long time approximations. In the case of a short time, keeping the first two terms in a series expansion of the square root gives

$$x_{ox} \approx \frac{B}{A}(t+\tau). \tag{2.20}$$

The growth rate is approximately linear in this case. As such, the ratio B/A is often referred to as the **linear rate constant**. For very long times, $t >> \tau$, and Eq. (2.18) reduces to

$$x_{ox} \approx \sqrt{B(t+\tau)}\,. \tag{2.21}$$

The oxidation rate is also dependent on the crystalline orientation of the wafer. In situations for which Eq. (2.20) is appropriate, the orientation dependence can be captured by changing the linear rate constant. The ratio of the linear rate constant for (111) silicon to (100) is given by

$$\frac{(B/A)_{(111)}}{(B/A)_{(100)}} \approx 1.68 \tag{2.22}$$

Thus, (111) silicon typically oxidizes 1.7 times faster than does (100) silicon. It is speculated that the increased reaction rate is due to the larger density of Si atoms in the (111) direction.

Example 2.2 illustrates the use of the Deal-Grove model.

Example 2.2

Growth time for wet etching

Find the time required to grow an oxide layer 800 nm thick on a (100) silicon wafer using wet oxidation at 1000°C.

Solution

Assuming a native oxide layer of 25 nm, we can use the suggested constants from Table 2.3. Solving (2.18) for t,

$$t = \frac{A^2}{4B}\left[\left(\frac{2}{A}x_{ox} + 1\right)^2 - 1\right] - \tau$$

$$t = \frac{0.424^2\ \mu\text{m}^2}{4(0.316)\ \mu\text{m}^2/\text{hr}}\left[\left(\frac{2}{0.424\ \mu\text{m}}(0.800\ \mu\text{m}) + 1\right)^2 - 1\right] - 0.0355\ \text{hr}$$

$$t = 3.06\ \text{hr}\ .$$

If we make the long time approximation of Eq. (2.20)

$$x_{ox} \approx \sqrt{B(t+\tau)}$$

Solving for t,

$$t \approx \frac{x_{ox}^{\ 2}}{B} - \tau$$

$$t \approx \frac{(0.800\ \mu\text{m})^2}{0.316\ \mu\text{m}^2/\text{hr}} - 0.0355\ \text{hr}$$

$$t \approx 1.99\ \text{hr}.$$

We see that the long time approximation is not a very good one in this case. And although three hours may seem like forever when waiting by an oxidation furnace, it does not qualify as a long time by the standards of Eq. (2.20). ◄

Oxide layer thicknesses can be measured using optical techniques that measure the spectrum of reflected light from the oxide layer. Due to the index of refraction of the oxide and its finite thickness, the light reflected from the top and bottom surfaces will be out of phase. Since visible light is in the wavelength range of 0.4-0.7 μm, which is the same order of magnitude as most oxide layers, the constructive and destructive interference of the reflected light will cause it be a different color depending on oxide thickness. Quick estimates of the thickness can be made by visually inspecting the oxide surface and comparing it to one of many available "color charts", such as in Appendix D.

2.4 Additive technique: Physical vapor deposition

Physical vapor deposition (PVD) refers to the vaporization of a purified, solid material and its subsequent condensation onto a substrate in order to form a thin film. Unlike oxidation, no new material is formed via chemical reaction in the PVD process. Rather, the material forming the thin film is physically transferred to the substrate. This material is often referred to as the **source material**.

The mechanism used to vaporize the source in PVD can be supplied by contact heating, by the collision of an electron beam, by the collision of positively charged ions, or even by focusing an intense, pulsed laser beam on the source material. As a result of the added energy, the atoms of the source material enter the gas phase and are transported to the substrate through a reduced pressure environment.

PVD is often called a direct line-of-sight impingement deposition technique. In such a technique we can visualize the source material as if it were being sprayed at the deposition surface much like spray paint. Where the spray "sees" a surface, it will be deposited. If something prohibits the spray from seeing a surface, a concept called *shadowing*, the material will not be deposited there. Sometimes shadowing is a thorn in the side of the microfabricator, but at other times s/he can use shadowing to create structures of predetermined shapes and sizes.

Physical vapor deposition includes a number of different techniques including thermal evaporation (resistive or electron beam), sputtering (DC, RF, magnetron, or reactive), molecular beam epitaxy, laser ablation, ion plating, and cluster beam technology. In this section we will focus on the two most common types of PVD, evaporation and sputtering.

2.4.1 Vacuum fundamentals

Physical vapor deposition must be accomplished in very low pressure environments so that the vaporized atoms encounter very few intermolecular collisions with other gas atoms while traveling towards the substrate. Also, without a vacuum, it is very difficult to create a vapor out of the source material in the first place. What's more, the vacuum helps keep contaminants from being deposited on the substrate. We see, then, that the requirement of a very low pressure environment is one of the most important aspects of PVD. As such, it is a good idea to spend some time learning about vacuums and how to create them.

A vacuum refers to a region of space that is at less than atmospheric pressure, usually significantly less than atmospheric pressure. When dealing with vacuum pressures, the customary unit is the torr. There are 760 torr in a standard atmosphere:

$$1 \text{ atm} = 1.01325 \times 10^5 \text{ Pa} = 760 \text{ torr}. \tag{2.23}$$

Vacuum pressures are typically divided into different regions of increasingly small pressure called **low vacuum** (LV), **high vacuum** (HV) and **ultra high vacuum** (UHV) regions. Table 2.5 gives the pressure ranges of these regions.

Table 2.5. Pressure ranges for various vacuum regions

Region	Pressure (torr)
Atmospheric	760
Low vacuum (LV)	Up to 10^{-3}
High vacuum (HV)	10^{-5} to 10^{-8}
Ultra-high vacuum (UHV)	10^{-9} to 10^{-12}

Vacuum pumps

Vacuums are created using pumps that either transfer the gas from the lower pressure vacuum space to some higher pressure region (gas transfer pump), or pumps that actually capture the gas from the vacuum space (gas capture pump). In general, transfer pumps are used for high gas loads, whereas capture pumps are used to achieve ultra-high vacuums.

A common mechanical pump used to create a vacuum is the rotary sliding vane pump. Such a pump operates by trapping gas between the rotary vanes and the pump body. This gas is then compressed by an eccentrically mounted rotor. (Fig. 2.15.) The pressure of the compressed gas is higher than atmospheric pressure so that the gas is released into the surroundings.

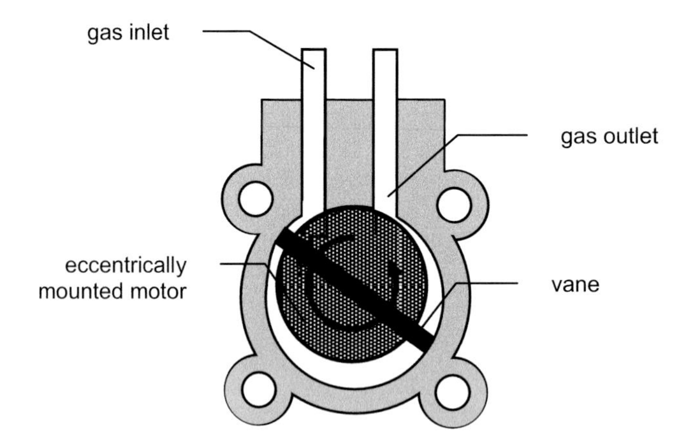

Fig. 2.15. A rotary vane pump

Rotary vane pumps are most commonly used as rough pumps, pumps used to initially lower the pressure of a vacuum chamber and to back (connected to the outlet of) other pumps. In addition to rotary vane pumps, other rough pump types include diaphragm, reciprocating piston, scroll, screw, claw, rotary piston, and rotary lobe pumps. Rough pumps are not capable of creating high vacuums and are only effective from atmospheric pressure down to 10^{-3} torr. Once a vacuum space is in this range, specialized pumps classified as high vacuum pumps are needed to achieve lower pressures.

High vacuum pumps generally work using completely different operating principles than rough pumps, which tend to be mechanical in nature. Common high vacuum pumps types include turbomolecular pumps (turbopumps), diffusion pumps and cryogenic pumps (cryopumps). Turbopumps operate by imparting momentum towards the pump outlet to trapped gas molecules. In such a pump a gas molecule will randomly enter the turbo pump and be trapped between a rotor and a stator. When the gas molecule eventually hits the spinning underside of the rotor, the rotor imparts momentum to the gas molecule, which then heads towards the exhaust. Diffusion pumps entrain gas molecules using a jet stream of hot oil. In diffusion pumps the downward moving oil jet basically knocks air molecules away from the vacuum chamber. Cryopumps trap gas molecules by condensing them on cryogenically cooled arrays.

Vacuum systems

An enclosed chamber called a vacuum chamber constitutes the environment in which PVD occurs. Such chambers are usually made of stainless steel or, in older systems, glass. To create and/or maintain a high vacuum in the vacuum chamber, a two-pump system of a rough pump and a high vacuum pump must be used, as no single pump of any type can both create and maintain a high vacuum space while operating between a high vacuum at its inlet port and atmospheric pressure at its exhaust. A generic vacuum system that may be employed in PVD is shown in Fig. 2.16.

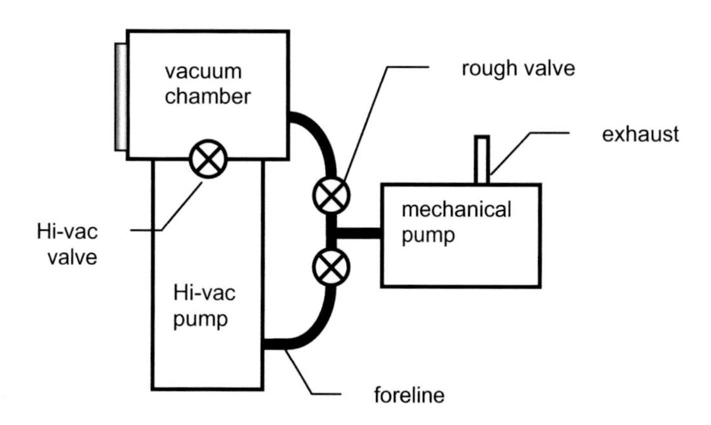

Fig. 2.16. Typical vacuum system setup in a PVD system

In Fig. 2.16 notice that although the exhaust of the rough pump is always the atmosphere, the inlet to the rough pump can be either the vacuum chamber or the exhaust of the high vacuum pump. The line connecting the exhaust of the high vacuum pump to the rough pump is called the foreline. The inlet to the high vacuum pump is the vacuum chamber itself. Operation of a vacuum system in order to first create a vacuum for PVD, perform PVD in the vacuum chamber, and then power down the system essentially involves alternately opening and closing the various valves in the correct order while running the appropriate pump(s).

Vacuum theory and relationships

Using a vacuum deposition process has many advantages essentially stemming from the fact that the number of molecules in a gas is directly related to pressure in a given volume as shown by the ideal gas equation,

$$PV = Nk_bT \tag{2.24}$$

where P is pressure, V is volume, N is the number of *molecules*, k_b is Boltzmann's constant and T is gas absolute temperature (K in the SI system).[4] And so, at the greatly reduced pressures of high vacuums there are correspondingly fewer molecules.

Another big advantage of using a vacuum process in order to deposit a thin film is the long *mean free path* of the desired material atoms. The mean free path represents the average distance a molecule travels before colliding with another molecule. In PVD the mean free path is of the same order of magnitude as the distance from the source to the substrate, and sometimes even longer. This means that the source atoms are unlikely to collide with other atoms on the way to the substrate, causing PVD's line of sight deposition characteristic. As such, a substrate that is placed in the line of sight of the source will receive most of the source atoms.

The kinetic theory of gases gives a very good estimate of the mean free path in a vacuum.[5] The mean free path of a molecule can be determined by:

$$\lambda = \frac{V}{\sqrt{2}N\sigma} = \frac{k_bT}{\sqrt{2}\sigma P} \tag{2.25}$$

where λ is the mean free path of the molecule and σ is the interaction cross section. The interaction cross section represents the likelihood of interaction between particles, and has dimensions of area.

Example 2.3 illustrates these relations.

---

[4] You may be familiar with a different form of the ideal gas equation: $PV = nR_uT$, in which n is the number of *moles* and R_u is the universal gas constant equal to 8.314 J/mol-K. In the form we use here we prefer actual number of molecules rather than moles, and Boltzmann's constant takes the place of R_u. In fact, one interpretation of Boltzmann's constant is the ideal gas constant on a per unit molecule basis.

[5] The kinetic theory models the molecules of a gas as infinitesimally small masses, therefore dictating that the only form of energy they can have is kinetic energy. Furthermore, the theory assumes that the collisions between molecules are all elastic, so that kinetic energy is conserved during the collision. The most well-known result of the kinetic theory is the ideal gas equation.

Example 2.3

Number of molecules and mean free path in air at high vacuum pressure

Estimate the number of air molecules in 1 cm^3 and the mean free path of air at room temperature at

(a) atmospheric pressure and
(b) 1×10^{-7} torr.

Take the interaction cross section to be $\sigma = 0.43$ nm^2.

Solution

(a) Let's take atmospheric pressure to be 760 torr and room temperature to be 20°C. Using the ideal gas equation and solving for N we have

$$N = \frac{PV}{k_b T} = \frac{(760\ \text{torr})(1\times10^{-6}\ \text{m}^3)}{(1.38\times10^{-23}\ \frac{\text{J}}{\text{K}})(20^o C + 273)\text{K}} \times \frac{133\ \text{Pa}}{\text{torr}} \times \frac{\text{J}}{\text{Pa}\cdot\text{m}^3}$$

$$= 2.50\times10^{19}$$

For the mean free path,

$$\lambda = \frac{k_b T}{\sqrt{2}\sigma P} = \frac{(1.38\times10^{-23}\ \frac{\text{J}}{\text{K}})(20°\text{C} + 273)\text{K}}{\sqrt{2}(0.43\times10^{-18}\ \text{m}^2)(760\ \text{torr})} \times \frac{\text{torr}}{133\ \text{Pa}} \times \frac{\text{Pa}\cdot\text{m}^3}{\text{J}}$$

$$= 6.58\times10^{-8}\ \text{m} = 65.8\ \text{nm}$$

At atmospheric pressure there are about 2.50×10^{19} molecules of air in 1 cm^3, and they tend to travel an average of a mere 66 nm before colliding with other molecules.

(b) Repeating for a pressure 1×10^{-7} torr (high vacuum region)

$$N = \frac{PV}{k_b T} = \frac{(1\times10^{-7}\ \text{torr})(1\times10^{-6}\ \text{m}^3)}{(1.38\times10^{-23}\ \frac{\text{J}}{\text{K}})(20^o C + 273)\text{K}} \times \frac{133\ \text{Pa}}{\text{torr}} \times \frac{\text{J}}{\text{Pa}\cdot\text{m}^3}$$

$$= 3.29\times10^{9}$$

Mean free path:

$$\lambda = \frac{k_b T}{\sqrt{2}\sigma P} = \frac{(1.38 \times 10^{-23} \frac{\text{J}}{\text{K}})(20^\circ \text{C} + 273)\text{K}}{\sqrt{2}(0.43 \times 10^{-18} \text{m}^2)(1 \times 10^{-7} \text{torr})} \times \frac{\text{torr}}{133 \text{ Pa}} \times \frac{\text{Pa} \cdot \text{m}^3}{\text{J}}$$

$$= 500 \text{ m}$$

We see that the number of molecules in 1 cm^3 of air has decreased by a factor of 10^{10} in the high vacuum compared to atmospheric pressure. If this high vacuum were to be used as the environment for PVD, the likelihood of impurities in the sample, especially reactive species such as oxygen, would also decrease by the same factor. Furthermore, with a mean free path of 500 m there is virtually no chance that a source atom will collide with an air molecule en route to the substrate. ◄

2.4.2 Thermal evaporation

As a physical vapor deposition process, thermal evaporation refers to the boiling off or sublimation of heated solid material in a vacuum and the subsequent condensation of that vaporized material onto a substrate. Evaporation is capable of producing very pure films at relatively fast rates on the order of μm/min.

In order to obtain a high deposition rate for the material, the vapor pressure (the pressure at which a substance vaporizes) of the source material must be above the background vacuum pressure. Though the vacuum chamber itself is usually kept at high vacuum, the local source pressure is typically in the range of 10^{-2}–10^{-1} torr. Hence, the materials used most frequently for evaporation are elements or simple oxides of elements whose vapor pressures are in the range from 1 to 10^{-2} torr at temperatures between 600°C and 1200°C. Common examples include aluminum, copper, nickel and zinc oxide. The vapor pressure requirement excludes the evaporation of heavy metals such as platinum, molybdenum, tantalum, and tungsten. In fact, evaporation crucibles, the containers that hold the source material, are often made from these hard-to-melt materials, tungsten being the most common.

The flux, or number of molecules of evaporant leaving a source surface per unit time and per unit area is given by

$$F = N_0 \exp\left(\frac{-\Phi_e}{k_b T}\right) \tag{2.26}$$

where F is the flux, Φ_e is the activation energy[6] (usually expressed in units of eV) and N_0 is a temperature dependant parameter. The utility of Eq. (2.26) is that it will tell you the relative ease or difficulty of evaporating one material versus another. It can be recast into another form:

$$F = \frac{P_v(T)}{\sqrt{2\pi M k_b T}}, \tag{2.27}$$

where $P_v(T)$ is the vapor pressure of the evaporant and M is the molecular weight of the evaporant.

We have already seen that the mean free path of a gas tends to become very large at small pressures. This has a strong correlation to the fraction of evaporant atoms that collide with residual gas atoms during evaporation. The collision rate is inversely proportional to a quantity known as the Knudsen number, given by

$$Kn = \lambda / d \tag{2.28}$$

where Kn is the Knudsen number, d is the source to substrate distance and λ is the mean free path of the residual gas. For Knudsen numbers larger than one, the mean free path of molecules is larger than the distance evaporated molecules must travel to create a thin film, and thus collision is less likely. Ideally this number is much greater than one. Table 2.6 gives the Knudsen number at various pressures for typical source-to-substrate distances of 25–70 cm. As seen in the table, high vacuum pressures are most favorable for evaporation.

Table 2.6. Typical Knudsen numbers for various vacuum pressures

Pressure (torr)	$Kn = \lambda / d$
10^{-1}	~0.01
10^{-4}	~1
10^{-5}	~10
10^{-7}	~1000
10^{-9}	~100,000

[6] We have seen the term "activation energy" once before. In general activation energy refers to the minimum amount of energy required to activate atoms or molecules to a condition in which it is equally likely that they will undergo some change, such as transport or chemical reaction, as it is that they will return to their original state. Here it is the energy required to evaporate a single molecule of the source material.

Types of evaporation

There are two basic methods to create a vapor out of the source material in evaporation. One method is by **resistive heating** in which the source material is evaporated by passing a large electrical current through a highly refractory metal structure that contains the source (such as a tungsten "boat") or through a filament. The size of the boat or filament limits the current that can be used, however, and contaminants within the boat or filament can find their way into the deposited film.

In **electron beam (e-beam) evaporation** an electron beam "gun" emits and accelerates electrons from a filament. The emitted e-beam eventually impacts the center of a crucible containing the evaporant material. The electron beam is directed to the crucible via a magnetic field, usually through an angle of 270°. (Fig. 2.17.) The crucible needs to be contained in a water-cooled hearth so that the electron beam only locally melts the material and not the crucible itself, which could contaminate the deposited film. This process usually occurs at energies less than 10 keV in order to reduce the chances that X-rays are produced, which can also damage the film.

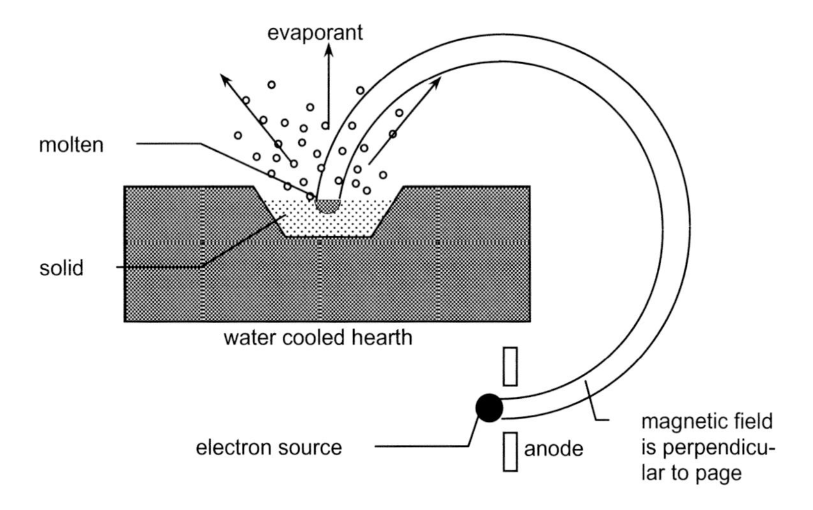

Fig. 2.17. Electron beam evaporation configuration (Adapted from Madou)

Shadowing

When we shine a flashlight, the light beam illuminates certain surfaces, but casts shadows upon others. An evaporant beam behaves much in the same way. In fact, when an evaporant is prevented from being deposited on a certain surface due to the evaporant stream's inability to "see" the surface, the phenomenon is appropriately called **shadowing**.

We have already introduced the flux F of the source material in evaporation to be the amount of material leaving a surface per unit surface area per unit time. Another quantity of interest is the arrival rate, or the amount of material incident on a surface per unit surface area per unit time. Due to the line-of-sight nature of evaporation the arrival rate is dependent on the geometry of the evaporation set-up. In reference to Fig. 2.18 the arrival rate is given by

$$A = \frac{\cos\beta\cos\theta}{d^2}F \tag{2.29}$$

where A is the arrival rate, β is the angle the substrate makes with its normal, θ is the angle the source makes with its normal and d is the distance from the source to the substrate. Equation (2.29) essentially captures how well the deposition surface can "see" the source. From Eq. (2.29) we see that where β and/or θ are large, the source does not see the substrate well and shadowing occurs. When the light-of-sight of the flux is normal to the substrate and source, however, the arrival rate achieves a maximum. Furthermore, we see that the large source to substrate distances result in lower arrival rates. It should be noted that Eq. (2.29) applies to small evaporation sources only.

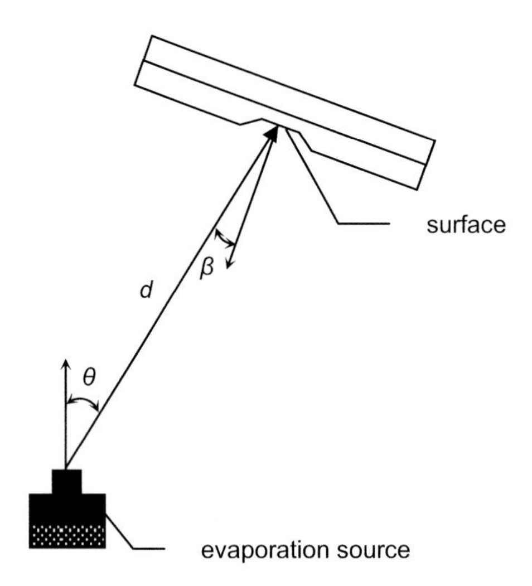

Fig. 2.18. Geometry of small source evaporation (Adapted from Madou)

This directional dependence makes it difficult to obtain uniform coatings over large surfaces areas, and particularly over topographical steps and surface features on a wafer. The ability of a technique to create films of uniform surface features is called *step coverage*. One way to gage step coverage is to compare the thickness of films at different locations on a surface. As the thickness of a film resulting from evaporation is proportional to the arrival rate, Eq. (2.29) can be used to make such a comparison. In reference to Fig. 2.19 (a), Eq. (2.29) predicts that the ratio of film thickness t_1 to film thickness t_2 is given by

$$\frac{t_1}{t_2} = \frac{\cos \beta_1}{\cos \beta_2}. \tag{2.30}$$

In this case $\beta_1 = 0$ and $\beta_2 = 60^\circ$, resulting in $t_1/t_2 = 2$. That is, the film thickness at location (1) is twice the thickness at location (2). For cases with large angles such as in Fig. 2.19 (b), shadowing can be extreme. We will see in Chapter 5 that this non-uniformity can result in unintended stress in the film, which can lead to the cracks or the film peeling away from the substrate.

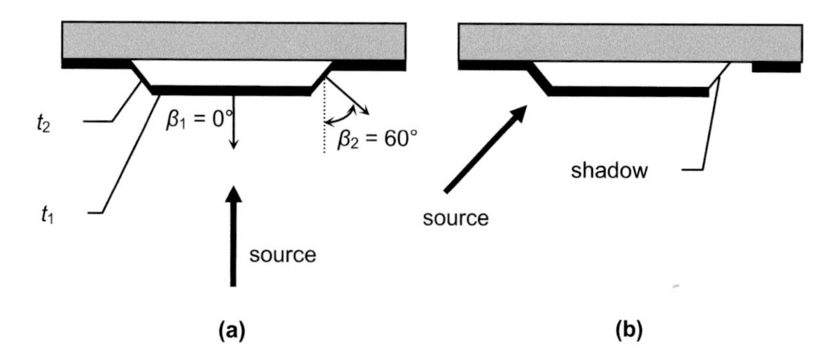

Fig. 2.19. Shadowing: (a) The film thickness on the horizontal surfaces are twice that of the sloped surfaces. (b) An extreme case of shadowing. (Adapted from Madou)

There are two major ways major to reduce shadowing. One way is continuously rotate the substrate in planetary holders during deposition to vary the angle β. Another way is to heat the substrate to 300-400°C in order to increase the ability of the deposited material to move around the surface. In a typical evaporation process both techniques are employed simultaneously.

Though shadowing can often be a problem during evaporation, the clever microfabricator can use shadowing to keep evaporated material from being deposited in undesirable locations.

2.4.3 Sputtering

In evaporation an energy source causes a material to vaporize and subsequently to be deposited on a surface. After the required energy is delivered to the source material, we can visualize the process as a gentle mist of evaporated material en route to a substrate. In sputtering, on the other hand, a target material is vaporized by bombarding it with positively charged ions, usually argon. In sputtering the high energy of the bombarding ions violently knocks the source material into the chamber. In stark contrast to a gentle mist, sputtering can be likened to throwing large massive rocks into a lake with the intent of splashing the water. (Fig. 2.20.)

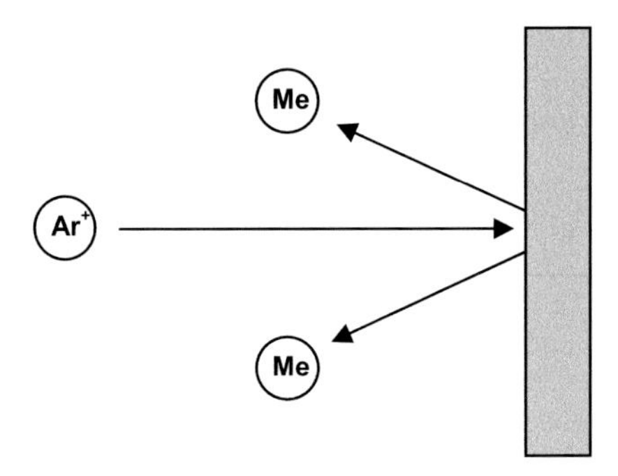

Fig. 2.20. Sputtering of metal by ionized argon

A more technical definition of sputtering is the ejection of material from a solid or molten source (the target) by kinetic energy transfer from an ionized particle. In almost all cases argon atoms are ionized and then accelerated through an electrical potential difference before bombarding the target. The ejected material "splashes" away from the cathode in a linear fashion, subsequently condensing on all surfaces in line-of-sight of the target.

The bombarding ions tend to impart much larger energies to the sputtered material compared to thermally evaporated atoms, resulting in denser film structures, better adhesion and larger grain[7] sizes than with evaporated films. Furthermore, because sputtering is not the result of thermally melting a material, virtually any material can be sputtered, and at lower temperatures than evaporation. As a result, sputtering does not require as high a vacuum as evaporation, typically occurring in the range of 10^{-2} to 10^{-1} torr. This is three to five orders of magnitude higher than evaporation pressures. However, sputtering tends to take a significantly longer period of time to occur than thermal evaporation, a relatively quick process.

The number of target atoms that are emitted per incident ion is called the *sputtering yield*. Sputtering yields range from 0.1 to 20, but tend to be around 0.5-2.5 for the most commonly sputtered materials. Typical sput-

---

[7] The silicon in a silicon substrate is an exception to most solid structures in that it is composed of one big single-crystal structure. By contrast, most deposited materials are *polycrystalline*; that is, they are made of a large number of single crystals, or *grains*, that are held together by thin layers of amorphous solid. Typical grain sizes vary from a just few nanometers to several millimeters. Indeed, when one sputters a thin film of silicon, it is usually referred to as *polysilicon*.

tering yields for some common materials are given in Table 2.7 for an argon ion kinetic energy of 600 eV.

Table 2.7. Sputtering yields for various materials at 600 eV

Material	Symbol	Sputtering yield
Aluminum	Al	1.2
Carbon	C	0.2
Gold	Au	2.8
Nickel	Ni	1.5
Silicon	Si	0.5
Silver	Ag	3.4
Tungsten	W	0.6

The amount of energy required to liberate one target atom via sputtering is 100 to 1000 times the activation energy needed for thermal evaporation. This, combined with the relatively low yields for many materials, leads to sputtering being a very energy inefficient process. Only about 0.25% of the input energy goes into the actual sputtering while the majority goes into target heating and substrate heating. Hence, deposition rates in sputtering are relatively low.

There are many methods used to accomplish the sputtering of a target. The original method is called DC sputtering (direct current sputtering) because the target is kept at a constant negative electric potential, serving as the cathode. As such, DC sputtering is limited to materials that are electrically conductive. RF sputtering (radio frequency sputtering), on the other hand, employs a time-varying target potential, thereby allowing non-conductive materials to be sputtered as well. In reactive sputtering a gas that chemically reacts with the target is introduced into the system and the products of reaction create the deposited film. To help overcome the relatively low deposition rates associated with sputtering, the use of magnets behind the target is sometimes employed in a process called magnetron sputtering. Magnetron sputtering can be used with either DC or RF sputtering. Each of these methods is described in more detail below.

DC sputtering

In DC sputtering the target material serves as a negatively charged surface, or cathode, used to accelerate positively charged ions towards it. As such, the two basic requirements for DC sputtering are that the source material be electrically conductive and that is has the ability to emit electrons. Typically the metallic vacuum chamber walls serve as the anode, but sometimes a grounded or positively biased electrode is used.

In the process of DC sputtering, an inert gas, typically argon, is introduced into a vacuum chamber. The gas must first be ionized before sputtering can begin. (An ionized gas is usually referred to as a *plasma*.) At first random events will cause small numbers of positive argon ions to form. The Ar^+ ions sufficiently close to the target will be accelerated towards the cathode, while the electrons near the anode are accelerated towards the anode. As the Ar^+ ions collide with the target, the target material, electrons and X-rays are all ejected. The target itself is heated as well. The ejected electrons are accelerated away from the electrode and into the argon gas, causing more ionization of argon. The sputtering process eventually becomes self-sustaining, with a sputtering plasma containing a near-equilibrium number of positively ionized particles and negatively charged electrons. Figure 2.21 shows a DC sputtering configuration.

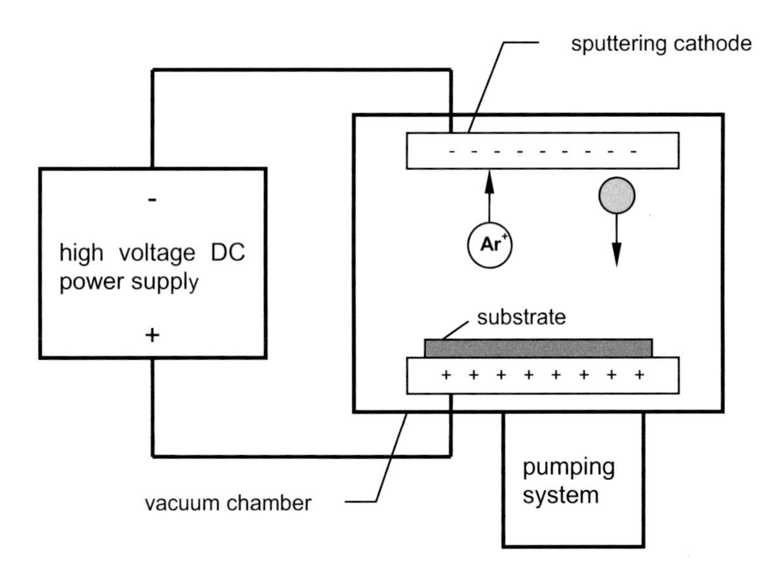

Fig. 2.21. A typical DC sputtering configuration

RF Sputtering

RF sputtering removes the requirement that the target material be electrically conductive. In RF sputtering the electric potential applied to the target alternates from positive to negative at high enough frequencies, typically greater than 50 kHz, so that electrons can directly ionize the gas atoms. This scheme works because the walls of the vacuum chamber and the target form one big electrical capacitor, and the ionized gas in the region next to the target and the target itself form another small capacitor contained in the larger one. The two capacitors thus have some capacitance in common, allowing for the transfer of energy between them in what is known as *capacitive coupling*. A result of this coupling, however, is that sputtering occurs both at the target and at the walls of the chamber. Luckily, the relative amount of sputtering occurring at the walls compared to the target is correlated to the ratio of wall to target areas. Since the area of the target is much smaller than the wall area, the majority of sputtering is of the target material.

Reactive Sputtering

Reactive sputtering is a method to produce a compound film from a metal or metal alloy target. For example, an aluminum oxide film can be deposited in reactive sputtering by making use of an aluminum target. In reactive sputtering a reactive gas, such as oxygen or nitrogen, must also be added to the inert gas (argon). The products of a chemical reaction of the target material and the gas form the deposited material. (The occurrence of a chemical reaction is a trait reactive sputtering has in common with **chemical vapor deposition**, discussed in the next section.). Formation of the reactive compounds may occur not only on the intended surface, but also on the target and within the gas itself.

The amount of reactive gas is very important issue in reactive sputtering, since excessive reactive gas tends to reduce the already low sputtering rates of most materials. A fine balance must be made between reactive gas flow and sputtering rate. Furthermore, the behavior of the plasma in reactive sputtering is quite complicated, as the reactive gas ionizes along with the inert gas. Hence, drastically different deposition rates may result for a metal oxide compared to the metal by itself. A number of approaches used to avoid the reduction in sputtering rate for reactive sputtering are detailed in the literature.

Magnetron sputtering

Sputtering processes do not typically result in high deposition rates. However, the use of magnetic discharge confinement can drastically change this. Magnets carefully arranged behind the target generate fields that trap electrons close to the target. Electrons ejected from the target feel both an electric force due the negative potential of the target and a magnetic force due to the magnets placed behind the target given by the Lorentz force

$$\vec{F} = q(\vec{E} + \vec{v} \times \vec{B})\,. \tag{2.31}$$

With properly placed magnets of the correct strength, electrons can only go so far away from the target before turning around and hitting the target again, creating an electron "hopping" behavior. The path length of the electrons increases near the cathode improving the ionization of the gas near the cathode. This local increase in ionization produces more ions that can then be accelerated toward the cathode, which in turn provides higher sputtering rates. Figure 2.22 shows a typical magnetron sputtering geometry and the resulting electron path.

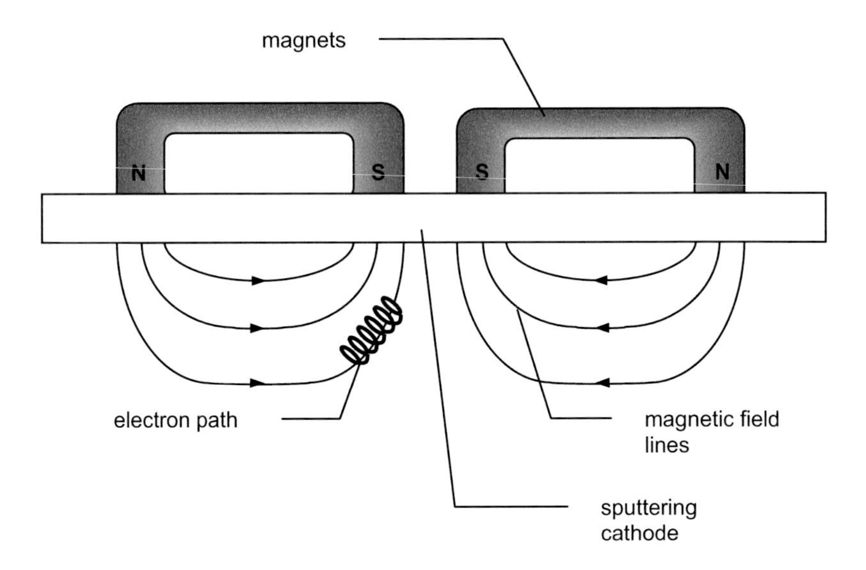

Fig. 2.22. Magnetron principle

2.5 Other additive techniques

2.5.1 Chemical vapor deposition

In chemical vapor deposition, or simply CVD, a chemical reaction takes place in order to deposit high-purity thin films onto a surface. In CVD processes, a surface is first exposed to one or more volatile **precursors**, vapors containing the to-be-deposited material in a different chemical form. A chemical reaction then takes place, depositing the desired material onto the surface, and also creating gaseous byproducts which are then removed via gas flow. (Fig. 2.23.) A common example of CVD is the deposition of silicon films using a silane (SiH_4) precursor. The silane decomposes forming a thin film of silicon on the surface with gaseous hydrogen as a byproduct.

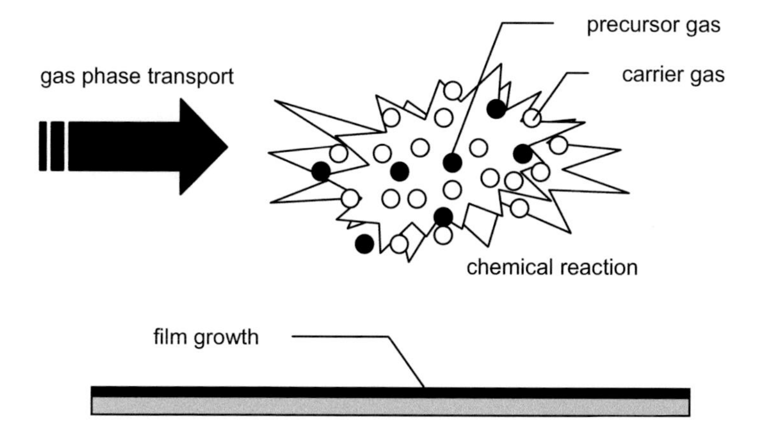

Fig. 2.23. Basic chemical vapor deposition process

Often CVD processes result in the formation of volatile and/or hazardous byproducts, such as HCl in the deposition of silicon nitride (Si_3N_4) using dichlorosilane (SiH_2Cl_2) and ammonia (NH_3) precursors. This hazardous waste is one of the disadvantages of the CVD technique as compared to physical vapor deposition methods. However, CVD is not a line-of-sight deposition method, and thus offers excellent step coverage and greatly improved uniformity over PVD. However, temperatures ranging from 500°-850°C are often required for CVD, making it impossible to use with silicon surfaces already deposited with certain metals such as aluminum or gold, as eutectics[8] may form.

There are several methods of enhancing the deposition rates during CVD. These include employing low pressures (LPCVD - low pressure CVD) as well as the use of plasmas (PECVD – plasma enhanced CVD).

2.5.2 Electrodeposition

Also called electroplating, **electrodeposition** refers to the electrochemical process of depositing metal ions in solution onto a substrate. It is commonly used to deposit copper and magnetic materials in MEMS devices.

Surface quality of electrodeposited films tends to be worse than films deposited via physical vapor deposition, exhibiting a higher degree of roughness. Uniformity of the films can also be an issue, but can be improved by careful control of applied electric current.

2.5.3 Spin casting

Some materials, particularly polymers, can be added to a substrate by dissolving them in solution, applying them to a wafer, and then spinning the wafer to distribute the solution across the surface via centrifugal force. Afterwards the wafer is baked to remove the solvent, leaving behind the thin film. The process is also commonly called simply **spinning**.

Gelatinous networks of colloidal suspensions containing solid polymer particles, called sol-gels, are often spin cast. Spin casting is also the standard technique for applying photoresist to wafers. It is discussed in more detail in the next chapter.

2.5.4 Wafer bonding

Perhaps the ultimate additive technique is to add an entire wafer to another one. This technique is used in the fabrication of some MEMS devices, but more commonly it is used to create a protective enclosure for a MEMS, or to **package** it.

---

[8] The term eutectic refers to an alloy of two or more metals whose melting point is lower than that of any other alloy composed of the same constituents in different proportions. In short, by combining silicon with the right proportions of aluminum or gold, the resulting interface can potentially start melting at significantly lower temperatures than any of the materials by themselves.

Silicon wafers can be bonded to one another via a high temperature (~1000°C) anneal, fusing the wafers together. Silicon wafers can also be bonded to 7740 Pyrex glass at temperatures of about 500°C if a voltage is simultaneously applied across the wafer stack. Such a process is called *anodic bonding*, with typical voltages on the order of 400-700 V. The Pyrex is required for the bonding since its coefficient of thermal expansion closely matches that of silicon. Otherwise the glass could shatter upon cooling.

Materials such as adhesives and low melting point solders are sometimes used to bond wafers in cases where higher temperature methods cannot be employed.

Essay: Silicon Ingot Manufacturing

Patrick Ferro
Department of Mechanical Engineering, Gonzaga University

For reliable operation of electronic devices, the chips upon which they are fabricated must be free of atomic-level defects. The precursor to defect-free devices is single-crystal silicon, grown as an ingot or boule.

Single-crystal silicon ingots are grown via a controlled, rotating withdrawal process known as the Czcochralski technique. Sometimes practitioners abbreviate Czochralski-grown silicon as 'CZ silicon'.

The equipment used to grow silicon ingots includes a resistance furnace which supplies radiant heat to a rotating quartz crucible. The resistance furnace and rotating crucible are contained within a vacuum chamber that has at least one interlock chamber and a capability for inert gas backfill.

The process starts by melting chunks of very pure silicon (impurities are measured at parts per million level) in a quartz crucible, along with controlled amounts of dopant elements. The crucible is slowly rotated in one direction at a specific rotational velocity (e.g. 10 rpm) and slowly heated until the silicon within it is molten. The rotating, molten silicon-filled crucible is stabilized at a particular temperature above the liquidus, or temperature above which crystals can no longer coexist with the melt in a homogeneous state. A counter-rotating crystal seed is then slowly lowered to touch the surface of the melt.

As the seed makes contact with the surface of the molten silicon, a temperature gradient is created. At this critical point in the process, the intention is to initiate and sustain the growth of a single grain, or crystal, of silicon. In practical terms, the operator will begin slowly withdrawing the rotating seed away from the counter-rotating melt, to pull a thin 'neck' of silicon up and out of the melt. The withdrawal rate is process-dependent and is designed to sustain the growth of a neck that is free of defects, called dislocations, within the crystal structure. The neck diameter is similar to that of a pencil, and the withdrawal rate is measured in only millimeters per minute. The neck-growth stage of the process continues until a neck of approximately 10 cm has been grown.

The process continues by temporarily decreasing the withdrawal rate to increase the diameter of the ingot. Once the diameter is at its desired dimension, the withdrawal rate is stabilized and the ingot is allowed to grow. The withdrawal rate is on the order of millimeters per minute and the seed and crucible rotation rates are on the order of 10 to 20 rpm. The transition between the neck and the ingot body is known as the 'shoulder'. Lines, or ridges, at a regular spacing can be seen in the radial direction on the shoulder and on the body of an ingot due to the crystallinity of the silicon. If dislocations develop in the silicon during the body-growth phase of the process, the appearance of these lines is affected. If dislocations are observed during the process, the partially-grown ingot will be scrapped and remelted. In a silicon ingot manufacturing plant, productivity is closely monitored because of the long processing times involved. For example, one good ingot may require more than twenty four hours of furnace time.

Reducing the onset of dislocation in ingots requires careful attention to changes in withdrawal rate and rotational speeds as well as minimizing the effects of vibrations, tramp element levels, vacuum and other process parameters. Because the requirements are so stringent, even secondary parameters such as the bubble morphology in the quartz crucibles may be carefully monitored.

A well-controlled process allows for the sustained growth of a body of single-crystal silicon. The growth of the body continues until the crucible is completely empty of molten silicon. A completed ingot, or boule, will have a tapered point at the bottom indicating that the entire crucible of silicon was pulled out. A partially grown ingot cannot be rapidly pulled out, since the shock of a sudden withdrawal would propagate through the length of the partially-grown ingot, rendering it unusable.

Solidification, in general and for all materials, proceeds in the direction of the highest thermal gradient. It is critically important during the growth of a silicon ingot to control and maintain a high thermal gradient at the solidification front. Furnace design, including the use of baffles, can help in the control of the thermal gradient. Also, the process parameters including withdrawal rate, rotational speed, and chamber pressure affect the thermal gradient.

The CZ process is relatively slow, and is nearly an art form. Some parts of the process may be automated, but still require a relatively high degree of involvement from skilled operators and process engineers. A silicon wafer manufacturing plant may have twenty or more CZ silicon furnaces to meet the continuous demand.

Dislocation-free and chemically pure silicon ingots allow for defect-free wafers, upon which both electronic devices and MEMS can be reliably fabricated.

References and suggested reading

Alciatore DG, Histand MB (2007) Introduction to Mechatronics and Measurement Systems, 3rd edn. McGraw Hill, Madison, WI

"Substrate Manufacture: Single Crystal Ingot Growth" U.S. Department of Labor Occupational Safety & Health Administration http://www.osha.gov/SLTC/semiconductors/substratemfg/snglcrystlingtgrowth.html

Franssila S (2004) Introduction to Microfabrication. Wiley, Chichester, West Sussex, England

Griffiths D (1999) Electrodynamics. In: Reeves A (ed) Introduction to Electrodynamics, 3rd edn. Prentice Hall, Upper Saddle River, NJ

Jaeger RC (2002) Introduction to Microelectronic Fabrication, 2nd edn. Prentice Hall, Upper Saddle River, NJ

Madou MJ (2002) Fundamentals of Microfabrication, The Art and Science of Miniaturization, 2nd edn. CRC Press, New York

Maluf M, Williams K (2004) An Introduction to Microelectromechanical Systems Engineering, 2nd edn. Artech House, Norwood, MA

Peeters E (1994) Process Development for 3D Silicon Microstructures with Application to Mechanical Sensor Design. Ph.D. thesis, Catholic University of Louvain, Belgium

Senturia S (2001) Microsystem Design. Kluwer Academic Publishers, Boston

Serway RA (1998) Principles of Physics, 2nd edn. Saunders College Pub., Fort Worth, TX

Vossen JL, Kern W (1978) Thin Film Processes, Academic Press, New York

"Sputtering Yields Reference Guide" *Angstrom Sciences* http://www.angstromsciences.com/reference-guides/sputtering-yields/index.html

Wolf S, Tauber RN (2000) Silicon Processing for the VLSI Era, Vol.1: Process Technology, 2nd edn. Lattice Press, Sunset Beach, CA

Questions and problems

2.1 Silicon is the most common material out of which substrates are made. Find at least one other material used for substrates. What are its advantages and/or disadvantages over silicon?
2.2 What are three reasons you might add a layer to a silicon substrate in making a MEMS?
2.3 Would you expect a silicon wafer doped with antimony to be a p-type or n-type wafer? Why? What about a silicon wafer doped with gallium?
2.4 Give three applications of a p-n junction.
2.5 What is shadowing? Is shadowing a good or bad thing? Why?
2.6 Why is a vacuum needed in physical vapor deposition?
2.7 Calculate the mean free path of an air molecule for a pressure of 10^{-6} torr. Also calculate the Knudsen number for an evaporation set-up for which the source to substrate distance is d = 50 cm. Is this an adequate pressure to use for this evaporation? Why or why not?
2.8 Determine the Miller indices for plane shown in the figure.

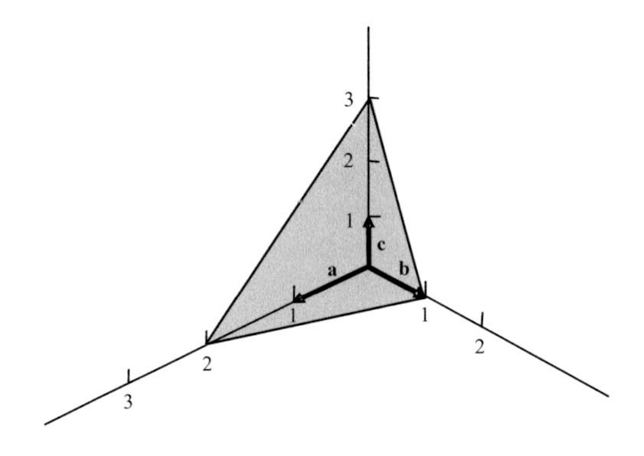

2.9 The lattice constant for Si, a cubic material, is a = 5.43 Å. Determine the distance between adjacent planes for both the [100] and [111] directions.
2.10 The angle between two crystal directions in a cubic material is given by

$$\cos\theta = \frac{h_1 h_2 + k_1 k_2 + l_1 l_2}{\sqrt{\left(h_1^2 + k_1^2 + l_1^2\right)\left(h_2^2 + k_2^2 + l_2^2\right)}}$$

The direction $[h\ k\ l]$ of the line produced by the intersection of planes $(h_1\ k_1\ l_1)$ and $(h_2\ k_2\ l_2)$, is given by

$$h = k_1 l_2 - k_2 l_1$$
$$k = h_1 l_2 - h_2 l_1$$
$$l = h_1 k_2 - h_2 k_1$$

Find the angle between the crystal directions [111] and [110]. Find the direction of the line produced by the intersection of these two planes.

2.11 An oxide layer is to be grown on a (111) silicon wafer that is initially 300 μm thick. After one hour of wet oxidation the oxide layer is estimated to be 300 nm thick. What is the new thickness of the wafer?

2.12 A (100) silicon wafer has an initial 100 nm of oxide on its surface.

a. How long will it take to grow an additional 600 nm of oxide using wet oxygen method at 1100°C?
b. Graph Eq. (2.18), i.e. plot time vs. oxide thickness and then look up the time required to grow 600 nm of oxide. Comment on the result.
c. What is the color of the final oxide under vertical illumination with white light? (Hint: Many of the references for this chapter have "color charts" that indicate oxide thickness as a function of apparent color.)

2.13 A p-type Si-wafer with background doping concentration of 1.4×10^{15} cm^{-3} is doped by ion implantation with a dose of phosphorus atoms of 10^{16} cm^{-2}, located on the surface of the wafer. Next thermal diffusion is used for the drive-in of phosphorous atoms into the p-type water (the anneal step). The wafer is then annealed at 1000°C for 3 hours.

a. What is the diffusion constant of phosphorous atoms at this anneal temperature?
b. What is the junction depth after the drive-in anneal?
c. What is the surface concentration after the drive-in anneal?

Chapter 3. Creating and transferring patterns—Photolithography

3.1 Introduction

We have learned much about the substrate and ways of adding layers to it. We next turn our attention to creating and transferring patterns onto the substrate and/or these layers. In so doing we lay the foundation of creating actual structures with specific geometries. The primary tool used to accomplish this is **photolithography**, which is essentially an optical printing process.

Figure 3.1 illustrates the steps in transferring a pattern using photolithography. Figure 3.1 (a) shows a wafer covered with a thin film, which in turn is covered with a layer of photosensitive material called **photoresist**. When exposed to light, photoresist becomes either less soluble in the case of a **negative resist**, or it becomes more soluble in the case of a **positive resist**. By use of a transparent plate with selective opaque regions, called a **mask**, only certain regions of the resist are exposed to light. Figure 3.1 (a) shows a mask in proximity to the wafer, whereas Fig. 3.1 (b) illustrates the exposure step. Next, a chemical called a **developer** is brought in contact with the exposed photoresist layer. The developer chemically attacks the photoresist, but it attacks the exposed and unexposed regions at much different rates. Which region is attacked more aggressively depends on whether the resist is **negative** or **positive**, but either way a pattern of photoresist is left behind on the wafer. In Fig. 3.1 a negative resist is used, so that the unexposed regions are attacked, leaving a window in the resist to the layer below. (Fig. 3.1 (c).) The shape of this window is the same as the opaque region of the mask, in this case a simple square as shown in a top view of the mask in Fig. 3.1 (d).

T.M. Adams, R.A. Layton, *Introductory MEMS: Fabrication and Applications*,
DOI 10.1007/978-0-387-09511-0_3, © Springer Science+Business Media, LLC 2010

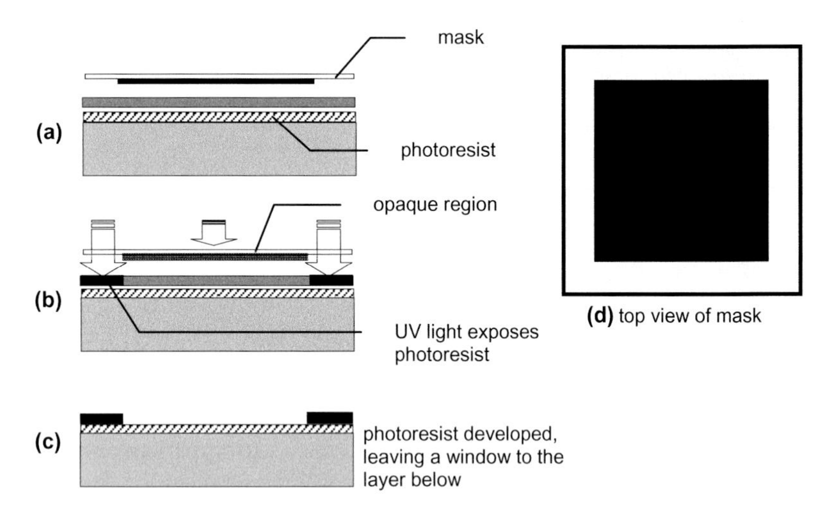

Fig. 3.1. Steps in a simple photolithographic process

In all microfabrication techniques cleanliness is of paramount importance. This is particularly true in photolithography, as a single unwanted speck of dust can have dire consequences on the photolithographic process, and thus the success of a design. As such, we will first discuss **clean rooms**, spaces in which most microfabrication processes take place, as well as some material cleaning techniques. We then turn our attention to the details of the aforementioned photolithography materials and processes.

3.2 Keeping it clean

Cleanliness is extremely important when fabricating micro-sized devices. When the devices we are creating have characteristic dimensions on the order of microns to hundreds of microns, an unwanted piece of matter with similar characteristic dimensions can spell disaster. The main culprit tends to be dust. Dust particles in the air can settle on substrates causing defects in the device structure. Cleanliness is probably most crucial during the photolithography steps, as when dust particles adhere to the surface of a photomask, they behave as unwanted opaque patterns on the mask. Thus, unintended patterns are transferred to the wafer.

The major way to combat this problem is to fabricate MEMS in a **clean room**, a space intentionally kept at a certain level of cleanliness. HEPA (high efficiency particulate air) filters are commonly used to control the contaminant content in clean rooms. The temperature and humidity of clean rooms are also tightly controlled, and the environmental pressure is typically kept higher than surrounding rooms.

Clean rooms are classified according to how many (or few, really) particles of a certain size can be found within a certain volume. In a class 1000 clean room, for example, only 1000 or fewer particles of size one-half micron or larger exist within a cubic foot. For a class 100, only 100 or fewer particles of this size are present per cubic foot. For a class 10, it is ten particles, and so on. Particles of smaller than one-half micron tend to exist in larger quantities, whereas larger particles exist in smaller quantities. Figure 3.2 gives the particle size distribution for the various clean room classes.[1]

Extra special consideration goes into the construction and maintenance of a clean room. Clean rooms are rarely constructed near smokestacks, for example, and clean room floors must be conductive in order to account for electrostatic discharge. Only proper attire, popularly referred to as a "bunny suit," is allowed in a clean room. The suits do not protect the user from mishaps, but rather they protect work pieces from the user, as the main source of airborne dust is human skin. (Fig. 3.3.) Only certain types of furniture are allowed, as is only specially designed stationary, which is to be used with pens rather than pencils. Certainly eating and drinking are not allowed in clean rooms, and the wearing of perfume, cologne and makeup is discouraged.

In addition to working in a clean environment, wafers themselves typically need to be cleaned in order to remove contaminants. One of the most widely used and effective methods for cleaning silicon is a two step process called an **RCA clean**, so named because it was developed by Werner Kern in 1965 while working at RCA Laboratories. In the first step a 1:1:5 to 1:1:7 by volume solution of NH_4OH : H_2O_2 : H_2O is used to remove organic contaminants and heavy metals from the surface. Next, a solution of HCl: H_2O_2: H_2O in a 1:1:5 to 1:2:8 volume ratio is used to remove aluminum, magnesium, and light alkali ions. Each step is carried out for ap-

[1] This classification is that of US FED STD 209E, which was actually cancelled in 2001. However, the terminology is still the most commonly encountered in this country. The International Organization for Standardization classification of clean rooms (ISO 14644-1) is based on the number of 0.1 μm-sized particles within one cubic meter, a class ISO 1 being fewer than 10, ISO 2 being fewer than 100 and so on.

proximately 20 minutes while gently heating to 75-85°C on a hot plate. Other chemical cleaning techniques, sometimes with such ominous names as "piranha clean", are also used to rid surfaces of contaminants.

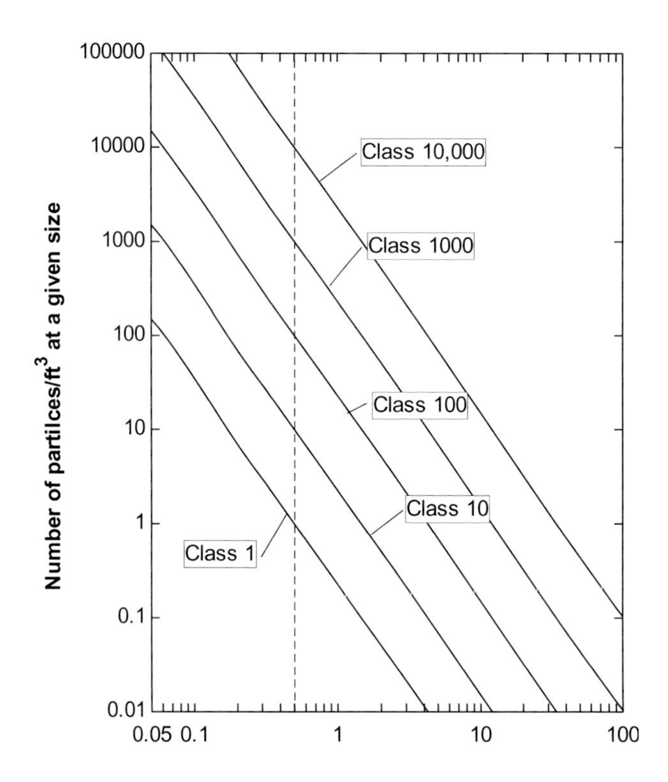

Fig. 3.2. Standards for clean room classification

Fig. 3.3. Only proper clean room attire must be worn while in a clean room.

3.3 Photoresist

The real stuff of photolithography is **photoresist**, sometimes simply called resist. Photoresist is made up of three components: a *base resin* (polymer) that gives the resist its structural properties, a *photoactive compound* (PAC), which is the light-sensitive component, and a *solvent*. The photoactive compound in resist is usually sensitive to a narrow band of ultraviolet (UV) light, and sometimes the post-exposure products generated by the reaction of UV light with the PAC are included as a resist component.

3.3.1 Positive resist

When a **positive resist** is exposed to UV light, it becomes more soluble to the developer. Hence, the unexposed regions of the resist are left behind after development, and the developed resist pattern is identical to the mask pattern. Exposure to light degrades the PAC allowing the resin to become readily soluble in alkalis such as NaOH or KOH, both of which are often used as developers. Positive resist tends to be very sensitive to the UV light with a wavelength of 365 nm, called the I-line of the mercury spectrum. Typically a one minute exposure is necessary for a 1-μm thick film using a Hg-Xe arc lamp as the UV source. Typical resist thicknesses are from 1 to 3 μm thick layers when using positive resist.

3.3.2 Negative resist

In a **negative resist**, exposure to UV radiation decreases solubility. Therefore the exposed regions of the resist are left behind after development, and the developed resist pattern is the *negative* of the mask pattern. (Hence the name negative resist.) The solubility decrease is accomplished either by an increase in molecular weight of the resist through UV exposure, or by a photochemical transformation of the resist to form new insoluble products. Negative resists are roughly ten times more sensitive than even the best positive resists and therefore require much shorter exposure times. They are also typically sensitive to UV light with a wavelength of 405 nm - the H-line - rather than the I-line of the typical positive resist.

Figure 3.4 shows the results of pattern transfer using positive and negative resists. In both cases the same mask in used.

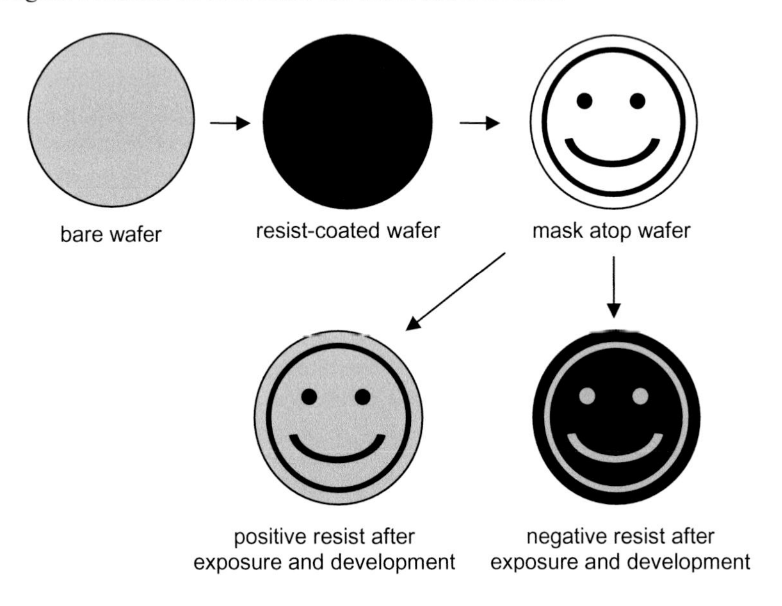

Fig. 3.4. Pattern transfer using positive and negative resists

3.4 Working with resist

3.4.1 Applying photoresist

When applying photoresist to a surface, there are three main requirements. First, the resist should adhere well to the surface; second, the thickness of the resist layer should be as close to uniform across the wafer as possible; and third, the applied thickness should be predictable from wafer to wafer.

To tackle the first requirement, a couple of strategies are employed. In many cases, the layer beneath the applied photoresist will be an oxide layer. SiO_2 tends to adsorb water from the atmosphere (molecular water), and form silanol groups (Si-OH) as well, both of which can contribute to poor resist adhesion. The molecular water can be effectively reduced by a *pre-bake* in which the wafer is placed in an oven at 200-250°C for roughly 30 minutes. Readsorption of water can begin in 30-60 minutes after the pre-bake, necessitating that the wafers be promptly coated. After baking, substrates are often treated with an adhesion promoter, such as hexamethyl disilazane (HMDS) for SiO_2 surfaces. Adhesion promoters undergo chemical reactions with the substrate to help remove any remaining molecular water from the surface.

Photoresist can either be spun or sprayed onto a wafer surface. In order to address uniform and predictable resist thicknesses, photoresists are usually spun, however. To apply the resist, the wafer is first mounted onto a vacuum chuck. Resist in a liquid form is applied to the center of the wafer, and the wafer is then spun at angular speeds ranging from 1500 to 8000 rpm. Typical spin times range from 10 to 60 seconds.

The resulting thickness of the film depends on the viscosity of the resist (η), the polymer concentration in the resist (C), and the angular spin speed (ω). An empirical relation often used to predict the film thickness is given by

$$T = \frac{KC^{\beta}\eta^{\gamma}}{\omega^{\alpha}} \tag{3.1}$$

where T is the film thickness and K, α, β, and γ are experimentally determined constants that vary from system to system. Typical values of α, β and γ are ½, 1 and 1, respectively.

Different devices require different resist thicknesses. For integrated circuit (IC) fabrication, a film thickness of the order of a micron or smaller may suffice. MEMS devices with large aspect ratio surface features may require much thicker layers. In some cases one may need to use a casting technique and/or the application of thick, dry sheets of photoresist to achieve the required thickness.

Although ultraclean conditions should be maintained during the entire procedure, the coating step poses the largest danger from a dust contamination point of view. This is because the spinning creates a slight vacuum along the plane of the wafer thereby promoting the delivery of any airborne particles of dust to its surface.

After the resist has been applied, a *post-apply bake* is typically performed at 75-100°C for approximately 10 to 20 minutes. This step effectively removes most of the remaining solvent and relieves any stress resulting from the spinning process. This step also further improves adhesion of the film to the substrate.

3.4.2 Exposure and pattern transfer

Once the photoresist has been adequately prepared, the wafer is ready for mounting in either a *contact aligner* (or standard *mask aligner*) or a *projection printer* in order to transfer the pattern from the mask. Once mounted, the mask and the wafer onto which the pattern is to be transferred first need to be carefully aligned with each other in order to transfer the pattern to the correct location on the wafer. (We will talk more about alignment later.) At this point the photoresist is ready to be exposed.

We have seen that photoresists are manufactured to absorb ultraviolet light of specific wavelengths. In most cases, the UV light is monochromatic, or occurring at only one wavelength. Most modern optical transfer equipment uses a high-pressure mercury-xenon vapor lamp to produce the UV light. (Incidentally, the clean rooms in which photolithography is performed have a characteristic yellowish hue to them as a result of filtering out the UV portion of the spectrum in the ambient lighting. This is to prevent unintended exposure of photoresist.) Ultraviolet light is usually divided into four regions. In order of decreasing wavelength these are called near UV, UV, deep UV, and extreme UV. Table 3.1 lists one classification of ultraviolet light based on wavelength, though the exact wavelength range of each of these regions is somewhat arbitrary.

Table 3.1 Ultraviolet light regions

Region name	Wavelength (nm)
Near UV	330-450
UV	260-330
Deep UV	200-260
Extreme UV	10-14

The emitted radiation from an Hg or Hg-Xe lamp has peaks at certain wavelengths and gives off relatively little radiation at others. (Fig. 3.5.) These peaks are commonly referred to as *spectral lines*, some of which are in the visible range. Many of these lines are identified by letters, including the E-line (546 nm), the G-line (436 nm), the H-line (405 nm), and I-line (365nm).

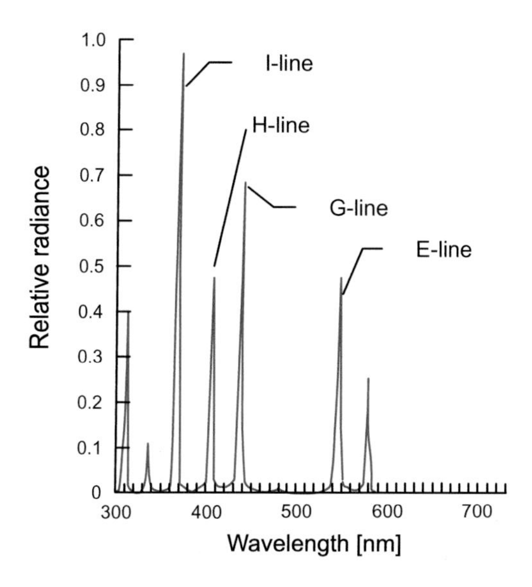

Fig. 3.5. Emission spectrum for an Hg-Xe source

In general smaller wavelengths result in a better **resolution**, which in this context refers to the smallest distinguishable feature size of a transferred pattern. For this reason the I-line is widely used in the semiconductor industry, as small feature sizes are a must. When dealing with short wavelengths, however, less energy is available to expose the photoresist, much of it being absorbed by the optical equipment. And so, for very short wavelengths one must be sure to use a resist with sufficient sensitivity or employ another light source capable of supplying higher energies at the required wavelength. Excimer lasers are often used for this purpose.

During exposure the term *aerial image* (image in air) is often used to refer to the image of the unmodified mask pattern projected onto the surface of the photoresist by an optical system. The reproduction of the aerial image in the resist layer itself as a spatial variation of chemical species is called the *latent image*. During exposure, the resist absorbs a portion of the entering light energy in order to form this latent image. As the latent image is formed, the component in the resist that absorbs the light also bleaches, making it transparent. Typically as the resist absorbs more than 40% of the incident energy image profile suffers degradation. However, too small an absorbency results in unacceptably long exposure times. The goal naturally is to have the latent image resemble as well as possible the original mask pattern.

A certain minimum amount of optical energy is required to expose the resist so that it completely develops away (positive) or remains (negative). The term **dose to clear** (D_p for positive resist and D_g^o for negative) denotes this amount of energy. Specifically, the dose to clear is the amount of energy *per unit surface area* required to expose a layer of photoresist down to the layer beneath it. The dose to clear is a function of the photoresist itself, information about which is supplied by the resist manufacturer. The actual amount of energy per unit surface area seen by the wafer during exposure is simply termed the **dose**, D. In practice typical doses are on the order of 10-100 mJ/cm^2. The energy delivered to the surface during exposure comes in the form of **light intensity**, I, which has dimensions of power per unit surface area. And so, to find the dose one need only multiply the intensity by the exposure time.

$$D = It_{exposure}, \tag{3.2}$$

where $t_{exposure}$ is the exposure time. Careful adjustment of the actual dose relative to the dose to clear is needed to create the sharpest images within the resist layer.

After the resist is exposed it undergoes another bake, a *post-exposure bake*. The purpose of this is to further densify the resist so as to reduce the dissolution rate of its undissolved regions. In addition, it improves resist adhesion and prevents its undercutting during development. It also toughens the resist. Another important benefit of the post-exposure bake is that it reduces striations caused by standing waves by redistributing the photo active compound which has been destroyed by optical radiation. The hard bake is usually done at a temperature of 100-120°C.

Contact printing

There are various methods used to transfer the mask pattern to the wafer during exposure. In **contact printing** (also called hard contact) the mask is in physical contact with the photoresist layer, and is strongly pressed against it. (Fig. 3.6 (a).) The method has the advantage of transferring patterns fairly true to the original mask as well as being relatively simple. The major drawback of this method is that the masks can only be used a few times before the patterns essentially rub off, thereby propagating defects with future use. This is especially true with masks that make use of gelatin, a material which tends to degrade quickly. Metal coated masks tend to last a bit longer with contact printing, but particular attention must be paid to keeping the masks clean, as dust and small particles of resist will adhere to the metal surface. As a result of all of these considerations, contact printing tends to be relatively low yield process.

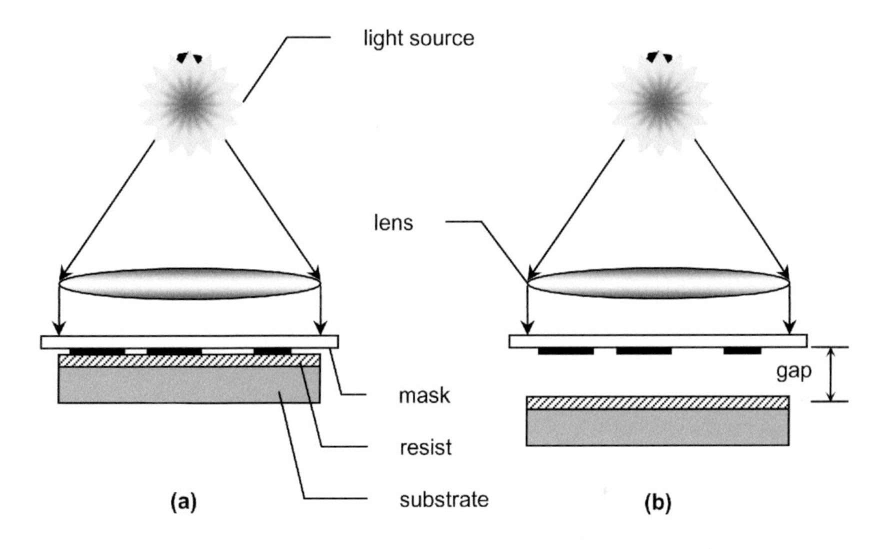

Fig. 3.6. Shadow printing methods: (a) Contact printing; (b) Proximity printing

Proximity printing

Proximity printing was developed to reduce problems encountered during contact printing. In proximity printing the mask is *not* pressed against the wafer, but rather, a small gap of 10 to 50 μm is left between the mask pattern and the wafer. (Fig. 3.6 (b).) Masks therefore last for a much longer time with proximity printing. The gap introduces diffraction, however, resulting in poorer resolution than in contact printing.

Projection printing

Both contact and proximity printing are called **shadow printing** methods because both rely on the opaque and transparent regions of the mask to directly transfer the mask pattern to the resist. In essence, the combination of the light source and the mask cast a shadow on the resist. In **projection printing**, by contrast (no pun intended), the mask image finds its way to the wafer by means of a projection system containing numerous optical components. Once mounted in the projection system, mask lifetime is virtually unlimited, banning operator mishandling that is. One big advantage of projection printing over shadow printing methods is its ability to change the size of the projected image via the lenses contained in the system. Typically the mask image pattern is reduced by a factor of 4× to 10× on the wafer.

Figure 3.7 shows a generic projection system. It consists of a light source, a condenser lens, the mask, the objective lens, and finally, the resist coated wafer. Though pictured as a single lens, the condenser lens may actually be a system of glass lenses (refractive elements) and/or mirrors (reflective elements). The combination of the light source and the condenser lens is called the illumination system, the purpose of which is to deliver light to the mask uniformly and with good intensity. After light leaves the illumination system, it passes through the clear areas of the mask and diffracts on its way to the objective lens. The objective lens then picks up a portion of the diffraction pattern and projects an image onto the wafer. With proper spectral characteristics, directionality and so forth, the projected image well resembles the mask pattern. Achieving this requires much fine tuning, however, and expensive optical components as well, making projection printing an expensive endeavor. The light is normally monochromatic but can be polychromatic.

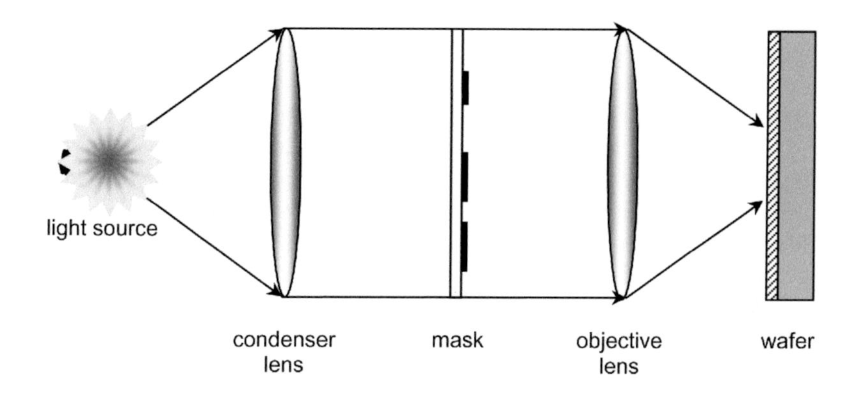

Fig. 3.7. A generic lithographic projection system

Just as the condenser lens may actually be many optical components, the objective lens is often a combination of twenty or more lenses, the purpose of which is to reduce any geometrical aberrations to a point where their effect on the image is negligible. Producing high quality images therefore results in lenses that are very expensive and heavy.

When using a single lens for imaging, the simple thin-lens Gaussian formula is often used:

$$\frac{1}{s}+\frac{1}{s'}=\frac{1}{f}, \tag{3.3}$$

where s is the object distance from the lens, s' is the image distance, and f is the focal length of the thin lens. In the end, the fundamental limit to how well optical techniques will work in transmitting a high quality image to the resist is dictated by the wavelength of light; i.e., the diffraction limit. The use of smaller wavelengths such as deep UV (200 nm) can extend this slightly. In electron beam lithography, on the other hand, beams of electrons are used in place of UV light, and the practical diffraction limitation is averted. But that is another story.

3.4.3 Development and post-treatment

After the resist film has been exposed, the next step is to develop that exposed film. This step represents a chemical reaction in which a chemical called a **developer** reacts with only selected regions of the resist, removing them from the wafer. Which areas get removed depends on what type of resist was used. In positive resists, the exposed areas become more soluble and are removed by the developer, leaving behind a positive image of the mask pattern. Negative resists become less soluble when exposed so that the surrounding unexposed material is removed by the developer, leaving behind a negative image of the mask pattern in the resist layer.

Pattern transfer can occur using either a "wet" technique or a "dry" technique. When the printed features are not too small, resist removal is usually accomplished using a wet technique. This simply involves dipping the wafer into an organic solvent. This method can cause resist swelling because of absorption. The effect is significant for negative resist but minimal for positive resist. For very small features, a dry etching technique is used. This technique will be discussed later.

After development, there still remains some unwanted exposed/unexposed resist on the wafer. A mild oxygen plasma treatment removes this resist. After this step, another postbake (also called a hard bake) is done to further harden the resist.

Resist stripping

After the patterned photoresist has served its purpose in subsequent processing steps, the final step is its removal, or **stripping**. Two methods are available, *wet stripping* and *dry stripping*. The "wet" in wet stripping refers to the use of aqueous or organic solutions, whereas the "dry" in dry stripping indicates the lack of such solutions.

Positive resists are usually removed with a wet stripping process using a chemical solvent such as acetone or methylethylketone. Many of these are highly flammable; what's more, effective use often requires application at roughly 80°C, which is close to the temperature at which they form ignitable mixtures with oxygen. Hence for safety reasons, many proprietary stripping solutions are used.

Negative resists are tougher to remove. One approach is to immerse the wafer in a 1:1 mixture of concentrated H_2SO_4 and H_2O_2 at 150°C. This mixture is often referred to as "piranha" etch because of it virulent qualities. (You may recall from section 3.2 that a "piranha clean" is sometimes used to rid surfaces of contaminants.) In another method hot chlorinated hydrocarbons are used to swell the polymer, together with acids to loosen its adhesion to the substrate. Many proprietary mixtures are used here as well, and include such solvent mixtures containing trichloroethylene, methylene chloride, and dichlorobenzene, combined with formic acid or phenol. Obviously a drawback of wet stripping of negative resist required handling and disposal of large amounts of corrosive chemicals.

In dry stripping photoresist is removed using an ionized gas, or plasma. The process is commonly called **plasma ashing**, and has many advantages over wet chemical methods. The plasma oxidizes the resist, producing mostly water, carbon monoxide, and carbon dioxide, which are rapidly desorbed. Very little residue remains at the completion of the process leaving the wafers ready for the next process step with a minimal amount of clean-up and rinsing required.

3.5 Masks

We have already seen that a **photomask**, or simply a **mask**[2], essentially behaves as a stencil during the photolithography process, allowing light to make its way to a layer of photoresist only in selective regions. Hence, masks are characteristically made of flat, transparent materials with opaque regions. Glass is therefore a typical mask material.

There are several types of photomasks used in MEMS. The most common is the **binary mask** made up of opaque and transparent regions such that light is either completely blocked or completely transmitted through the mask at certain locations. The masks we have seen thus far all fall within the category of binary mask, and unless otherwise noted in the text we will assume that a mask is binary.

Other mask types exist, however. One type called a **phase shift mask** contains spatial variations in intensity and phase transmittance of light. These variations tend to increase resolution by allowing features to be printed more closely. **Polarization phase shift masks** improve on ordinary phase shift masks by altering the polarization of the light as it passes through the mask. The incident illumination may be either unpolarized or polarized depending on the mask design. Either way the final goal is the same, to improve resolution.

When a binary mask is used with near UV light, it is typically made from a nearly optically flat glass plate. Masks for use with deep UV typically employ quartz instead. To create the opaque regions, the mask plates are selectively coated with either a high resolution emulsion (gelatin) or a

[2] It should be noted that the word mask has many contexts in MEMS. For example, the oxide remaining on the wafer in Fig. 1.3 (g) and (h) is sometimes referred to as a **hard mask**, as it is used to protect the silicon directly beneath it during chemical etching. Another type of mask is the **field mask** made of photoresist in Fig. 2.9, which is used for selective doping of a wafer. In this chapter, however, we use the term mask to refer to the separate physical entity containing the pattern to be transferred to a layer of resist.

thin film of metal, usually chromium. Typical metal thickness range from 1000Å to 2000 Å. Sometimes a master mask is created first after which replica masks are created from the master for use in the actual printing process.

Having a high quality mask with good resolution is often the bottleneck for being able to create high quality MEMS with small feature sizes. The question naturally arises, then, of how the masks themselves are made. Somewhat ironically, a lithographic process itself is used to create masks.

In one method a device called an *optical pattern generator* exposes a tiny adjustable-size rectangle on a coated plate. The plate is then repositioned and exposed once more. The process is repeated until the desired pattern is created, much like a hi-tech *Etch A Sketch*® at the micro-scale. With emulsion masks the generator creates the pattern directly in the emulsion layer, as emulsions are photosensitive. For metal masks, the entire mask is first covered with metal that in turn is covered with a layer of photoresist. The pattern is generated in the resist layer. The resist is developed, selectively exposing the metal layer below. The metal is then etched and the resist stripped completing the mask. Emulsion masks are cheaper to make, though metal masks tend to resist surface damage better.

In order to generate the highest quality masks *electron beam lithography* is employed. In its simplest sense, electron beam lithography, or simply e-beam lithography, replaces the optical beam of an optical pattern generator with an electron beam. Just as electron microscopes bypass the wavelength restrictions of optical microscopes to allow one to see smaller and smaller things, e-beam lithography bypasses the optical wavelength restrictions of optical pattern generators, allowing for masks with virtually unrestricted shapes with very small feature sizes to be created. E-beam lithography requires the use of a different type of resist material, called electron beam resist, which undergoes chemical changes when penetrated by the e-beam. The e-beam is typically a raster scan, which means it creates an image line by line by being blinked on and off, as in the cathode ray tube in older televisions.

Producing masks via optical pattern generators and e-beam lithography are commercially available services obtainable through vendors. Depending on the quality of the mask and the required resolution, masks generated thusly can range in price from relatively inexpensive to exceedingly expensive, with corresponding turn around times.

An inexpensive option in creating a mask is to photographically produce a negative image containing the desired mask pattern. Patterns are typically designed using a computer aided design (CAD) software program and then printed on paper. The paper is photographed in order to obtain the negative. An even cheaper option is to print directly onto a transparency and use that as a mask. Naturally the quality of the masks produced using these options depends heavily on the resolution of the printer used.

3.6 Resolution

Ideally the pattern on a mask is perfectly transferred to a photoresist layer so that after development, the photoresist left on the wafer looks just like the mask pattern. In reality several things prevent this from happening.

3.6.1 Resolution in contact and proximity printing

In Fig. 3.8 we see a rather boring mask consisting of an opaque region of width b adjacent to a transparent region of width b and so on. The result of an idealized exposure and development of a positive resist layer during a proximity printing process is also shown. Now one may initially believe that as long as one is able to produce a mask with smaller and smaller widths b that those widths should translate into smaller and smaller features formed in the photoresist. However, this is not the case since as the width b in the mask grows smaller diffraction effects become increasingly important, impeding the ability to successfully transfer the pattern to the resist layer. The smallest width b that can be successfully transferred to the resist layer is the **resolution**[3] of the printing process. In terms of the quantities given in Fig. 3.8 the resolution R is given by

$$R = b_{\min} \cong \frac{3}{2}\sqrt{\lambda\left(s + \frac{z}{2}\right)} \tag{3.4}$$

where s is the gap distance between the bottom of the mask feature and the resist surface, λ is the vacuum exposure wavelength, and z is the resist thickness. Equation (3.4) comes from the consideration of the effects of diffraction and geometry. It is valid for both contact printing and proximity printing. In the case of contact printing, the gap distance s becomes zero.

[3] Just as the term mask has many connotations in MEMS, so does resolution. Specifically, resolution also refers to the smallest change in input that can be detected by a MEMS sensor. Other terms used for resolution within photolithography are line-width control and critical dimension.

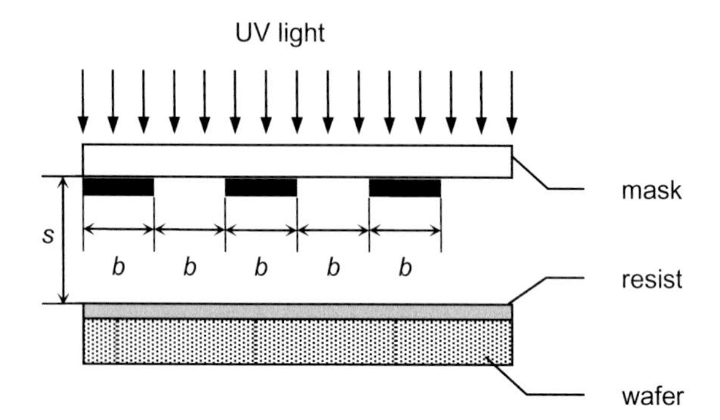

Fig. 3.8. Parameters affecting resolution in proximity printing

From Eq. (3.4) we see that, all other things being equal, contact printing results in better resolution than does proximity printing. The tradeoff comes in that contact printing tends to be more likely to damage the mask and thus shorten mask life.

3.6.2 Resolution in projection printing

In the case of projection printing resolution is also affected by the characteristics of the projection system. An experimental correlation for resolution is given by

$$R = \frac{k_1 \lambda}{NA} \tag{3.5}$$

where k_1 is an experimentally determined parameter for a given system that depends on the photoresist, projection system optics and process conditions, and NA is the **numerical aperture** of the optical system, a dimensionless number that characterizes the range of angles over which the system can accept or emit light. In terms of the simple system of Fig. 3.9 (remember the "lenses" in projection systems are often really the combination of many optical components) numerical aperture is given by

$$NA = n \sin\theta_{\max} = \frac{D}{2F} \tag{3.6}$$

where, n is the index of refraction of the imaging medium, usually air, D is the lens diameter and F is the ratio of the lens's focal length to its diameter (sometimes called the *effective F number*). For modern systems NA ranges from 0.16 to 0.93 and k_1 from 0.3 to 1.1.

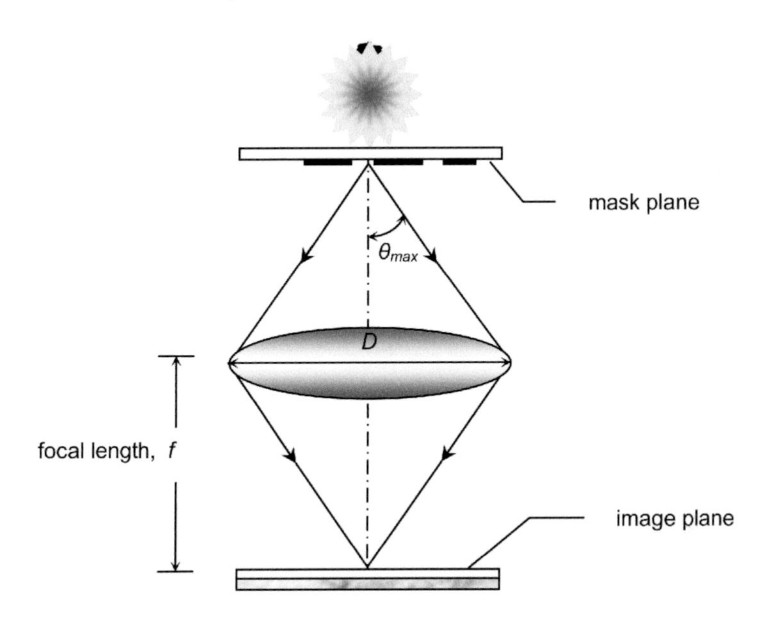

Fig. 3.9. Relationship of numerical aperture to other quantities

A characteristic that both Eqs. (3.4) and (3.5) have in common is that the wavelength of the optics used in photolithography directly affects the resolution. This is the reason for the push towards smaller and smaller wavelengths regardless of printing method. For projection printing Eq. (3.5) shows us that increasing numerical aperture also enhances resolution. Increasing NA, however, usually involves complex and therefore costly optics. Decreasing wavelength is usually favored but requires new resist materials.

Ideally, the latent image in projection printing would be perfectly reproduced throughout the entire resist thickness. In reality, however, the image is well-focused at one location within the resist with a certain tolerance on either side in which the image has an acceptable level of focus. This tolerance is called the **depth of focus**, an experimental correlation for which is given by

$$\pm\delta = \pm\frac{k_2\lambda}{(NA)^2}, \tag{3.7}$$

where k_2 is an experimentally determined constant depending on contrast. Eliminating NA from Eqs. (3.5) and (3.7) gives

$$\pm\delta = \pm\frac{k_2 R^2}{k_1^{\,2}\lambda}. \tag{3.8}$$

We see, then, that there is a tradeoff between resolution and depth of focus. That is, smaller wavelengths result in smaller (and thus better) resolution, but also in smaller (and thus worse) depth of focus. Even so, resolution usually takes precedence.

3.6.3 Sensitivity and resist profiles

Optical exposure causes a photochemical reaction to take place in a layer of resist. Not all the radiation is incident at the same angle, however, and some of the energy scatters as it moves through the resist towards the wafer. Furthermore, development is a time dependent process in which the dissolved photoresist cannot be swept away instantaneously. The result is that the aerial image formed on the resist surface does not translate directly into a perfect latent image throughout the entire resist thickness. Most noticeably this results in non-vertical sidewalls within the resist profiles as shown in Fig. 3.10 (a) and (b) for positive and negative resists, respectively. Also shown in Fig. 3.10 are exposure response curves. These curves give the amount of resist remaining on a wafer after exposure and development as a function of the dose, D. Since positive resist becomes more soluble after exposure whereas negative becomes less soluble, the slopes of the curves have opposite signs.

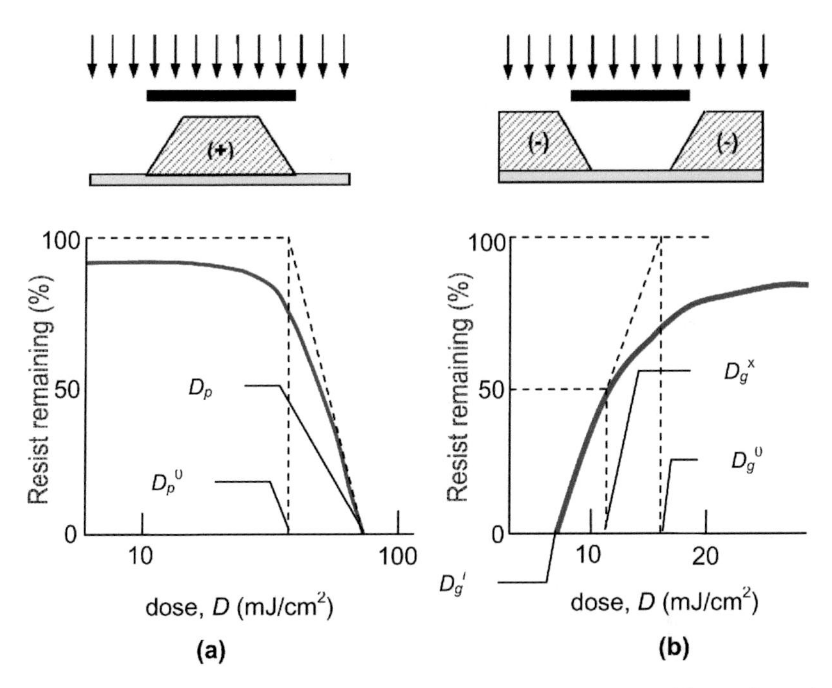

Fig. 3.10. Resist profiles and exposure response curves for (a) positive resist and (b) negative resist

For a positive resist in which the exposed regions become more soluble, the dose required to completely clear out the resist down to the underlying layer after development—that is, the dose to clear—is defined as the **lithographic sensitivity**, D_p. The reference dose D_p^0 in Fig. 3.10 (a) is that at which the developer first begins to attack the irradiated film. The slope of the curve formed by these two doses helps us define the **resist contrast**, γ_p as

$$\gamma_p = \frac{1}{\ln D_p - \ln D_p^0} = \left[\ln\left(\frac{D_p}{D_p^0}\right)\right]^{-1} \tag{3.9}$$

The resist contrast describes the sharpness of the pattern in the resist formed during the lithographic process. Larger values of γ_p imply sharper images. A typical value of γ_p for positive resist is between 2 and 4.

For negative resist a critical dose D_g^i is needed to start the polymerization processes within the resist in order to decrease solubility. For a dose of D_g^0 the entire layer is polymerized. The contrast is found in a similar fashion as for positive resist:

$$\gamma_n = \frac{1}{\ln D_g^0 - \ln D_g^i} = \left[\ln\left(\frac{D_g^0}{D_g^i} \right) \right]^{-1} \tag{3.10}$$

A typical value for of γ_n is 1.5.

The sensitivity for a negative resist, however, is not defined as the dose to clear as for a positive resist. Rather, the sensitivity is defined as the dose that gives optimal resolution after development. This typically corresponds to resist layers of 50-70% of the original thickness remaining after development. The sensitivity is shown as D_g^x in Fig. 3.10 (b).

The lithographic sensitivities defined above give insight as to what doses are required to produce sharp images in exposed and developed resist layers. A different but related concept is *intrinsic resist sensitivity.* This quantity, sometimes called the photochemical quantum efficiency, is defined as the number of photoinduced events divided by the number of photons required to accomplish that number of events.

$$\Phi = \frac{\text{Number of photo induced events}}{\text{Number of photons absorbed}} \tag{3.11}$$

Φ ranges from 0.5-1 for negative resists and from 0.2-0.3 for positive resists.

In short, intrinsic resist sensitivity represents how sensitive a resist is to exposure, whereas lithographic sensitivity measures how sensitive a resist is to exposure and subsequent development in order to produce the best patterns.

3.6.4 Modeling of resist profiles

Figure 3.11 shows the resulting profile for a positive resist after exposure and development. A semi-empirical relation for the slope of the sidewalls for such a layer created via projection printing is given by

$$\frac{dz}{dx} = \frac{2NA}{\lambda(a + \alpha t) D_p \left[1 - kk_2\right]^2} \tag{3.12}$$

where a and α are constants characterizing the resist absorbance, t is the resist thickness, k is a parameter depending on the coherence of the light source and k_2 is a process dependent contrast (≈ 0.5), the same constant that appears in Eqs. (3.7) and (3.8) in regards to depth of focus. Note that for a given resist (a and α) we see that reducing wavelength, resist absorption and resist thickness, and/or lithography sensitivity all result in steeper slopes and therefore higher wall angles. Increasing the numerical aperture of the projection system will also result in steeper sidewalls, though as previously discussed this tends to be an expensive endeavor.

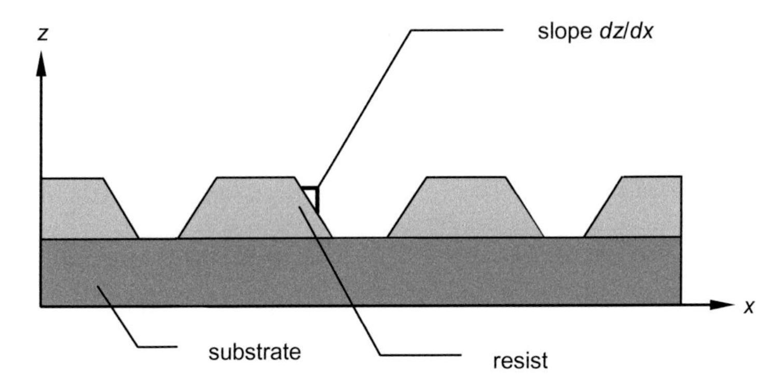

Fig. 3.11. Side view of a positive resist sidewall profile

Though at first sight the idea of non-vertical sidewalls seems like a huge problem, as always crafty MEMS designers have found ways to exploit the phenomenon. The active control of resist layer profiles, especially of positive resists, can help one selectively deposit layers of other materials in a process known as **lift-off**. Lift-off is usually used to deposit metals and is explored in more detail in Chapter 4.

3.6.5 Photolithography resolution enhancement technology

The push for smaller feature sizes in MEMS devices necessitates better resolution during photolithography steps. Many technologies are actively being developed to address this issue.

One strategy is to improve resist performance. For example, chemically amplified resists that contain the appropriate catalysts can be used to multiply the number of chemical reactions taking place, thereby making the resist more soluble in the developer. This requires a post-exposure bake and works well in the deep UV region.

Another resist performance enhancement technique is to use image reversal. Because positive resists do not swell during development, several process variations aim to reverse the tone of the image so that the resist can act as a high resolution negative resist. That is, a positive resist is made to behave like a negative resist in producing the negative of a pattern, yet retain its original higher contrast. After the initial mask and exposure steps, reversal is accomplished by means of a bake step, followed by flooding the resist with UV.

Diffraction effects in particular represent a fundamental limit on the resolution of lines in optical lithography. Hence, another strategy involves improved mask technology to overcome this limitation.

During diffraction, the light intensity from each of two adjacent illuminated slits is diffracted into the obscured regions which surround them. As the slits are moved closed together, the individual intensities cannot be resolved at the image plane. Resolution can be improved by controlling the phase as well as the amplitude of the light at the image plane. The phase-shift mask, invented by Marc Levenson, addresses this problem. By placing a transparent layer across one of the openings of sufficient thickness to provide a 180° phase shift to light passing through it, the intensity pattern at the image plane can be resolved. The result is that features on the mask can be moved closer together and the images at the wafer plane can still be resolved. The use of phase-shifting masks of this type has the potential of extending the resolution of modern optical systems by a factor of two, although the technique is limited to highly repetitive patterns. More recently, however, a number of modifications have been proposed which allow the method to be applied to arbitrary mask patterns.

3.6.6 Mask alignment

The fabrication of MEMS devices is inherently a layered process, building one structural layer on top of another. As such, several photolithography steps are often required. If each subsequent layer is not placed just right on top of the layer beneath it, however, the resulting structure can be a disaster, regardless of how carefully the individual steps are controlled or how good the resolution may be.

And so in processes requiring multiple masks, it is of paramount importance to align patterns on a mask with those already on the wafer. This alignment procedure is referred to as registration, or simply **mask alignment**. Mask alignment is usually accomplished by building a set of *alignment marks* into each and every mask that is used, and then using those marks to line up subsequent layers. Figure 3.12 illustrates this process.

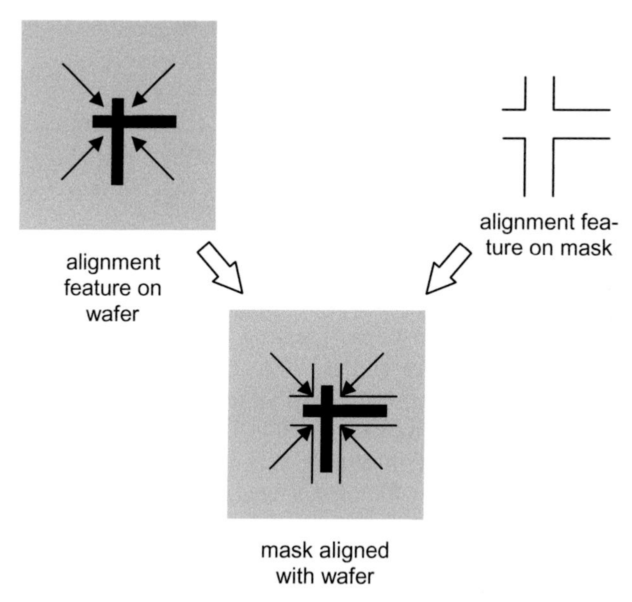

Fig. 3.12. Mask alignment accomplished through the use of alignment marks. Arrows are often used to make alignment marks easier to find.

3.7 Permanent resists

In this chapter we have focused on the use of photolithography as a tool for transferring patterns that are ultimately needed for some auxiliary step, such as protecting the silicon directly beneath it during chemical etching, or the selective doping of a wafer. The resist layers are temporary and are stripped off the wafer after the subsequent processing takes place. However, sometimes a patterned resist can become a permanent part of the MEMS device. Such resists are called **permanent resists**.

The polymer polyimide is a common example of a permanent resist. Polyimide has many properties making it attractive for such use. It is both chemically and thermally stable, withstanding temperatures up to 500°C, and it has a dielectric constant and dielectric strength comparable to those of SiO_2, but with an electrical resistivity about ten thousand times lower. It is often used as a low-dielectric insulation layer and also as a structural material for creating flexible hinges in mechanical structures.

As a resist, polyimide comes in liquid solution form and is typically applied to wafers via spin-coating. The film is then cured by heating to approximately 300°C for several hours.

Essay: Photolithography—Past, Present and Future

Jerome Wagner
Department of Physics and Optical Engineering, Rose-Hulman Institute of Technology

Lithography is a technique used for creating 3-D relief patterns on a substrate. The word lithography comes from a combination of two Greek words: graphia, meaning to write, and lithos, meaning stone. Literally translated, lithography is "writing a pattern in stone". The lithography process, originally chemical in nature, was invented by Bavarian author Alois Senefelder in 1796. At the time it was used to print text or artwork onto paper or other suitable materials.

In today's modern world, the stone is replaced by a CD, DVD, semiconducting wafer, or other suitable substrates, and the chemical process is replaced by an optical/chemical process which produces a mask image in a photosensitive recording medium (photoresist), hence the name photolithography. Photolithography has been used for decades in the fabrication of integrated circuits. The same basic lithography process is also used to fabricate MEMS devices. The use of optical radiation in theses processes greatly enhances the speed at which patterns can be written, which leads to greater *throughput*, or the number of devices that can be fabricated per unit time.

The optical region of the electromagnetic spectrum which includes the infrared, visible, and ultraviolet, was originally used for exposure. Higher resolution (printing smaller mask features) requires sources having shorter wavelengths. This suggests using wavelengths in the ultraviolet. Originally, a high pressure mercury vapor lamp provided three useable wavelengths: 436 nm, 405 nm, and 365 nm, the latter being heavily used by the semiconductor industry in the fabrication of integrated circuits. To further increase resolution, it became necessary to develop sources of shorter wavelengths. In 1970 an excimer laser was invented by Basov, Danilychev and Popov at the Lebedev Institute in Moscow. A modern excimer (excited dimer) or, more generally, an exciplex (excited complex) laser typically uses a combination of an inert gas (argon, krypton, or xenon) and a reactive gas (fluorine or chlorine). Such a combination has a bound excited state but a repulsive ground state. Upon dissociation, either by spontaneous or stimulated emission, and, depending on the materials used, radiation wavelengths have been produced ranging in wavelength from 126 nm (Ar_2) to 351 nm (XeF). These lasers provide significantly more power than UV sources and overcome the problem of energy absorption in projection systems at shorter wavelengths.

In addition to ultraviolet exposure of photoresists, other sources are available, none of which has been developed to the point that it is ready for implementation in manufacturing. One method, still within the optical region is called extreme ultraviolet lithography (EUVL) which is a next generation lithography technology (NGL) using the 13.5 nm wavelength. Since all materials absorbs EUV, this lithography must take place in a vacuum. Also, the system optics absorbs approximately 96% of the available EUV radiation requiring that the source be significantly bright enough leaving enough energy to expose the resist. A large investment will be required to develop this technology with the hope of achieving a manufacturing capability by 2013.

Optical lithography is limited by diffraction. Using the x-ray region of the electromagnetic spectrum was motivated by the idea that diffraction effects could be essentially neutralized by using radiation of extremely short wavelengths. However, at these wavelengths, there are no materials available for constructing image-formation mirrors or lenses. Therefore, the use of proximity printing rather than projection printing is used for x-ray lithography. Photoresists can also be exposed by electron beams producing the same results as UV exposure. The principal difference between the two methods is that while photons are absorbed, depositing all of their energy at once, electrons deposit their energy gradually, and scatter within the photoresist during the process. Special resists have been developed dedicated specifically to electron-beam exposure. Ion beams can also be used for exposure. Ions scatter very little in solids which can result in very high resolution. However, because ions are significantly more massive than electrons, they cannot be deflected at the same speeds as electrons limiting large-field exposures.

The first integrated circuits using the method of contact printing began to appear in the early 1960s. Since that time, optical lithography has remained the only manufacturing approach used for high-volume IC production. In 1965, Gordon E. Moore, a co-founder of Intel, described an important trend in the history of computer hardware: that the number of transistors that can be inexpensively placed on an integrated circuit is increasing exponentially, doubling approximately every two years. This is known as Moore's Law. For Moore's law to continue to hold, smaller and smaller dimensions will be required. This brings into question the resolution limits of optical lithography which, when achieved, will lead to the implementation of the next generation lithography. Optical imaging, under the most favorable conditions (λ=193 nm with NA=1.35) will provide approximately 80 nm features. However, smaller features in the 10 nm to 30 nm range are desired. Resolution could be achieved by reducing the wavelength from 193 nm (ArF exciplex laser) to 157 nm (F_2 excimer laser). For practical reasons, this reduction has not occurred, mostly related to the costs and time required to develop 157 nm lens materials. Other possibilities for the production of smaller features include nanoimprint lithography (NIL) and double-patterning lithography (DPL). In NIL, embossing, pressing a relief pattern from a hard master into a softer substrate material, is a well known and long-used printing technique. This technique is known for both its high resolution and low cost but is subject to high defect levels. In DPL, two separate patterning steps intersperse features to cut the pitch resolution in half. However, given the extra processing and doubling the patterning steps to create one final pattern, the lithography cost essentially doubles. Another problem is that of overlay errors and the effect they have on critical dimension errors.

Another aspect of optical lithography is simulation. This involves the accurate description of semiconductor optical lithography using mathematical equations. In the early 1970's Rick Dill of IBM began the process of transforming the lithography method from an art to a science. Highly sophisticated simulation programs such as SAMPLE and PROLITH/2 have been developed and are used as a research tool, a development tool, a manufacturing tool, and a learning tool. These programs allow the researcher to develop lithographic models to test our understanding of the physics and chemistry of lithographic imaging by comparing model predictions with experimental observations. Simulation programs are now being used as a guide to design the next generation of lithography tools, materials, and processes.

References and suggested reading

Franssila S (2004) Introduction to Microfabrication. Wiley, Chichester, West Sussex, England

Levenson MD (2001) Generic Phase Shift Mask, US Patent 6251549

Lin BJ (1992) The Attenuated Phase-Shifting Mask. Solid State Technology January 1992:42

Madou MJ (2002) Fundamentals of Microfabrication, The Art and Science of Miniaturization, 2nd edn. CRC Press, New York

Senturia S (2001) Microsystem Design, Kluwer Academic Publishers, Boston

Questions and problems

3.1 What are the main components that make up a photoresist?
3.2 What is a two-component resist? What are the two components?
3.3 What does the concept of dose involve?
3.4 What effect does the dose have on the molecular weight of a negative resist? (Does it cause it to decrease or increase?)
3.5 Identify one problem that can arise when positive resist is applied to a thermally grown oxide wafer (silicon substrate). How can this problem be resolved?
3.6 Give one advantage of contact printing over proximity printing. Give one disadvantage of contact printing compared to proximity printing.
3.7 A pattern is to be printed using deep UV radiation for the exposure. Should the mask material be made of glass or quartz? Why?
3.8 Pattern resolution is increased by decreasing exposure wavelength. If a projection printer is used, will the objective lens need to be redesigned? Why or why not?
3.9 In a clean room rated at class 1000, on the average, how many particles per cubic meter are there? Assume a particle size of 1 μm.
3.10 After exposing a resist, why is it a good idea to bake the resist?
3.11 When positive resist is exposed, what happens to its solubility?
3.12 With respect to transferring a pattern to the resist, what is meant by resolution?
3.13 What kind of cleaning process is best suited for a wafer before it is oxidized?
3.14 The initial puddle of resist dispensed on a 10 cm silicon substrate by a spin-coater is 5.0 cm in diameter and 0.3 cm high. After spinning at 3500 rpm, the entire wafer is coated with resist 1 μm thick. How much of the original resist has been wasted?
3.15 The heat generated by the UV source must not be allowed to affect the pattern embossed on the mask. Suppose that a pattern is to be printed on a wafer having a diameter of 15 cm. This pattern is to overlay an existing pattern already on the wafer. It is essential that the two patterns be aligned correctly. The linear coefficient of thermal expansion of quartz used for the mask plate is $\alpha = 5.5 \text{ X } 10^{-7}$ / °C. Alignment accuracy across the wafer must be maintained from one layer to the next within 0.5 μm. What is the allowable range of mask plate temperature during alignment?
3.16 For a positive resist (AZ 1350J), the sensitivity is D_p = 90mJ/cm^2 where, D_p^0 = 45 mJ/cm^2. Determine the resist contrast.

3.17 For the negative resist (Kodak 747), the sensitivity is $D_p = 7$ mJ/cm^2 where $D_p^0 = 12$ mJ/cm^2. If the sensitivity of a negative resist is defined as the energy required to retain 50% of the original resist thickness in the exposed region, determine the resist contrast.

3.18 From Eq. (3.4) we see that proximity printing always results in worse resolution (larger minimum realizable feature size) than does contact printing, since $s > 0$. One way to gain better resolution is to use a thinner resist thickness, z. Find the limit of $R_{proximity}/R_{contact}$ as resist thickness goes to zero.

3.19 A 1 μm thick resist is spun onto a wafer. A contact print is made using the h-line wavelength (405 nm). Determine the resolution for this process. Repeat the calculation for $\lambda = 248$ nm.

3.20 A 1 μm thick resist is spun onto a wafer. A proximity print is made using the H-line wavelength (405 nm). The distance between the mask and the resist is 10 μm. Determine the resolution for this process. Repeat the calculation for $\lambda = 248$ nm.

3.21 A G-line stepper (436 nm) has a numerical aperture of 0.54. The parameter $k_1 = 0.8$.

a. Calculate the resolution of the projection printer. Express the wavelength in microns.

b. We now switch to an I-line stepper (365 nm) which has the same resolution as the G-line stepper. What numerical aperture should the I-line lens have?

3.22 With a given spinner and photoresist formulation, a spin speed of 4000 rpm gives a resist thickness of 0.7 μm.

a. Suggest a simple way to increase the thickness to 0.8 μm, and

b. to 1.0μm.

Chapter 4. Creating structures—Micromachining

4.1 Introduction

We have explored the nature of the silicon substrate, seen how to tailor its composition by doping it, how to deposit thin film layers, and also how to create and transfer patterns using photolithography. We next turn our attention to using the substrate, thin film deposition, and pattern transfer processes to create the actual structures that make up a MEMS device. This shaping of materials into structural elements is commonly known as **micromachining**.

Micromachining comes in two basic flavors. In **bulk micromachining** one selectively takes material away from a substrate to create the desired structures. It is much like creating a doughnut by starting with a lump of dough and then removing the unwanted material in the middle to form the doughnut hole. In **surface micromachining** we add material on top of the substrate and then selectively take away some of that added material. In this case the substrate does not form a structural element. Rather, it plays a role similar to the large, flat, green piece of a popular interlocking building block toy on top of which many other blocks are added. In a MEMS device fabricated using surface micromachining, the added **thin films** are analogous to the building blocks and form the actual structures. Many MEMS devices utilize both bulk and surface micromachining in their fabrication.

Some of the processes and tools used in micromachining, such as the use of **masks** and **wet and dry etching**, have already been discussed in previous chapters. More detail is given in the present chapter, as are introductions to additional processes and issues, such as **release** and the problem of **stiction**. How one properly orders these techniques to produce a finished device, the art of **process flow**, concludes the chapter.

T.M. Adams, R.A. Layton, *Introductory MEMS: Fabrication and Applications*,
DOI 10.1007/978-0-387-09511-0_4, © Springer Science+Business Media, LLC 2010

4.2 Bulk micromachining processes

The process of bulk micromachining refers to creating structures by removing material from the substrate. This can occur in many ways. The most prevalent method is to chemically eat away the material in a process called **chemical etching**, or simply **etching**. Solutions containing the required components to achieve the desired chemical reaction(s) are appropriately called **etchants**. Historically most etchants have been in liquid form, and the process has therefore also been called **wet chemical etching**. This also distinguishes the process from those in which the reactive components are contained in a gas or plasma, which is called **dry etching**.

The overall goal in bulk micromachining is to remove material from the substrate itself. The amount of material removed per unit time during etching is called the **etch rate**. It is usually measured in dimensions of length (signifying the depth into the substrate) per unit time. In general, higher etch rates are desirable. This allows for batch fabrication to proceed more quickly.

More recently, methods that remove material by physical rather than chemical means have been gaining more widespread use. These processes include micromilling, laser ablation, electrodischarge machining and ultrasonic machining. These are not usually batch fabrication techniques, and therefore traditional etching processes still form the larger part the microfabrication world. However, in some cases the economics of physical methods can win over traditional micromachining. Combining methods for the fabrication of certain devices, such as laser drilling through-holes in otherwise chemically etched pressure sensors, is also gaining popularity.

4.2.1 Wet chemical etching

As an example of wet etching, consider Fig. 4.1, which shows a two step process used to etch a pit into a (100) silicon wafer. In the first step a silicon dioxide layer is etched through a window of a photoresist mask. The resist layer is then stripped, leaving a window through the oxide layer to the silicon substrate below. The oxide layer now forms a mask for etching the substrate. (This is often referred to as a hard mask.) Using a different etchant, the exposed portion of the silicon wafer is etched next, forming the desired pit in the substrate.

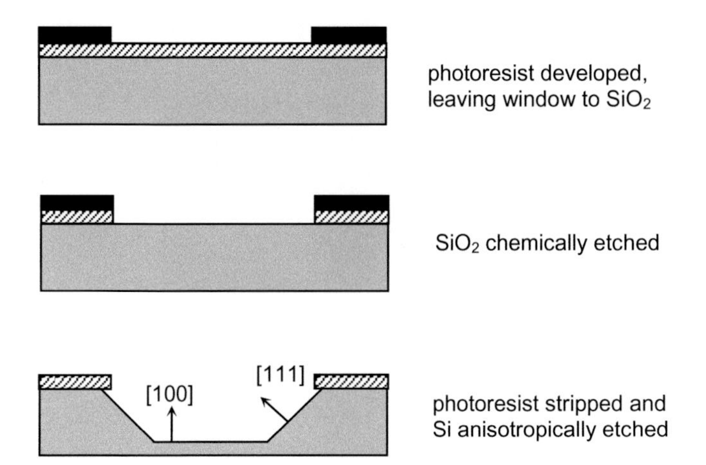

Fig. 4.1.Two step wet etching process to produce a pit in a (100) silicon wafer

It is important to realize that the etchants used in the above described process react with everything with which they come into contact, at least to a small degree. For the process to work, however, the etchants must react very slowly with some materials, while reacting very quickly with others. For example, we would like the etchant used to etch the oxide layer to react very quickly with the oxide itself, but very slowly with the photoresist layer as well as with the silicon substrate below. In this way we can ensure that the resulting pattern in the oxide layer closely resembles the photoresist pattern, and that the reaction stops when the substrate is reached. Likewise, the etchant used for creating the pit in the substrate should react quickly with the silicon, but very slowly with the oxide.

The relative etch rate of an etchant solution with one material compared to another is called **selectivity**. In essence we are playing games with selectivity in order to bring about the desired structures.

You may have noticed in Fig. 4.1 that the resulting window in the oxide layer does not resemble the photoresist mask perfectly. Specifically, the opening in the oxide is slightly wider than that of the photoresist. This is because the chemical reaction has proceeded in the horizontal direction to some degree as well as in the thickness direction of the oxide layer. This phenomenon is called **undercutting**.

Often we wish to minimize the degree of undercutting in order to keep the transferred pattern as true as possible. At other times, however, we rely on undercutting to actually create structures. A good example is the undercutting of a silicon dioxide layer over a silicon substrate to form a suspended structure. Figure 4.2 shows an example of a MEMS cantilever created in this way.

Fig. 4.2. Scanning electron microscope image of a silicon dioxide cantilever formed by undercutting (Courtesy: S. Mohana Sundaram and A. Ghosh, Department of Physics, Indian Institute of Science, Bangalore)

You may also have noticed in two preceding figures that the silicon wafer pits have slanted sidewalls. This is because the etchant used not only is selective towards the silicon material, but to certain crystallographic directions as well. When an etching process proceeds at different rates in different directions, it is called an **anisotropic etch**. For the etchant and crystalline orientation of the wafer shown here, the resulting shape of the pit resembles an inverted pyramid with sidewalls corresponding to the {111} planes. When the etching process proceeds at the same rate in all directions, we have an **isotropic etch**. The resulting shape resembles the rounded pit shown in Fig. 4.3.

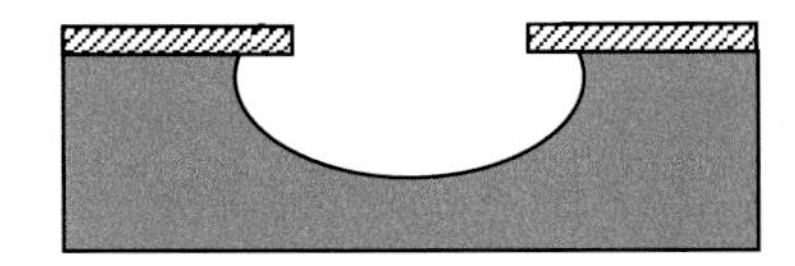

Fig. 4.3. Side view of an isotropically etched pit

Table 4.1 gives the etch rates of various commonly used etchants with respect to silicon and silicon dioxide. Etch rates for various materials are available from numerous sources including handbooks and online resources. They can differ significantly based on differences in conditions, such as concentration and temperature. From Table 4.1 we see that the first two etchants listed are highly selective towards SiO_2 over Si, whereas the last three are selective towards Si. Furthermore, basic etchants tend to etch silicon anisotropically, whereas acidic etchants are isotropic. Since most photoresists are soluble in basic solutions, it is therefore usually necessary to form a silicon dioxide hard mask to etch silicon anisotropically.

Table 4.1. Applications and properties of different etchants (Data taken from Madou)

Etchant	Application	Etch Rate (s)	Notes
48% HF	SiO_2	20-2000 nm/min 0.3 Å/min for Si	Isotropic
Buffered oxide etch (BOE) (28 mL HF/113 g NH_4F/170 mL H_2O)	SiO_2	100-500 nm/min (25°C)	Isotropic
Poly etch $HF/HNO_3/HC_2H_3O_2$ 8/75/17 (v/v/v)	Si	5 μm/min (25°C)	Isotropic
KOH (44 g/100 mL)	Si	1.4 μm/min (80°C) 28 Å/min SiO_2	Anisotropic IC incompatible
Tetramethylammonium hydroxide (TMAH) (22 wt%)	Si	10 μm/min (90°C) SiO_2 virtually unreactive	Anisotropic IC incompatible

In wet chemical etching, an etchant solution is brought into contact with the surface to be etched. The reactive component in the etchant solution first finds it way to the surface. Next, one or more chemical reactions take place involving the surface material and the etchant. Last, the products of the reaction(s) move away from the surface, and new etchant moves in to take the place of the products. This process continues until actively stopped in some way.

Sometimes the chemical reaction proceeds very quickly compared to the rate at which reactants and products are transported to and from the surface. In this case the etching process is called *diffusion limited*, referring to the mass diffusion of species to and from the surface. (Mass diffusion was discussed in Chapter 2 in the context of a dopant diffusing within a substrate. Here, of course, we are referring to individual chemical species diffusing within an aqueous solution.) One can enhance the etch rate under these conditions simply by stirring the solution. Sometimes the process of mixing the solution is automated, as in the case of devices called "wafer rockers".

In other cases the chemical reaction between the substrate and etchant is slow compared to mass diffusion. When this is the case, the etch process is said to be *reaction limited*, which is to say that the rate of the chemical reaction itself is what makes the etching process fast or slow. Simply put, if the chemical reaction is a fast one, then the etching is also fast; if the chemical reaction is slow, then so is the etching process. Ways of increasing the etch rate for the reaction limited case include increasing temperature, increasing etchant concentration, or changing the etchant composition. All of these strategies are naturally aimed at increasing the rate of the chemical reaction.

In general, reaction rate limited etching is preferred to diffusion limited, as it is more readily controlled and reproduced.

Isotropic etching

In isotropic wet etching the etch rate is the same in all directions. Isotropic etchants are typically acidic with the reactions carried out at room temperature. The isotropy is due to the fast chemical kinetics (fast reactions) of the etchant/material combination. Hence, isotropic etching is tends to be diffusion limited. Isotropic etchants also have the fastest etch rates, on the order of microns to tens of microns per minute.

One of the most prevalent isotropic etchants used with silicon is *HNA* ($HF/HNO_3/HC_2H_3O_2$) also called *poly etch* for its frequent use with polycrystalline silicon[1]. The overall reaction is given by

$$HNO_3(aq) + Si(s) + 6HF(aq) \rightarrow H_2SiF_6(aq) + HNO_2(aq) + H_2O(l) + H_2(g) \quad (4.1)$$

The etching process actually occurs in several steps. In the first step, nitric acid oxidizes the silicon.

$$HNO_3(aq) + H_2O(l) + Si(s) \rightarrow SiO_2(s) + HNO_2(aq) + H_2(g) \quad (4.2)$$

In the second step, the newly formed silicon dioxide is etched by the hydrofluoric acid.

$$SiO_2(s) + 6HF(aq) \rightarrow H_2SiF_6(aq) + 2H_2O(l) \quad (4.3)$$

In the isotropic etching of glass or silicon dioxide, the oxidation step is not required. Hence, there is no need for nitric acid, and a buffered oxide etch (BOE) consisting of aqueous HF and NH_4F can be used. The reaction proceeds directly from the second step of Eq. (4.3).

[1] Polycrystalline silicon is often referred to as *polysilicon*, or simply poly. It is a common material used for structural layers in surface micromachining.

In isotropic etching the amount of undercutting of a mask will be on the order of the depth to which the layer is etched. For example, the opening at the top of an oxide layer created using a photoresist mask with initial dimensions of 200 μm × 200 μm will have dimensions of 210 μm × 210 μm after etching to a depth of 10 μm.

Anisotropic etching

Unlike isotropic etching, anisotropic etching occurs at different rates in different directions, specifically along different crystalline planes. The etchants are typically alkaline instead of acidic, and the reactions are carried out at slightly elevated temperatures, usually between 70°C and 90°C. Also unlike isotropic etching, the process is reaction limited, resulting in slower etch rates on the order of 1 μm/min. The anisotropy itself is due to different crystal planes etching at different rates.

In the anisotropic etching of silicon, silicon dioxide is formed first. Then, the oxide is reacted with a strong base, producing $Si(OH)_4$ (aq) or more likely, $H_2SiO_4^{2-}$(aq). The {111} planes etch the slowest, and the {100} planes etch the fastest. In fact, the etch rate for the {100} planes is 100 times greater than that for {111} planes when using KOH as the etchant!

Common anisotropic etchants of silicon and their properties are summarized in Table 4.2

Table 4.2. Selected anisotropic etchants of silicon.

Etchant	Temperature	Si etch rate (μm/min)	{111}/{100} selectivity	SiO_2 etch rate (nm/min)
KOH (40-50 wt%)	70°-90°C	0.5-3	100:1	10
EDP (750ml Ethylenediamine 120g Pyrochatechol, 100 ml water)	115°C	0.75	35:1	0.2
TMAH (Tetramethylammonium hydroxide 22 wt%)	90°C	0.5-1.5	50:1	0.1

Several theories have been proposed as to the reason for the differing etch rates. In the model put forth by Siedel *et al.* it is postulated that the lower reaction rate for the {111} planes is caused by the larger activation energy required to break bonds behind the etch plane. This is due to the larger bond density of silicon atoms behind the {111} plane. For this orientation, each Si atom is bonded with three others below it. There is one unfilled valence or "dangling bond" above the plane. (Fig. 4.4 (a).) However, the {100} and {110} planes both have two dangling bonds above the plane, which can bind two OH^- ions from the etchant. (Fig. 4.4 (b).) The presence of the two bonded OH^- ions serves to weaken the two bonds behind the etch plane, making the etching reaction (breaking of Si-Si bonds) more energetically favorable.

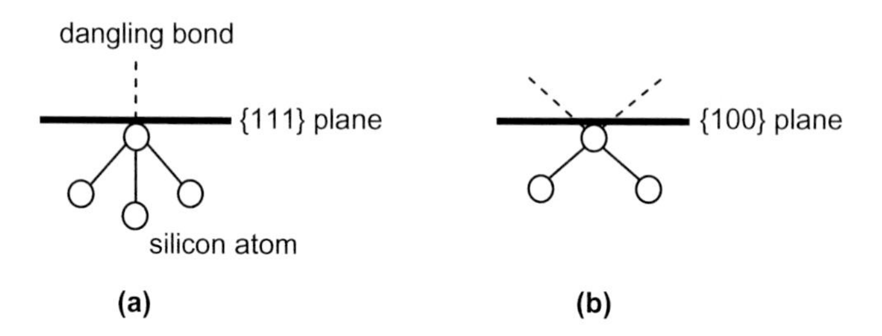

Fig. 4.4. (a) There is one dangling bond above the {111} plane. (b) There are two dangling bonds above both {100} and {110) planes, making them easier to etch.

In the Elwenspoek *et al.* model, surface roughness is suggested as the reason for the discrimination of etch rate. Specifically, the {111} plane is atomically flat, thereby creating a barrier to nucleation and leading to slower etch rates. From this perspective, slower growing crystal planes are also ones that etch more slowly.

MEMS designers use anisotropic etching techniques in order to create the uniquely shaped trenches often seen in MEMS devices. The most common of these is the inverted pyramid shape seen in the anisotropic etching of (100) silicon, the sidewalls of which make angles of 54.7° with the plane of the wafer. (Fig. 4.5.) If the window through the hard mask is made small enough, the etch will continue only until the intersection of the {111} planes is reached. The crystalline structure serves as its own etch stop in this case. However, this technique only works when the {111} planes form concave corners. Even the slightest deformity on a convex corner of the intersection of {111} planes will expose much faster etching planes, which leads to massive undercutting of the hard mask. Sometimes this is most undesirable and the designer must make use of cleverly designed sets of masks for "corner compensation". However, this undercutting is often planned right into the process, as is the case in creating suspended structures over etched pits. (Fig. 4.6.)

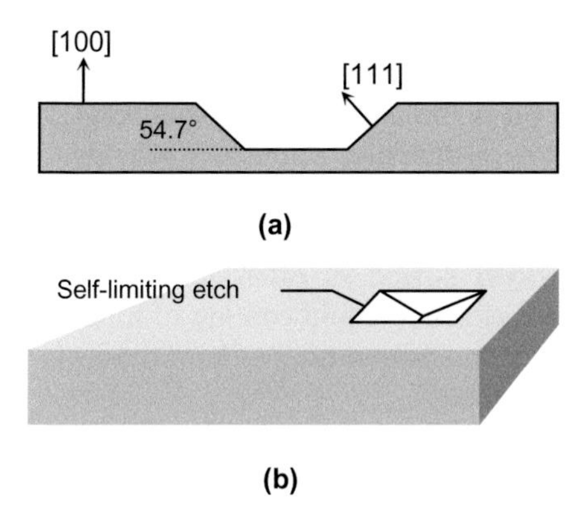

Fig. 4.5. Anisotropic etching of (100) silicon. (a) Anisotropic etching exposes {111} planes, creating sidewalls at 54.7° angles. (b) Intersection of various {111} planes creating a self-limited etch.

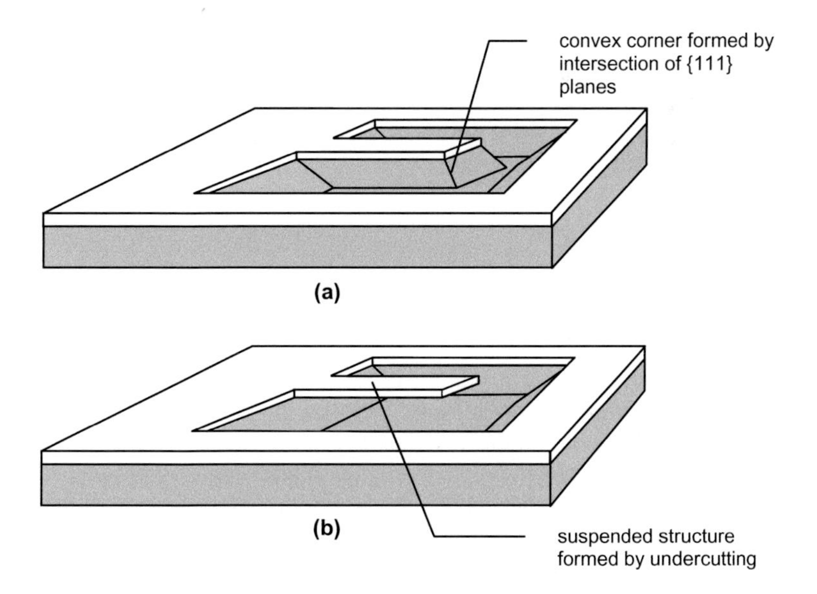

Fig. 4.6. Purposely exposing convex corners formed by {111} planes (a) can be used to undercut hard masks, thereby creating suspended structures (b).

For (110) silicon anisotropic etching results in an opening in the plane of the wafer that resembles a lopsided hexagon. The sidewalls make different angles with the horizontal than in (100) silicon. (Fig. 4.7.) Four of these sidewalls are vertical, a feature that otherwise usually requires more expensive and involved techniques compared to wet chemical etching. The bottom of the etched pit is initially a {110} plane, and therefore flat. For long etch times, however, the two slanted {111} planes intersect each other, creating a self-limiting etch. This self-limiting etch is pronounced for mask openings with small aspect ratios, but not large ones. Hence, anisotropic etching of (110) silicon is often used to create long narrow trenches like those used in active liquid cooling of microelectronic components. As with (100) silicon, undercutting of hard masks is sometimes used to create suspended structures.

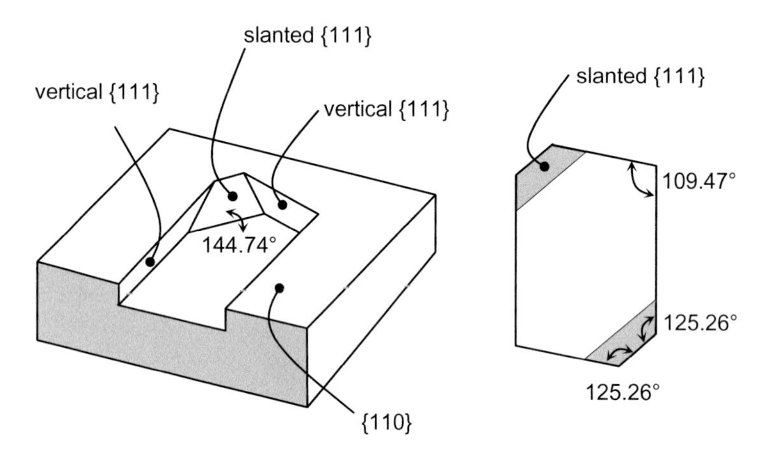

Fig. 4.7. Anisotropic etching of (110) silicon exposes various {111} planes, producing four vertical sidewalls and two slanted sidewalls.

For (111) silicon, the surface plane of the substrate is the one that etches the slowest. Since no other planes are exposed, a bare (111) wafer placed in an anisotropic etchant simply etches very, very slowly with no result other than the wafer becoming thinner. If a "starter hole" is pre-etched in the surface using some other technique, however, other planes will be exposed. Anisotropic etching can now proceed. This is most often used to create pits or trenches underneath a hard mask on the surface. (Fig. 4.8 (a).) When the sidewalls of the pre-etched shape are protected, deep trenches well below the surface can be formed. (Fig. 4.8 (b).) This technique is sometimes used to create microfluidic channels.

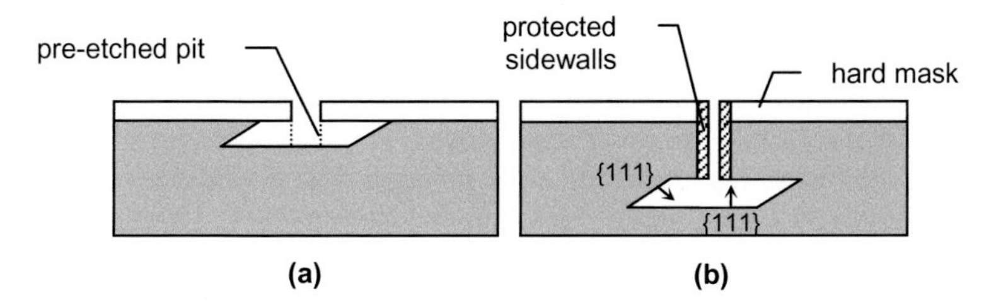

Fig. 4.8. Anisotropic etching of (111) silicon: (a) A pit must be pre-etched before anisotropic etching can proceed. (b) Protecting the sidewalls of the pre-etched shape can be used to produce deep trenches.

Etch stop

Getting the chemical reaction that makes up an etching process to stop can pose a difficult problem within microfabrication. One technique is to time the etch, simply pulling a wafer out of the wet etching solution when the desired depth has been reached. Variability of etch rates, however, makes this technique less than precise. The resolution for thickness is generally on the orders of microns. That is to say, if you attempt to leave a membrane in a silicon wafer less than 10 μm thick using this technique, it could result in a membrane 0 μm thick!

In an *insulator etch stop* the etching process stops when the etchant reaches an insulating layer. In the case of a silicon wafer, the insulator is usually an oxide layer grown on one side of the wafer. Due to the selectivity of the etchant, the etching essentially ceases upon reaching the silicon dioxide. In the case of an insulator etch stop, the insulator makes up part of the structure of the MEMS.

Other etch stop techniques involve varying etchant composition and temperature. Much more precise, however, is to make use of a wafer that has been previously doped at a certain depth. This can allow for a very precise control over etch depth, which has in large part led to microfabrication becoming a high yield process.

Anisotropic etchants do no attack heavily doped Si layers (often designated p^+), and severe drops in etch rate result when the p-type layer is reached. As boron is most frequently used to create p-type silicon, this process is often called a *boron etch stop*. According to the anisotropic etching model of Siedel *et al.* the p-type layer is electron deficient, and there are no electrons available to react with water at the surface. (According to this model, the reduction of water is the rate-limiting step in the etching process.) One disadvantage of a boron etch stop is that the high level of p-type doping, on the order of $5\times10^{18}/cm^3$, is not compatible with CMOS standards for integrated circuit fabrication.

One technique that is compatible with CMOS standards is called *electrochemical etch stop*. In this particular etch stop an n-type layer is grown on a p-type Si substrate, creating a p-n junction. The required level of doping is very light compared to the boron etch stop method. During etching, an electric potential is applied across the p-n junction creating a reverse-bias diode. The diode keeps current from flowing across the junction, and etching of the p-type substrate can occur readily. As soon as the etching reaches the n-type layer, however, the diode vanishes and current starts flowing. The newly freed electrons oxidize the Si at the surface, forming SiO_2. Since silicon dioxide etches very slowly in an alkaline medium, the etching stops almost immediately.

4.2.2 Dry etching

In contrast to wet etching, which employs etchants in aqueous solution form, **dry etching** delivers the chemically reactive species to the etching surface in gaseous form, or within an ionized gas, called a *plasma*. In dry etching the gas or plasma bombards the etching surface, producing a sputtering-like effect. Hence, dry etching actually consists of a combination of physical and chemical etching mechanisms. This bombardment also gives the etching process a directional component, and dry etching techniques have become the standard for creating near vertical sidewalls with large aspect ratios.

An example of an etchant used in dry etching is xenon diflouride (XeF_2). XeF_2 is a highly selective etchant that can etch silicon or polysilicon without etching metals, silicon dioxide, or many other structural layers. This selectivity makes XeF_2 vapor a valuable etchant for the release step in surface micromachining processes in which polysilicon is the sacrificial layer.

Plasma etching

In **plasma etching**, the chemically reactive gas is ignited by an RF (radio frequency) electric field, usually in the range of 10-15 MHz. The reactive chemical species is contained within a plasma inside a vacuum chamber where the surface to be etched also resides. The plasma provides the necessary energy for, or "excites" the reactive gas in order to etch the wafer.

One variant of plasma etching called **plasma ashing** is commonly used to remove photoresist from wafers after completing photolithography steps. In this process an oxygen plasma converts the polymer that makes up the photoresist into carbon dioxide and water vapor, which are then removed by the vacuum pump of the ashing system.

Reactive ion etching (RIE)

In the limit of low pressures, plasma etching can take on much higher degrees of directionality, especially when the ions are directed normal to the surface. This process has become known as **reactive ion etching**, or **RIE**. RIE is capable of creating features with high aspect ratios and nearly vertical sidewalls of high surface quality. (Fig. 4.9.)

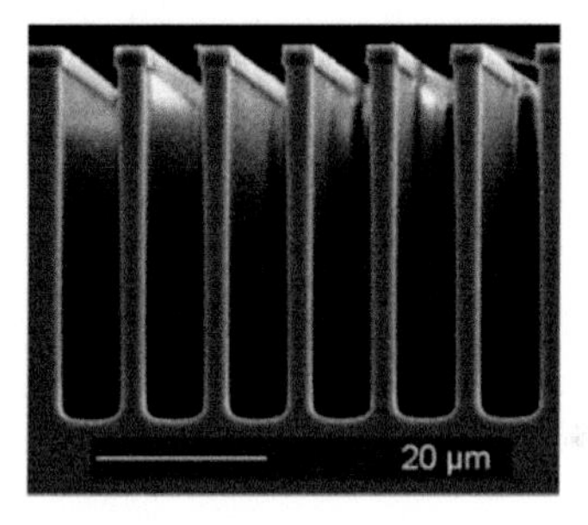

Fig. 4.9. MEMS device created using RIE. The device has a high-aspect ratio and near-perfect vertical sidewalls. (Courtesy Intellisense Corporation)

RIE is typically done by bombarding wafers with heavy ions such as Ar ions in the presence of an energetic plasma. (Fig. 4.10.) Since the ions are directed toward horizontal surfaces, they typically do not hit the sidewalls during etching. It is by this mechanism that vertical walls are produced.

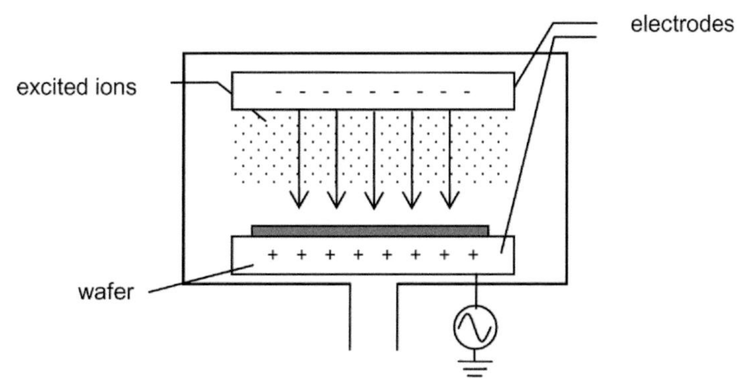

Fig. 4.10. A schematic of an RIE chamber. Electrodes produce an RF electric field that excites ions and directs them toward the surface of the wafers.

Several parameters control the effectiveness of the RIE process. These include the RF power level and frequency, plasma composition, gas flow rates, and reactor pressure. These parameters are particularly critical in achieving desired selectivities. Etch rates tend to be very high, on the order of 10 μm/min.

Table 4.3 provides a list of common materials etched using dry etching techniques and their corresponding types of reactive gases.

Table 4.3. List of materials and corresponding types of reactive gases (etchants) used in dry etching systems

Material	Reactive gas
Silicon (Crystalline or polysilicon	Chlorine-base: Cl_2, CCl_2F_2 Fluorine-base: CF_4, SF_6, NF_3
SiO_2	Fluorine-base: CF_4, SF_6, NF_3
Al	Chlorine-base: Cl_2, CCl_4, $SiCl_4$, BCl_3
Si_3N_4	Fluorine-base: CF_4, SF_6, NF_3
Photoresist	O_2 (Ashing)

4.3 Surface micromachining

Surface micromaching (SMM) is a process that uses thin-film deposition, patterning via photolithography, and chemical etching to build mechanical structures on top of a substrate, typically a silicon wafer. It is a layered fabrication process in which some layers form structural elements and others are etched away. The layers are referred to as **structural layers** and **sacrificial layers**, respectively, and form the building blocks of the process. The typical process of SMM is first to deposit a layer, then pattern it, and finally chemically etch away unwanted material. This set of steps can be repeated several times in order to create complicated structures, often with moving parts.

The SMM process was developed in the early 1980s at the University of California at Berkeley in order to produce polysilicon mechanical structures with dimensions of a few microns. Since then several sophisticated polysilicon SMM processes have been established, among which are Sandia National Lab's SUMMIT (Sandia's Ultra-planar Multi-level MEMS Technology) and MEMS CAP's polyMUMPs (Multi User MEMS Processes). The SUMMIT process can produce a five level MEMS device incorporating four depositions of polysilicon layers. The PolyMUMPs includes three layers of polysilicon in addition to a metal layer. Fig. 4.11

shows an example of a MEMS device fabricated using PolyMUMPs SMM.

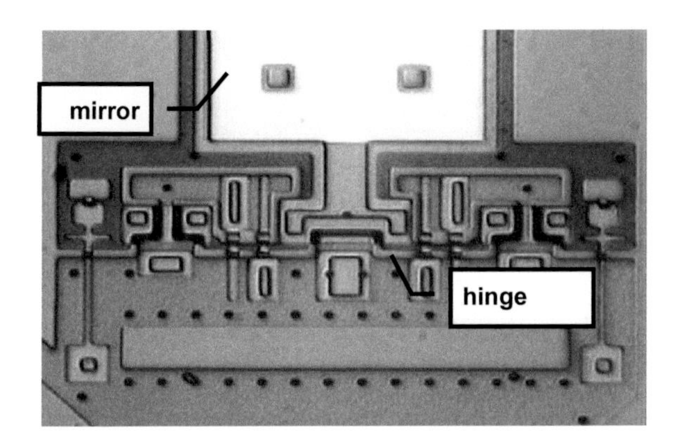

Fig. 4.11. Photo of a PolyMUMPs surface-micromachined micro-mirror. The hinge design allows for out-of-plane motion of the mirror.

SMM processes have a major advantage of being compatible with CMOS (complementary metal oxide silicon) technology used in microelectronics fabrication. Surface micromachined MEMS can therefore often be easily integrated with their control electronics on the same chip. Furthermore, many SMM processes have developed their own sets of standards, allowing for efficient and relatively inexpensive fabrication.

4.3.1 Surface micromachining processes

Surface micromachining requires three to four different materials in addition to the substrate. These include the sacrificial material, the structural/mechanical material, and in some cases electrical isolators and/or insulation materials. For a given MEMS device each of these materials may need to possess certain properties. For example, the structural material may be required to exhibit certain electrical and mechanical properties such as being a good electrical conductor while exhibiting low levels of mechanical stress.

As an example of a simple device that can easily be created using SMM, consider a cantilever as shown in Fig. 4.12. The figure shows the essential steps in the fabrication process. In this example silicon dioxide serves as the sacrificial layer and a metal is the structural material.

The oxide layer is first photolithographically patterned and then etched using HF as the etchant, resulting in an oxide layer as shown in Fig. 4.12

(a). Next, deposition of the metal structural layer takes place. This step is followed by another photolithography step and then a metal etch as shown in Fig. 4.12 (b). In the final step the sacrificial layer is removed, here using BOE (buffered oxide etchant). This removal of the sacrificial material is called **release**.

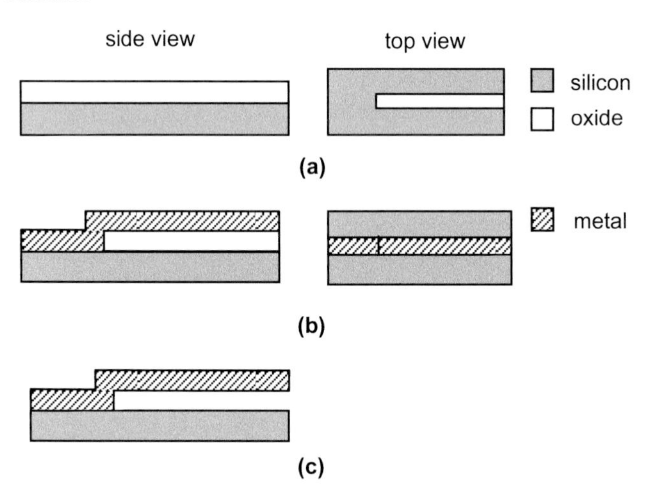

Fig. 4.12. Process steps in fabrication of a cantilever structure using surface micromachining: (a) Deposition of oxide as a sacrificial layer, (b) metal as a structural layer and (c) the release step.

It is in the release step that movable parts are created in surface micromachining. As a result release is often the most critical of the SMM processes. In order to complete this step successfully the sacrificial layer must have a much higher etch rate than the structural layers and the substrate, as well as any isolation/insulation layers that may be present. That is, the etchant should quickly remove the sacrificial layer while leaving other layers intact and undamaged. If we define the etch rate of the structural, sacrificial and substrate/isolation materials to be, R_s, R_m and R_i, respectively, this requirement amounts to

$$R_s >> R_m > R_i \,. \tag{4.4}$$

For example, buffered oxide etch (BOE) is used to remove SiO_2 sacrificial layers when polysilicon is the structural material. In this case oxide is etched at a rate of about 100 nm/min while polysilicon is etched at a rate of only 0.04 nm/min.

Typical structural/sacrificial material pairs along with the associated etchants used for release are provided in Table 4.4.

Table 4.4. Materials and etchants for typical SMM

Structural material	Sacrificial Material	Etchant
Si/Polysilicon	SiO_2	Buffered oxide etch (BOE) (HF-NH_4F ~ 1:5)
Al	Photoresist	Oxygen plasma
Polyimide	Phosphosilicate glass (PSG)	HF
Si_3N_4	Polysilicon	XeF_2

In surface micromachining methods best results are obtained when structural materials are deposited with good step coverage. Hence chemical vapor deposition (CVD) methods are preferred over PVD. When PVD methods are employed, sputtering is preferred over evaporation.

4.3.2 Problems with surface micromachining

Wet versus dry etching

The chemical etching steps in surface micromachining may be accomplished by wet or dry etching. There are advantages and disadvantages of each technique, and care must be taken when choosing one for a given process.

The main advantage of wet chemical etching is that it has been used for over 40 years by the semiconductor industry, thereby providing the MEMS designer with a wealth of data and experience. Furthermore, wet etchants have very high selectivities for different materials in addition to their ability to remove surface contaminants. However, wet etchants used in SMM tend to be isotropic, etching uniformly in all directions. Thus wet etching always undercuts the masking layer.

Dry etching techniques used in SMM include both plasma and reactive ion etching (RIE), both of which exhibit much higher aspect ratios than wet etching techniques with no undercutting and better resolution as well. Selectivity tends to be worse than in wet etching, however, and etching parameters (RF power level and frequency, plasma composition, pressure, etc.) must be tightly controlled.

Stiction

One of the most insidious problems encountered in surface micromachining occurs during release when using a wet etchant. During and/or after the release step, structures or mechanical parts may adhere to the substrate due to the surface tension of trapped liquids. This sticking of structural layers to the substrate is known as "**stiction**", coming from the combination of the words stick and friction. Stiction is a primary example of an unfavorable scaling encountered in MEMS, in this case due to the dominance of surface tension forces at small scales.

There are several solutions that have been used by researchers and MEMS manufacturers to reduce stiction. One approach calls for coating the substrate surface with a thin hydrophobic layer, thereby repelling liquid from the surface. Another popular technique is to dry surfaces using supercritical CO_2.[2] This removes fluids without allowing surface tension to form. Still other techniques utilize "stand-off bumps" on the underside of moving parts. These bumps act as pillars, propping up movable parts wherever surface tension may from. However, the most effective method for avoiding stiction is simply to use dry etching techniques for release where possible.

4.3.3 Lift-off

Lift-off is a technique that is most often lumped with the additive techniques of Chapter 2. It has much in common with surface microcmachining, however, in that a material is selectively deposited on a wafer by making use of a temporary layer which is ultimately removed.

Lift-off somewhat resembles a reverse etching technique. First photoresist is spun on a wafer and exposed, creating the desired pattern. (Fig. 4.13 (a).) Care is taken to ensure that the resist has either straight side walls, or more desirably, a reentrant shape. Next a material is deposited through the photoresist mask via a line-of-sight method, such as evaporation. (Fig. 4.13 (b).) Since a line-of-sight method is used, some shadowing takes place, leaving at least part of the photoresist sidewalls without any deposited material on them. When the photoresist is stripped, it therefore leaves behind only the material deposited through the opening, lifting off the material on top of it. (Fig. 4.13 (c).) Naturally the thickness of the deposited material needs to be thin compared to the resist thickness for the technique to work. Lift-off is most often used to deposit metals, especially those that are hard to etch using plasmas.

[2] Every fluid has a unique temperature and pressure, called the critical point, above which the distinction between gas and liquid disappears. A supercritical fluid is one which exists at temperatures and pressures above this point. Since surface tension forces form at the interface between a liquid and a gas, supercritical fluids cannot produce surface tension.

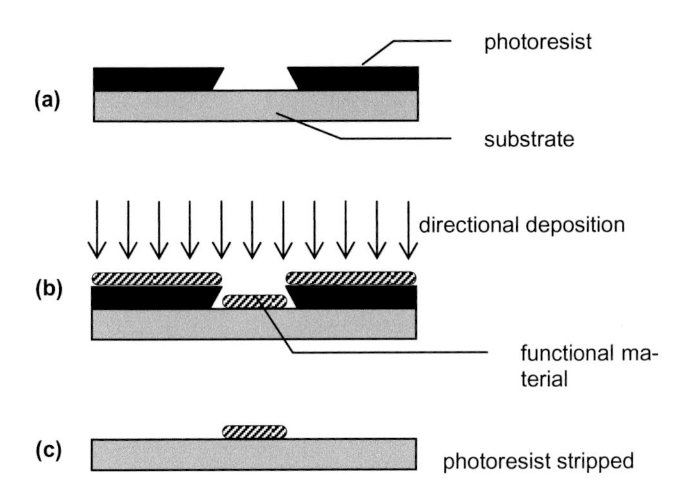

Fig. 4.13. Steps in a lift-off process: (a) Photoresist on a substrate; (b) Material deposited via a line-of sight method; (c) Photoresist stripped, lifting off unwanted material

4.4 Process integration

So far we have discussed many techniques for processing materials to create MEMS devices, including oxidation, PVD, patterning via photolithography, and bulk and surface micromachining. In this section we turn our attention to describing how to put all these steps together in order to create a MEMS device.

This can be a difficult task, as not only must we be able to choose the materials and processes to fabricate a MEMS, but we must also be able to put them *in the correct order*. A list of all these necessary fabrication steps in sequential order is called a **process flow**.

Process flows exist at many different levels of detail. The required process flow for actually creating a MEMS in a microfabrication facility should be highly detailed, with specific materials, processes, dimensions, times, etc., all given in sequential order. Such a process flow essentially serves as a step-by-step set of instructions to fabricate a device. When one is first figuring out how to make a particular MEMS device, however, many of these steps can be left out. As an example of the latter, consider Fig. 4.14, in which the process flow steps are given for a typical surface micromachined MEMS.

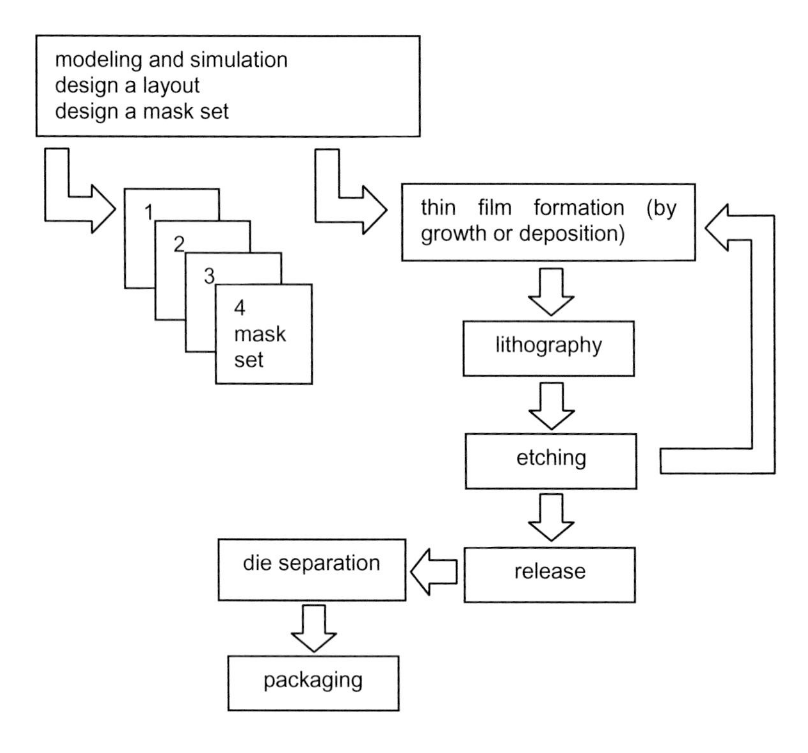

Fig. 4.14. Typical process steps for a generic surface micromachining process

On the right of Fig. 4.14 you should see several process steps that are familiar to you. The order in which they appear should also come as no surprise. That is, thin film formation typically precedes lithography, which in turn is followed by etching. Where creating process flows gets tricky is in repeating these steps. To successfully create a functioning MEMS device, we must often build it layer by layer, using different masks for each. Indeed, designing a proper set of masks to be used in the correct order forms a large part of developing most process flows.

After the mechanical parts are released, Fig. 4.14 shows two additional steps which heretofore have not been discussed. For batch fabrication processes in which many devices are made at once on the same wafer, the wafer must be cut apart in order to separate the individual devices. This step is called **die separation**. Die separation is typically accomplished by physically sawing or scribing the wafer. MEMS devices must also be **packaged** in order to provide them with the appropriate electrical connections and also to protect them from the environment in which they are used. In many cases the packaging must provide both protection from the environment and some limited access to it, as in the case of a pressure sensor or an inkjet print head. Issues like these have made packaging a difficult engineering problem, resulting in packaging forming the largest cost of producing many devices. In fact some MEMS professionals claim that one receives MEMS devices for free and pays for their packaging.

Though Fig. 4.14 shows die separation occurring before packaging, many process flows call for at least some degree of packaging to occur at the wafer level, which can have many advantages. This is not always possible, and which step comes first depends strongly on the device to be fabricated. And so we see that although general guidelines can be given, the process flow for every MEMS device will be unique.

4.4.1 A surface micromachining example

Since process flows can vary so much from device to device, learning to create one is probably best taught by example. As such an example, consider creating a process flow and a corresponding mask set that can be used to produce a wheel that is free to rotate about a hub as shown in Fig. 4.15.

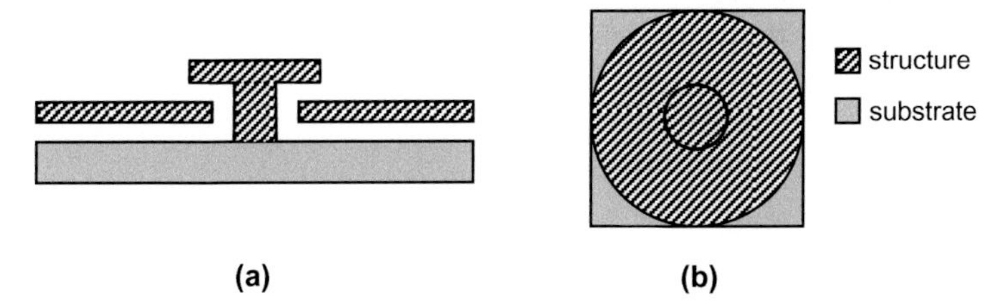

Fig. 4.15. A MEMS wheel and hub (a) Side view (b) Top view

For this example we will use surface micromachining techniques and assume that all photoresist is positive.[3] At first we will be interested in the major fabrication steps only. Details such as mask design, material choices, etchants, etch rates, necessary thickness, etc. will be filled in later.

From the last section we already know that the process flow will consist primarily of thin film deposition, photolithography, and etching steps. By studying Fig. 4.15 we see that there will need to be at least two structural

[3] Remember that for positive photoresist the exposed regions are dissolved during development, resulting in a resist pattern that matches the mask. Therefore, patterns etched through such resist layers also match the mask pattern used in the photolithography step.

layers, one for the wheel and one for the hub. Each of these requires a sacrificial layer. Our first attempt at a process flow might therefore look like this:

1. Deposit sacrificial layer
2. Deposit structural material for wheel
3. Do lithography
4. Etch sacrificial material (release wheel)
5. Deposit more sacrificial material
6. Deposit structural material for hub
7. Do lithography
8. Etch sacrificial material (release hub)

When we critically review the steps above, we will find that this process flow is flawed. Mainly, the wheel is completely patterned and released before the hub has been fabricated. This means that once the wheel is released it will come right off the wafer! This can be remedied if we have only one release step instead of two, releasing the hub and wheel at the same time.

Another problem with the above process flow is that it does not allow for the fact that the hub material extends all the way from the substrate, through the center of the wheel, and also above it. To fix this we will need to create a hole through the second sacrificial layer, the first structural layer (the wheel layer) and the first sacrificial layer before depositing the second structural layer.[4] With these ideas in mind, our second attempt at a process flow may look like this:

1. Deposit sacrificial layer
2. Deposit structural material for wheel
3. Do lithography using mask 1
4. Etch wheel material through photoresist
5. Deposit more sacrificial material
6. Do lithography using mask 2
7. Etch to create hole for hub
8. Deposit structural material for hub
9. Do lithography using mask 3
10. Etch to get hub
11. Etch sacrificial material (release)

Fig. 4.16 shows what the device looks like after each of the fabrication steps listed above.

[4] Such through-holes are often called **vias** in microfabrication.

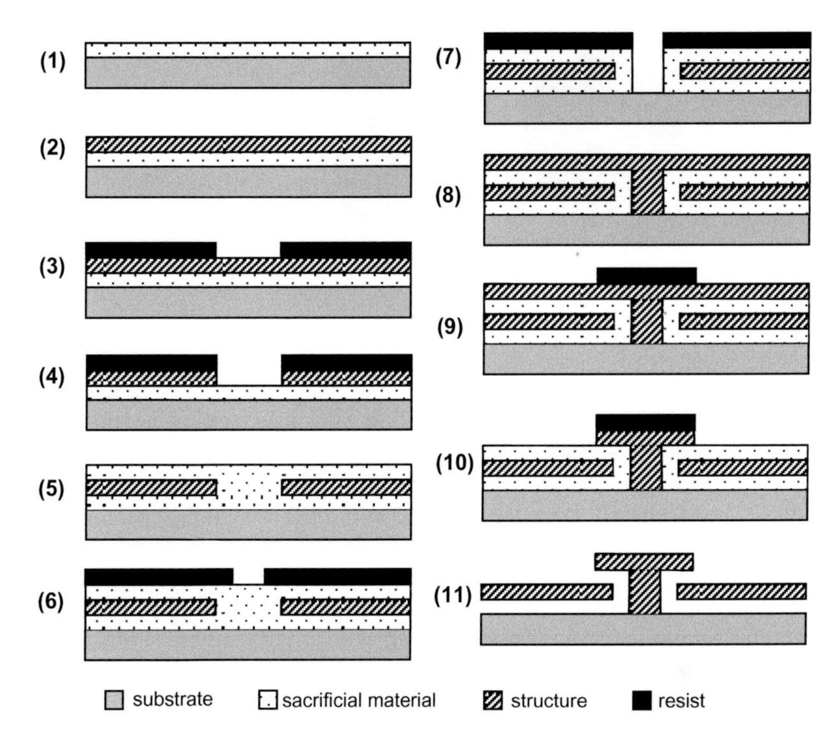

Fig. 4.16. Fabrication steps for MEMS wheel and hub device using SMM

From the improved process flow we see that we now need to think about mask design. Three different masks are required for the device: one to pattern the wheel, one for the through-hole, and one for the hub, in that order. The mask designs for these three tasks are given in Fig. 4.17. Since we are using positive photoresist throughout, the clear regions in the masks correspond to the regions where material is etched away after the photolithography step, whereas the opaque regions are left behind. (To help remember this many people use the mnemonic "Positive resist protects".)

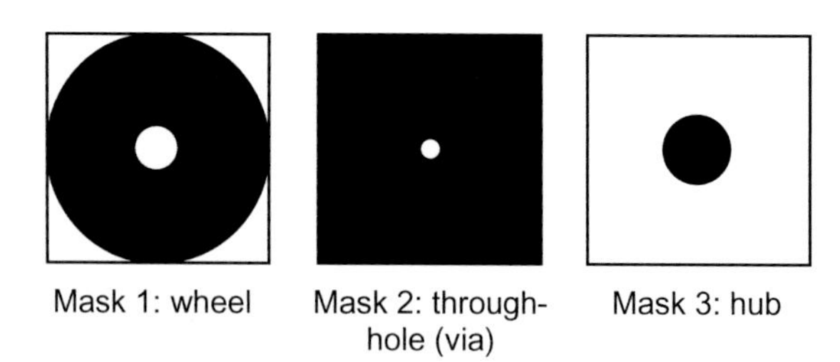

Fig. 4.17. Three-mask set used in wheel/hub fabrication

We are now in a position to start thinking about which materials to use for the fabrication. Since the wheel and hub are released at once, both structures can be made from the same material. We can consult Table 4.4 for candidates for the structure/sacrificial materials. Let's chose polysilicon as the structural material and silicon dioxide for the sacrificial material, using buffered HF as the etchant for the sacrificial material. Next we consult Table 4.1 in search of an etchant for use with polysilicon. KOH looks like a good candidate, as it etches silicon but does not attack either SiO_2 or photoresist significantly. With these choices, we can add a bit more detail to the process flow:

Starting material: (100) Silicon wafer

1. RCA clean
2. Grow oxide - Wet oxidation for a thin sacrificial layer
3. Deposit polysilicon film - Deposit poly for wheel using CVD
4. Photolithography - Mask 1: pattern poly to get wheel
5. Etch - Remove poly using KOH
6. Strip - Strip photoresist
7. Oxidize - Grow more oxide to create space between wheel and hub
8. Photolithography - Mask 2: pattern though-hole (via) into oxide for hub
9. Etch - Etch via into SiO_2 using BOE
10. Strip - Strip photoresist
11. RCA clean
12. Deposit polysilicon film- Deposit poly for hub using CVD (will fill in through-hole)
13. Photolithography - Mask 3: pattern hub
14. Etch - Remove poly using KOH
15. Strip - Remove photoresist
16. Release structure - Etch sacrificial SiO_2 using BOE

17. Dry in a way to avoid stiction

This latest version of the process flow contains more information still, including cleaning steps and mentioning specific fabrication processes by name, such as depositing the polysilicon layers using chemical vapor deposition. The process flow is starting to resemble one that can be used in a fabrication facility. It is still not complete, however, since such details as times in the oxidation furnace, required thicknesses of the deposited layers, when and where to add alignment marks on the wafer, etc., are not yet included. Nonetheless, this example should have given you a good idea of how a process flow evolves.

Even after this flow is completed, it still may need to be revised after attempting to fabricate the wheel and hub. For instance, step 9 calls for etching a via through a thick layer of SiO_2 using BOE. Buffered oxide etch is an isotropic etchant in SiO_2, however, and some undercutting of the mask will occur. If too much undercutting takes place we could make the hole too big or even release the wheel prematurely. (Fig. 4.18.) If this happens, we may need to adjust the size of the hole in Mask 2, or even go to a dry etching technique that offers more directionality. There are always unknowns in microfabrication, and there is no replacing experience and some amount of experimentation.

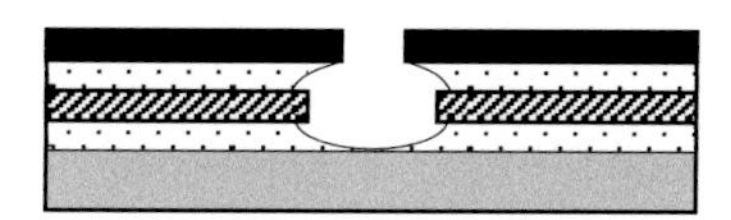

Fig. 4.18. Too much undercutting could necessitate changing the process flow.

4.4.2 Designing a good MEMS process flow

Now that we have seen the evolution of a process flow, we are in a better position to discuss some general guidelines in creating one. There are many issues affecting the choices in creating a process flow, and consequently many ways to design a MEMS device. There is no one right or wrong way to go about this task, but there are many things that should be considered well before attempting to fabricate a device for the first time.

Several rules of thumb have developed to aid in the design of a good process flows. For example, cleaning steps are generally needed before deposition steps and high temperature processes. Also, there are some generally held guidelines for the placement and use of alignment marks. Some good advice to follow includes:

1. *Use assymetric alignment marks.* This helps assess wafer orientation as well as alignment.
2. *Make sure alignment marks on masks do not obsure those on the wafer*
3. *Use several different kinds of marks.* Larger marks, for example, are bst used for general alignment, whereas the size of most other marks should be on the same order as the device's smallest feature size.
4. *Use more than one mark and place them across from each other on a surface.* In this way errors in rotational alignemnt can be minimized.

Long before this level of detail begins, however, one must answer some basic questions. One of the first is whether or not to integrate the MEMS device and any necessary electronics on the same chip; that is, the question of **system partitioning**. In certain cases it may be desirable to integrate these components. For example, one may want to incorporate the electronics that tell a car's airbag when to deploy on the same chip as a MEMS accelerometer. This option creates much less flexibility in design decisions, however, as the IC industry has very strict guidelines as to what types of materials can be used in microelectronics, and even what materials can be brought into fabrication facilities. Some metals used in MEMS, for instance, may diffuse into the substrate, interfering with the functionality of microelectronics components. Substrates must also be limited to semiconductor materials, whereas for MEMS devices by themselves many more substrate options exist. Furthermore, there is a large difference between device dimensions for MEMS and integrated circuits, as well as differences in etch times, growth times, etc. Thermal processing steps must also be limited.

One alternative is to use multi-module systems when possible. In a multi-module system the MEMS device and its associated electronics are each manufactured separately and then put together in the final package.

If a one-chip solution must be used, there are several considerations. High temperature mechanical processes can melt metals used in electronics, or further drive in existing dopants. When performing a microelectronics fabrication step, the material in existing mechanical structures may diffuse into the electronics. Furthermore, more than three layers of metal are typically needed in microelectronics, which can drastically affect the mechanical properties of mechanical structures. In order to minimize these effects, it is usually best to process all the electronics first and then all the mechanical parts, or *vice versa*.

System partitioning is just one of many concerns to be addressed in developing a MEMS fabrication procedure. What follows are some of the other major issues that MEMS designers think about when designing a MEMS process flow.

Process partitioning

Will a material used in one process bond with and/or affect the properties of materials in other processes? If so, the order of the process steps may matter significantly. For example, we have already seen that some deposited metals can diffuse into a silicon substrate and interfere with existing electronics for an integrated device.

So far these issues have not affected MEMS devices by themselves as much as for integrated devices. However, new sensing and actuating methods are being developed all the time within MEMS, bringing with them an ever growing array of materials into the fabrication laboratory.

Backside processing

Many MEMS devices are fabricated using both sides of the substrate. A good example is a bulk micromachined MEMS pressure sensor. (Fig. 4.19.) In this device a thin diaphragm is created by etching through the bottom of the wafer. Etching through the bottom of the wafer in this way is an example of **backside processing**. Piezoresistive sensing elements and metal interconnects are also needed for the device, but these are processed from the top.

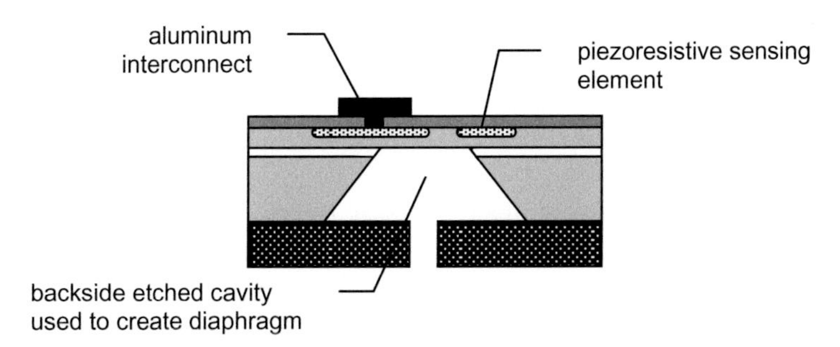

Fig. 4.19. Bulk micromachined pressure sensors are examples of devices requiring backside processing.

The use of backside processing can greatly simplify the fabrication of a device. There are certainly tradeoffs, however. For example, alignment will become much more of an issue when the substrate has to be flipped over several times during a process flow. Furthermore, processes such as oxidation and chemical vapor deposition affect both the front and back of a wafer. It may therefore become necessary to actively protect the backside from these processes if they interfere with backside processing steps that follow.

Even if a MEMS device does not require any backside processing *per se*, it may still be necessary to protect it. The presence of an oxide layer on the backside of a wafer can interfere with die-attachment contacts, for example. Residual stresses in backside thin films can also cause the substrate to bend unintentionally. (We will discuss stress and its consequences in detail in Chapter 5.)

Thermal Constraints

Materials currently on a device cannot always handle a high temperature step to come. For example, photoresist cannot stand up to the elevated temperatures associated with oxidation, diffusion or low pressure chemical vapor deposition (LPCVD), and it must therefore be removed before these processes. Furthermore, every time a wafer is heated dopants diffuse more, creating the need to adjust drive-in times.

Another tradeoff resulting from a thermal constraint involves the use of metals and insulators. Many devices that employ metal also require insulator materials such as dielectric films, which can be readily deposited using LPCVD, a high temperature process. Metals generally cannot be subjected to high temperature processes, however, lest they start flowing or diffusing. The deposition of insulator materials after metals must therefore be done using a lower temperature process such as PECVD, usually resulting in a lower quality film.

Device geometry

If you were only to review the three mask set given in Fig. 4.17, it would be very difficult to visualize the three dimensional wheel and hub device that they are intended to create. Probably the best solution to this visualization problem is to make use of the solid-modeling capabilities of CAD software developed specifically for MEMS.

One may also draw several different cross sections of the MEMS device from many angles. This approach has the advantage of being able to predict where unwanted structural features may form (usually at the edges of the device) as well as the position of the **stringers**, pieces of unetched material unintentionally left behind after an etching step. These features can interfere with subsequent fabrication steps, particularly photolithography, deposition of structural layers, and bonding during packaging. Well designed process flows aim to avoid the formation of these features altogether.

Even in the absence of stringers, just the topography of a MEMS device after a certain fabrication step may interfere with subsequent steps. In some cases it is necessary to even out this topography in a process called **planarization**. Planarization processes include chemical mechanical polishing (CMP) and spin-casting polymer films, particularly photoresist. Planarization is often performed in surface micromachining before the addition of sacrificial layers in order to facilitate the release step.

Mechanical stability

The fabrication of thin films can result in residual stresses within them, which in turn affects the mechanical behavior of structural elements. Such stresses can cause structures to bend, or in severe cases even to break. The formation of these stresses is material and process dependent. Oxide layers tend to induce compressive stress on silicon substrates, for example, whereas nitride films likely result in tensile stress. Metal layers can result in either depending on deposition conditions and thermal history.

In surface micromachining it is the release step that causes structures to experience the most rapid changes in their state of stress, thereby incurring the greatest likelihood of fracture. Silicon dioxide sacrificial layers can exacerbate this effect, and sometimes different materials must be used. Experimentation with thin film deposition can be used to offset these risks as well, as can doing simulations of the fabrication process.

Process accuracy

Many processing steps are prone to variations that can lead to unintended outcomes. Expansion or shrinkage of photoresist during exposure or development, for example, can cause masks to be reproduced inaccurately on the wafer. Variation in the thickness across photoresist layers can be a problem as well.

PR (in any deposited thin films) will reduce yield. Undercutting in etching and mask alignment errors (random variations) needs to be considered.

Alignment issues can be particularly difficult. Alignment marks on wafers are not only used for photolithographic alignment as discussed in Chapter 3, but also to identify crystalline plane directions. A good practice is to bulk micromachine features on at least three corners of the wafer to serve as alignment marks. They are often etched by first patterning an oxide layer and then using it as a hard mask to etch the silicon substrate with TMAH. These marks must be visible after each thin film deposition process, and film materials and their thicknesses must therefore be considered carefully when developing a process flow.

Non-uniformity in thin film layers across a wafer can cause variations in feature sizes. This is typically the worst at wafer edges where etchants can more easily find their way to the material to be etched. Structures at the edge of wafers are therefore more likely to be over-etched, and consequently, to fail.

Die separation and packaging

When one separates a die by sawing or scribing the wafer it create lots of dust particles that can get into mechanical structures, preventing them from moving freely. Subsequently cleaning them, however, can leave liquid in small gaps, leading to stiction. For these reasons it is desirable to encapsulate the moving parts before separating whenever possible. For example, silicon wafers can be bonded to patterned glass wafers before die separation. Such a process is known as *wafer-level packaging*.

In certain applications like chemical, optical and thermal sensing, environmental access *and* vacuum packaging may both be required. Packaging in these cases becomes a formidable challenge indeed, especially when searching for a wafer-level packaging scheme. The last hill on a process flow is often the hardest to climb.

4.4.3 Last thoughts

Ultimately good process design must deal with three critical issues: device performance, manufacturability, and cost. Carefully considering the tradeoffs and finding the optimum balance among these three issues is the mark of a good designer, whether s/he be in the field of MEMS or not.

One approach has been to create a base design which incorporates the minimum required functionality of a device in order to minimize cost and maximize manufacturability. From there the base design is tweaked as needed in search of an optimum. Of course in the end it is the market that decides whether the final product makes the grade or not.

One should not discount the role of serendipity and inspired creativity, however, as designs that can win in all three critical issues are possible. A classic example of a process flow that does just that is the self-aligned gate metal-oxide-semiconductor field-effect transistor (MOSFET) in microelectronics.

Fig. 4.20 shows schematics of two MOSFETs fabricated using different techniques. The purpose of the devices and the details of their operation are not important for our discussion here. We need only know that a MOSFET requires heavily doped (n^+) regions in a p-type substrate, a thin oxide layer on top of these, and a deposited material on top of the oxide to serve as an electrode. We are primarily interested in how the differences in the process flows have made the device in Fig. 4.20 (b) much more desirable than that in Fig. 4.20 (a).

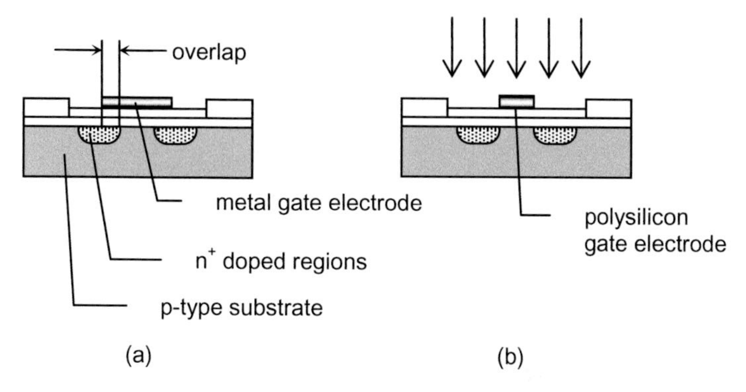

Fig. 4.20. MOSFETs created using different process flows: (a) The overlap between gate electrode and diffused regions causes unwanted capacitance. (b) The self-aligned gate demonstrates greatly enhanced device performance.

Older MOSFETs looked like that shown in Fig. 4.20 (a). In these devices the n^+ regions were diffused first, after which a metal was deposited and patterned for the gate. An overlap between the gate and the n^+ regions was purposely built into the process flow in order to minimize the likelihood of the type of photolithographic misalignment discussed in the last section. This overlap causes an unwanted increase in device capacitance, one which reduces transistor switching speed. And so we see an obvious tradeoff between manufacturability and device performance.

In the process flow for the device in Fig. 4.20 (b), a polysilicon gate electrode is deposited and patterned first, and then the doped regions are created using ion implantation using the polysilicon electrode as a mask. The implantation is followed by a high temperature drive-in.

Polysilicon is used for the device in Fig. 4.20 (b) instead of a metal since metals can't handle the high temperature drive-in step. The smaller electrical conductivity of polysilicon compared to metals causes a decrease in device performance. However, for the device in Fig. 4.20 (b) there is virtually no overlap between the electrode and the doped regions. (This is why it has become known as a self-aligned gate.) This results in a much smaller capacitance, which leads to much faster transistor switching speeds. The corresponding increase in device performance dwarfs any drawbacks of the polysilicon electrode material, and the process flow is somewhat simpler too. Rather than a tradeoff between performance and manufacturability we have a "win-win" situation. It has changed the manufacture of integrated circuits significantly.

Essay: Introduction to MEMS Packaging

Michael F. McInerney
Department of Physics and Optical Engineering, Rose-Hulman Institute of Technology

Background

The packaging of MEMS devices has largely been developed within the electronic device community. Techniques have evolved from the first transistors to modern integrated circuits and continue to evolve as packages get smaller and smaller. Size reduction necessitates automation with consequent capital investment that slows change of traditional packaging and puts pressure on MEMS packaging to conform to existing norms.

Function of packaging

Microscale devices are fragile and difficult to interface in their 'raw' state. Their intricate structures can be destroyed by corrosion and simple physical interference. They are too small to easily interface so packaging must both protect and provide access. The packaging of MEMS devices is further complicated by their great variety, often including moving parts and a need for direct access to the atmosphere.

The properties required of device packaging can be divided into mechanical support; the provision of access and protection; and thermal management. Mechanical support must minimize stress due to external loads. The device must not physically break while in normal use. This support must also allow access, usually in the form of conducting leads and not cause the development of internal stress in the package.

Mechanical support is traditionally provided by placing the device on a metal or ceramic *lead frame*, a thin layer of metal that connects the wiring from tiny electrical terminals on the semiconductor surface to the large-scale circuitry on electrical devices and circuit boards. This provides basic rigidity and electrical connection between the device and the outside world. Protection from the environment is obtained by placing the device and leadframe within a box or by encapsulation in a polymer. Many modern devices must also protect against electromagnetic interference.

The electrical connections must allow sufficient power to be delivered to the device, they must be thick enough not to melt under the current load.

The increased density of devices on a single chip has resulted in a power law (Rent's law) increase in the number of data lines required as a function of time. This increase is posing serious problems in micro packaging but they are mostly confined to processor and memory chips. MEMS generally require at present (2009) far fewer connections.

Miniaturization has decreased the power per unit 'activity' but has dramatically increased power density and associated heating. Packaging must remove this heat before it damages the device while at the same time not stressing the device due to differential thermal expansion and contraction.

Packaging levels

Electronic packaging is diverse but can be broadly divided into four levels, zero through three. These are, in turn, wafer, die, device and systems levels. Wafer-level refers to operations on the wafer before singulation (separation into single devices). The Die Level refers to operations performed immediately after the die is singulated. A completed die level package is ready for insertion into a circuit board for device level packaging. Level three is the 'Systems Level' and refers to the final product package such as a television or radio or airbag controller.

Packaging requirements peculiar to MEMS

MEMS packaging differs in significant ways from traditional electronic packaging. MEMS often have moving parts that must be protected and they often have to interact directly with a medium. Accelerometers are examples of the former and pressure sensors an example of the latter. An electronic chip must often be included in the package with them.

The figure below gives an example of a packaged MEMS, specifically a MEMS pressure detector. The only movement in the illustrated device is the flexing of the strain gauges as the pressure changes. But many MEMS have moving parts that require caps to protect them. The pressure device also makes use of caps and the placing and bonding of caps is an important part of MEMS packaging. Methods of bonding caps include epoxy adhesive, gold/silicon eutectic, anodic bonding, silicon fusion and welding.

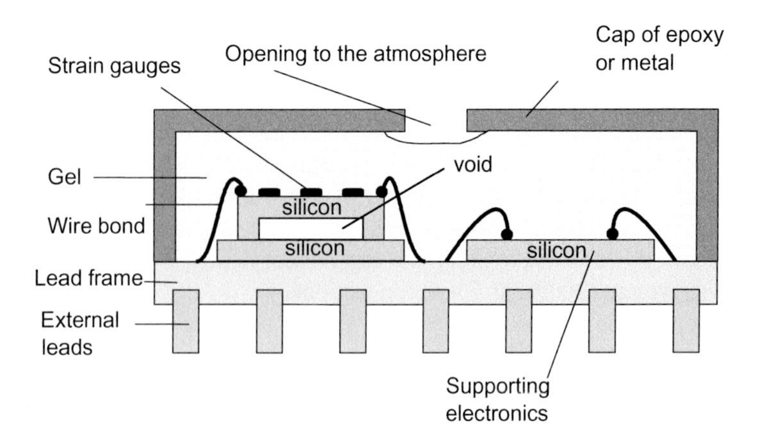

Schematic of a packaged MEMS pressure detector

The figure should make clear some of the packaging steps required. The MEMS must be made and a silicon cap formed over a void with the strain gauges in the cap. The silicon wafers must be bonded to the lead frame, correct power and data connections must be made with wire bonds or bumps, the hard cap with an opening must be placed over everything, and a gel dispensed inside so that corrosive elements are kept out.

References and suggested reading

Campbell SA (1996) The Science and Engineering of Microelectronic Fabrication. Oxford University Press, New York

Franssila S (2004) Introduction to Microfabrication. Wiley, Chichester, West Sussex, England

Jaeger RC (2002) Introduction to Microelectronic Fabrication, 2nd edn. Prentice Hall, Upper Saddle River, NJ

Madou MJ (2002) Fundamentals of Microfabrication, The Art and Science of Miniaturization, 2nd Ed. CRC Press, New York

Maluf M, Williams K (2004) An Introduction to Microelectromechanical Systems Engineering, 2nd edn. Artech House, Norwood, MA

Senturia S (2001) Microsystem Design. Kluwer Academic Publishers, Boston

Tuckerman DB, Pease RFW (1981) High-Performance Heat Sinking for VLSI. IEEE Electron Device Lett., EDL-2:126-29

Wang Z (2004) Design, Modeling and Fabrication of a TiNi MEMS Heat Engine. Master's thesis, Rose-Hulman Institute of Technology

Wolf S, Tauber RN (2000) Silicon Processing for the VLSI Era, Vol.1: Process Technology, 2nd edn. Lattice Press, Sunset Beach, CA

Questions and problems

4.1 You are planning to etch a pattern into a Si wafer using KOH as the etchant. Which is the more acceptable mask material, an oxide layer or photoresist? Why?

4.2 What is undercutting? Give one example of when it is considered bad and one example of when it can be useful.

4.3 What does "BOE" stand for, and what is its approximate composition?

4.4 What is the difference between isotropic etching and anisotropic etching?

4.5 What is selectivity?

4.6 Describe two models that attempt to explain the selective etch rates for the various crystalline planes in the anisotropic etching of silicon.

4.7 What is etch stop? Give two examples.

4.8 Give an advantage and a disadvantage of wet etching.

4.9 Give an advantage and a disadvantage of dry etching.

4.10 If you needed to fabricate a MEMS device with straight sidewalls with high aspect ratios, would you be more likely to use a wet etching or a dry etching technique? Why?

4.11 Explain the basic process of surface micromachining. How does it differ from bulk micromachining?

4.12 Why are chemical vapor deposition (CVD) techniques preferred over PVD methods in depositing structural layers in surface micromachining?

4.13 What is stiction? Give two methods commonly used to avoid it.

4.14 What is lift-off?

4.15 List four issues that may figure into the order of the process steps for fabricating a MEMS. Explain their relevance.

4.16 What is meant by backside processing?

4.17 Sketch the cross-sections resulting from anisotropically etching the silicon wafers shown with the given masks.

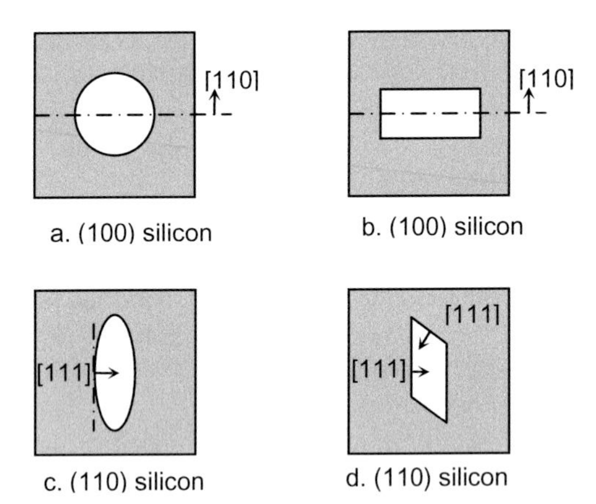

4.18 Starting with a (100) wafer that is 500 μm thick, calculate the time required to produce a 400 μm x 400 μm diaphragm that is 20 μm thick. The etchant is 33% KOH with an approximate etch rate of 10 μm/h for the {100} planes and an etch rate of 10 nm/h for the {111} planes.

4.19 For the last set of process steps given in the surface micromachining example in section 4.4.1, add information about alignment and the types of alignment marks to be used.

4.20 You are asked to make v-shaped grooves 60 μm deep in an oxidized (100) silicon wafer.
   a. How wide must the openings in the oxide mask be in order to achieve this result?
   b. The etch rate for the {111} plane is 10 nm/h. Will the degree of undercutting, due to etching into the {111} plane, be appreciable compared to the dimensions of the desired feature?

4.21 Develop a general level process flow that will produce a polysilicon cantilever beam whose deflection is sensed by a piezoresistor. A sketch is shown below. Be sure to list what chemicals and materials you have chosen for your design. (Note: The piezoresistor can be implemented using boron doping in polysilicon.)

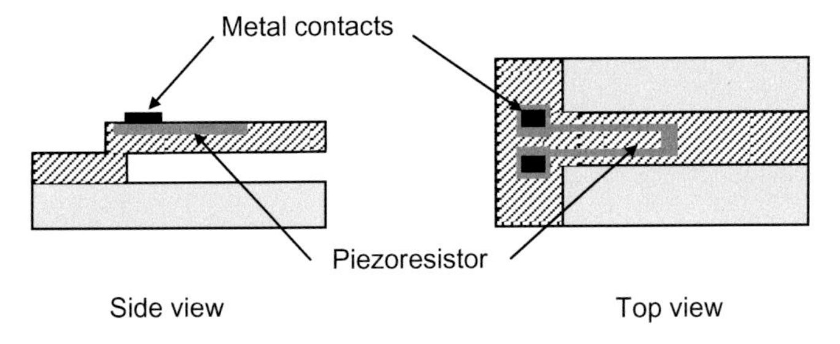

Chapter 5. Solid mechanics

5.1 Introduction

We have already examined silicon in terms of its crystalline structure and properties. However, as exemplified by the second M in MEMS, (the one that stands for "Mechanical") MEMS devices often have moving parts. Therefore an understanding of forces, displacements and other mechanical concepts is paramount when fabricating a MEMS device. In particular, applied forces, both intentional and unintentional, can lead to deformation of the elements composing a MEMS. Furthermore, the effects of thermal expansion can also lead to deformation, which in turn results in stresses in MEMS structures. The branch of physics relating all these mechanical parameters of solids is aptly known as **solid mechanics**.

The presentation of solid mechanics given here is intended to serve as a basis for understanding the bending, stretching and compression of components that can occur during the fabrication process, as well as to serve as a foundation for some of the actuation and sensing principles explored in the later chapters of the text. Due to the small scales encountered in MEMS, some of the solid mechanics concepts presented may be unfamiliar even to those readers who have had an introductory course on the subject. Nonetheless, only those concepts most important in regards to MEMS and microfabrication are developed. We neither assume nor require prior experience in the area.

5.2 Fundamentals of solid mechanics

Our study of solid mechanics will start with an introduction of the basic solid mechanics definitions of **stress**, **deformation** and **strain**. We then explore how to relate these quantities using **elastic theory** as a model. Next, a discussion of thermal effects is given through the introduction of the **thermal expansion coefficient**, which is a convenient way to link the

T.M. Adams, R.A. Layton, *Introductory MEMS: Fabrication and Applications*,
DOI 10.1007/978-0-387-09511-0_5, © Springer Science+Business Media, LLC 2010

thermal properties of MEMS materials to the mechanical behavior that often results from thermal processing.

5.2.1 Stress

Stress is necessarily a *surface phenomenon*. When speaking of stress, it must always be in relation to some surface area. The surface itself may be a real surface, such as a table top, or an imagined surface within a piece of solid material. Stress has the dimensions of force per unit area. A typical unit of stress is a N/m^2, or a pascal (Pa).

$$[Stress] \doteq \frac{[Force]}{[Area]} \tag{5.1}$$

There are different kinds of stresses based on the orientation of the force to the surface area in question. Forces per unit area oriented normal to that area are known as **normal stresses** and are usually denoted by the symbol σ. Those forces per unit area in the plane of the surface area are called **shear stresses**. These are given the symbol τ. Normal and shear stresses are both further classified depending on which surface and/or direction is of interest. The use of subscripts helps us identify the surfaces and directions.

Figure 5.1 shows an infinitely small arbitrary element within a solid piece of material. Shown on the element of Fig. 5.1 are three normal stresses and six shear stresses. The normal stresses include the normal stresses on the x, y and z surfaces, denoted by σ_x, σ_y, and σ_z, respectively. The shear stresses are denoted with two subscripts. The first subscript refers to the surface, whereas the second refers to the direction of the stress. For example, τ_{xy}, refers to the shear stress on the x plane in the y direction; τ_{yz} refers to the shear stress on the y plane in the z direction. Not shown in the figure are the stresses on the back, left and bottom faces of the element. The quantities are assumed positive in the direction they are drawn.

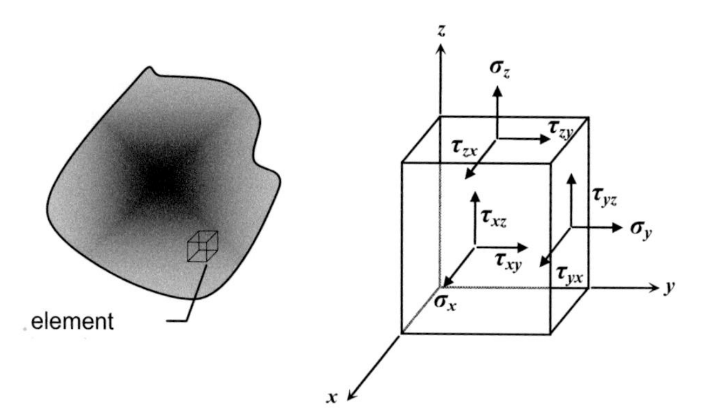

Fig. 5.1. Arbitrary stress element

Upon first inspection it may appear that there are a lot more stresses than one can readily keep track of. However, not all the stresses in an element are independent of each other. In the absence of body forces (e.g., gravity) the laws of conservation of linear and angular momentum require that the normal forces on either side of the element be equal, and that the shear stresses with subscripts interchanged also be equal. That is,

$$\tau_{xy} = \tau_{yx} \tag{5.2}$$

$$\tau_{yz} = \tau_{zy} \tag{5.3}$$

$$\tau_{xz} = \tau_{zx} \tag{5.4}$$

This leaves six independent components of stress, σ_x, σ_y, σ_z, τ_{xy}, τ_{yz}, and τ_{xz}. If this were not the case, then our infinitely small element would take on infinite acceleration!

5.2.2 Strain

Conceptually, strain is nothing more than differential deformation:

$$Strain = \frac{[Change\ in\ element\ length]}{[Original\ element\ length]}. \tag{5.5}$$

Hence strain is dimensionless and has no units. However, when most solid materials deform, they deform very little. And so a pseudo-unit called a

microstrain (μ-strain) is often used for strain. A microstrain is 10^{-6}. That is, a solid experiencing 1 μ-strain has deformed 1×10^{-6} times its original length.

In general, a material element can deform into seemingly arbitrary shapes. Figure 5.2 (a) shows a two dimensional element doing just that. Since the element's resulting shape after deformation may be fairly complex, it is useful to break up the deformation into two pieces. One piece corresponds to the change in length in one direction only and is dubbed **normal strain**. The other piece corresponds to the deformation that occurs with no change in volume (or area in the two dimensional case). This is the **shear strain**. Figure 5.2 (b) shows the two components of strain. (The *y*-component of axial strain is not shown in Fig. 5.2 (b).)

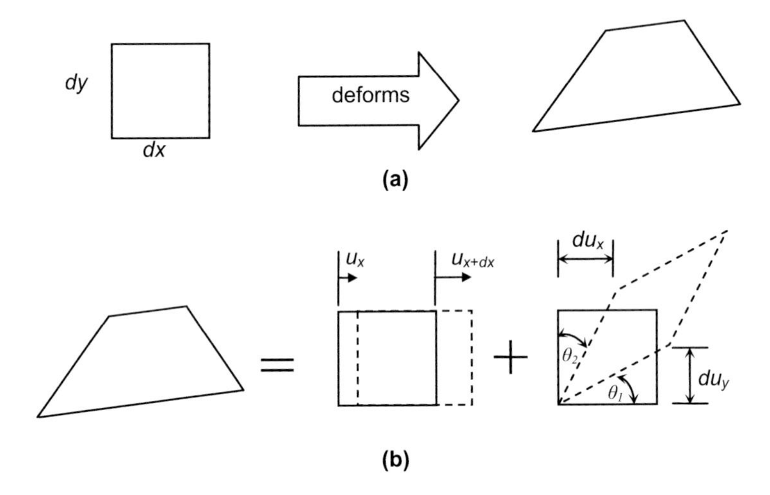

Fig. 5.2 (a) A two dimensional element deforming into an arbitrary shape; (b) Normal and shear components of strain

We now explore the formal mathematical definitions of strain. The definition of normal strain is the most intuitive, and so we begin there. If we let u be displacement, then the x direction normal strain is given by

$$\varepsilon_x = \frac{du_x}{dx} \tag{5.6}$$

We see that this corresponds to the change in x direction length of the element per unit length. Note that if displacement is not a function of direction, then there is no deformation and therefore no axial strain.

Shear strain's formal definition is given by

$$\gamma_{xy} = \left(\frac{du_x}{dy} + \frac{du_y}{dx} \right). \tag{5.7}$$

If the strain is small (and it almost always is) then the small angle approximations of trigonometry allow us to interpret shear strain as the sum of the angle θ_1 and θ_2 in Fig. 5.2 (b).

5.2.3 Elasticity

One question that naturally arises is if stress is related to strain. Indeed it is! If one applies a normal stress to thin rod of material and watches how the material deforms under the action of the stress, one can generate a plot similar to that of Fig. 5.3. In Fig. 5.3 only the normal stress is given as a function of normal strain in the same direction.

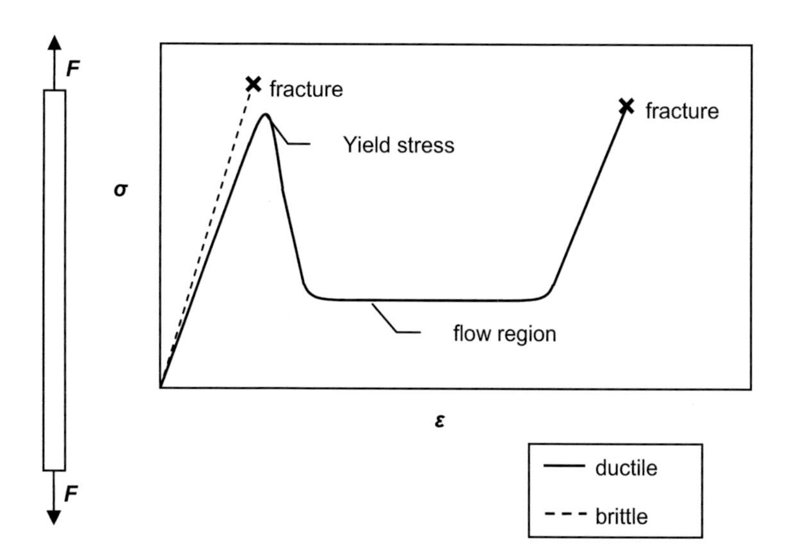

Fig. 5.3 Relation of normal stress to normal strain

The plots of both a **ductile** material and a **brittle material** are shown in Fig. 5.3. For a brittle material the stress and strain are linearly related right up until the point where the sample breaks. For a ductile material, the linear region is followed by some maximum stress called the **yield stress**, after which the stress starts decreasing. This is followed by the flow region, in which the stress is nearly constant; the hardening region where the stress again increases with strain; and finally fracture.

An important thing to note about Fig. 5.3 is that both brittle and ductile materials show a linear relationship between stress and strain at sufficiently small strains. When this is the case, materials behave elastically, and the appropriately named theory of **elasticity** holds true. This theory provides sufficient accuracy for most MEMS of interest. The slope of the linear region is known as **Young's modulus**, and is usually denoted by the symbol E:

$$\sigma = E\varepsilon \tag{5.8}$$

Equation (5.8) is known as Hooke's Law. A useful way to interpret Eq. (5.8) is as the stress-strain equivalent of the equation for a linear spring encountered in elementary physics:

$$F = kx \tag{5.9}$$

Rather than force F we now have normal force per unit area, the stress σ. Instead of displacement x we have deformation (strain) ε. Finally, Young's modulus takes the place of the spring constant k.

Fig. 5.3 shows only the relationship between normal stress and normal strain in one direction. When only these quantities are of importance Eq. (5.8) provides a sufficient model while in the elastic region. However, strains in one direction can cause strain in other directions. For example, if you stretch a rubber band you'll notice that it gets skinnier. (Fig. 5.4.) This additional strain is also related to stress. And so for the more general case, we need a few more equations.

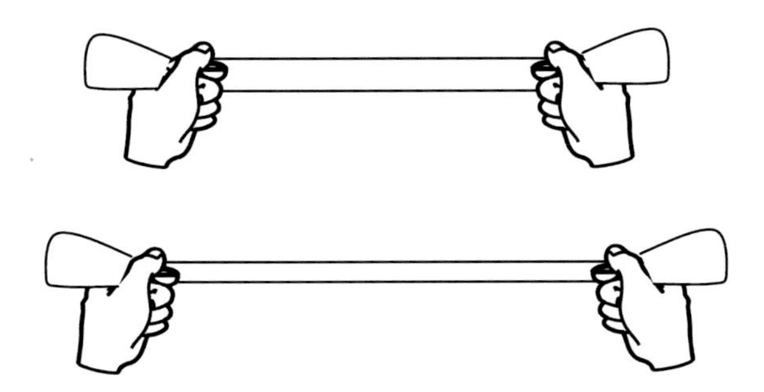

Fig. 5.4. Stretching a rubber band makes it skinnier.

A very simple equation exists for relating normal strain in one direction to the normal strain perpendicular to it:

$$\varepsilon_y = -\nu\varepsilon_x \tag{5.10}$$

The variable ν is called **Poisson's ratio**. It has a lower limit of zero for which no strain is induced, and an upper limit of 0.5, in which case the material behaves as an incompressible substance. The negative sign of Eq. (5.10) is a result of the rubber band getting skinnier and not thicker. This phenomenon is often called Poisson contraction.

Just as an applied normal stress causes normal strain in a material, applied shear stress will cause shear strain. For elastic materials this relationship is also linear, and is given by

$$\tau_{xy} = G\gamma_{xy} \tag{5.11}$$

where G is the shear modulus of the material. It can be shown that Young's modulus, Poisson's ratio and the shear modulus are all related to each other by

$$G = \frac{E}{2(1+\nu)} \tag{5.12}$$

Hence, knowledge of only two out of the three elastic material properties E, G and ν is sufficient to characterize an isotropic material.

Equations (5.8) and (5.11) suffice when pure normal stress/strain or pure shear stress/strain situations exist, respectively. Due to Poisson contraction, however, stress and strain, both normal and shear, can be important in all directions. This would certainly be true for the element undergoing three dimensional deformation as shown in Fig. 5.5.

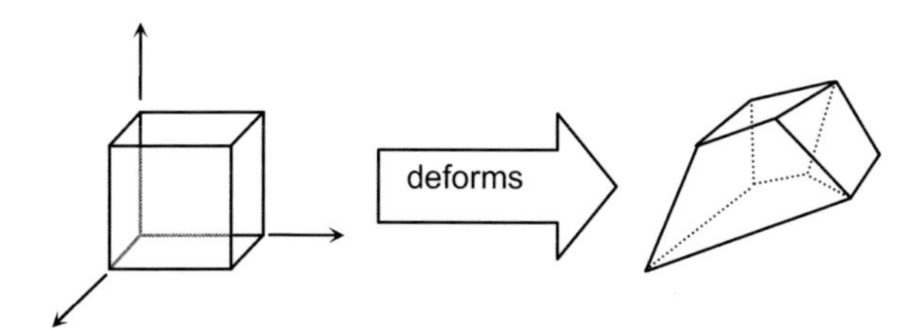

Fig. 5.5. General 3-D deformation

The x direction normal stress causes x direction normal strain. In a like fashion y direction normal stress causes y direction normal strain. Through Poisson contraction, however, the y direction normal strain causes an *additional* strain in the x direction. The same is true for the z direction. As a result, the general relation for x direction strain in terms of stress is given by

$$\varepsilon_x = \frac{\sigma_x}{E} - \nu\frac{\sigma_y}{E} - \nu\frac{\sigma_z}{E} = \frac{1}{E}\left[\sigma_x - \nu(\sigma_y + \sigma_z)\right]. \tag{5.13}$$

In the absence of any y or z direction stress, we see that Eq. (5.13) reduces to Eq. (5.8) for the x direction.

Applying the same idea to the y and z directions and including shear stress/strain we arrive at a set of equations collectively called the **generalized Hooke's Law** for three dimensional deformation:

$$\varepsilon_x = \frac{1}{E}\left[\sigma_x - \nu(\sigma_y + \sigma_z)\right] \tag{5.14}$$

$$\varepsilon_y = \frac{1}{E}\left[\sigma_y - \nu(\sigma_x + \sigma_z)\right] \tag{5.15}$$

$$\varepsilon_z = \frac{1}{E}\left[\sigma_z - \nu(\sigma_x + \sigma_x)\right] \tag{5.16}$$

$$\tau_{xy} = G\gamma_{xy} \tag{5.17}$$

$$\tau_{yz} = G\gamma_{yz} \tag{5.18}$$

$$\tau_{xz} = G\gamma_{xz} \tag{5.19}$$

5.2.4 Special cases

For general deformations Eqs. (5.14) through (5.19) must be solved simultaneously. Due to the complexity of these coupled equations, people must often attempt this numerically, using finite element analysis (FEA) software or other methods. There are some simple cases in which specific assumptions are valid, however, that result in Eqs. (5.14) through (5.19) having rather simple closed form solutions. Some of these are listed below without derivation.

Uniaxial stress/strain

When only normal stress in one direction is important, this is dubbed **uniaxial stress**. The basic form of Hooke's law is appropriate here:

$$\sigma = E\varepsilon \,. \tag{5.20}$$

Normal stress same in all directions with no shear stress

In some cases there is no shear stress present and all normal stresses are the same each directions. The hydrostatic pressure on a solid object submerged in an incompressible fluid is a good example. In this case, the stress/strain relation is given by

$$\sigma = \sigma_x = \sigma_y = \sigma_z = K\frac{\Delta V}{V} \tag{5.21}$$

where K is the bulk modulus given by

$$K = \frac{E}{3(1-2\nu)} \tag{5.22}$$

and the quantity $(\Delta V/V)$ is the volume strain.

Biaxial stress

In many two dimensional situations only normal stresses are important. If they can also be assumed equal, then we have **biaxial stress** given by

$$\sigma = \sigma_x = \sigma_y = \frac{E}{1-\nu}\cdot\varepsilon \,. \tag{5.23}$$

The quantity $E/(1\text{-}\nu)$ is known as the **biaxial modulus**. Biaxial stress is of particular importance in MEMS, as it is a valid model for most thin films.

5.2.5 Non-isotropic materials

The previous relations strictly hold for isotropic materials in which material properties are the same in all directions. For many materials (and crystalline silicon is one of them) material properties are different for different directions. For a general non-isotropic material, no single Young's modulus exists. In their place we can have up to 81 components of a forth rank **tensor**![1]

---

[1] You may have heard that a scalar has a magnitude only and that a vector has a magnitude and a direction. When you take the derivative of a vector, or sometimes when relating two vectors to each other, you need something that has all sorts of magnitudes and directions associated with it—a tensor. If you are not

Luckily, most non-isotropic materials exhibit some degree of symmetry and the number of independent components of this tensor, known as **stiffness coefficients**, goes way down. For any material the maximum number is 21. In the case of Si, a cubic material, only three independent components exist. The matrix representation of the stress-strain relation in terms of the stiffness coefficients is given below.

$$\begin{bmatrix} \sigma_x \\ \sigma_y \\ \sigma_z \\ \tau_{yz} \\ \tau_{zx} \\ \tau_{xy} \end{bmatrix} = \begin{bmatrix} C_{11} & C_{12} & C_{12} & 0 & 0 & 0 \\ C_{12} & C_{11} & C_{12} & 0 & 0 & 0 \\ C_{12} & C_{12} & C_{11} & 0 & 0 & 0 \\ 0 & 0 & 0 & C_{44} & 0 & 0 \\ 0 & 0 & 0 & 0 & C_{44} & 0 \\ 0 & 0 & 0 & 0 & 0 & C_{44} \end{bmatrix} \times \begin{bmatrix} \varepsilon_x \\ \varepsilon_y \\ \varepsilon_z \\ \gamma_{yz} \\ \gamma_{zx} \\ \gamma_{xy} \end{bmatrix} \tag{5.24}$$

where the symbol C_{ij} represents the stiffness coefficients. For crystalline silicon, the values are $C_{11} = 166$ GPa, $C_{12} = 64$ GPa and $C_{44} = 80$ GPa.

A much more compact form of such a relationship is given by

$$\sigma_i = C_{ij}\varepsilon_j \tag{5.25}$$

where the indices i and j vary from 1 to 6. When letting the indices vary, however, all the terms including the index j are added together for a single value of i. For example, σ_1 is given by

$$\sigma_1 = C_{11}\varepsilon_1 + C_{12}\varepsilon_2 + C_{13}\varepsilon_3 + C_{14}\varepsilon_4 + C_{15}\varepsilon_5 + C_{16}\varepsilon_6\,. \tag{5.26}$$

This is the case wherever an index is repeated, as is j in the present example. Summing values for a repeated index is called the **Einstein summation convention.**

In many cases this notation requires an index to vary only from 1-3, corresponding to the three rectilinear coordinate directions x, y and z. This is the true when relating some vectors to each other, for example. Stress and strain are both second rank tensors, however. Nonetheless, we still can get away with a single index for both σ and ε by letting the indices 4-6 represent the shear quantities corresponding to yz, zx and xy subscripts, respectively. As examples, σ_3 would be σ_z and ε_5 would be τ_{zx}.

---

familiar with tensors, one way to think about them is as something like a higher order vector. (Strictly speaking, scalars are tensors of rank zero, vectors are tensors of rank one, and what most of us call tensors start at rank two.)

There is virtually no hope of solving Eq. (5.24) in closed form in most situations, and advanced numerical techniques are usually employed for the task. Such techniques are beyond the scope of this text, and Eq. (5.24) is included mostly for the sake of completeness. However, we will revisit Eq. (5.24) in Chapter 10 in which we model piezoelectric transducers.

When to use what equation or equations in modeling the solid mechanics of a MEMS is based on which assumptions are valid, and, as all engineering skills, takes time and experience to master. The most important things for the new student is first to understand what is meant by the concepts of stress and strain, and then that elastic theory relates them to each other via one or more moduli.

5.2.6 Thermal strain

We have already seen that one way to deform a material is to apply a stress to it. Another way is to heat or cool it. Most things expand when heated and contract when cooled. For solids, a convenient way to express this relationship is by defining the **thermal expansion coefficient**:

$$\alpha_T \equiv \frac{d\varepsilon}{dT}. \tag{5.27}$$

Here α_T is the thermal expansion coefficient, ε is strain and T is temperature.

In general α_T is a function of temperature. However, for most materials we can assume α_T to be a constant, even over relatively large temperature changes. Hence we can usually approximate the thermal strain of a material as it goes from temperature T_0 to T by

$$\varepsilon(T) = \varepsilon(T_0) + \alpha_T(T - T_0) = \varepsilon(T_0) + \alpha_T \Delta T. \tag{5.28}$$

Furthermore, most materials, even crystalline materials, tend to expand or contract the same amount in all directions when heated or cooled. Only one thermal expansion coefficient is therefore sufficient to describe thermal strain for most materials.

An important point to realize is that thermal strain does not necessarily result in stress. In fact, if a piece of material is physically unconstrained and therefore free to expand in all directions, the resulting strain due to heating or cooling will result in no stress whatsoever. If the material is physically constrained upon the temperature change, however, then stresses will develop within it. This idea is explored in Problem 5.11. It is also a common theme in thin film deposition processes, and therefore, in many MEMS devices.

5.3 Properties of thin films

We have already seen that one of the most common MEMS fabrication processes involves the deposition of a thin layer of something on a thick layer of something else. These layers, of course, are called thin films.

Due to details of the deposition process, residual stresses often appear in thin films. These stresses can be either compressive or tensile. Sometimes the MEMS fabricator can use the resulting stresses to his/her advantage. At other times, however, the resulting stresses can cause significant problems in a MEMS device. It is therefore necessary to understand the mechanical properties of thin films in particular.

5.3.1 Adhesion

Thin films are not particularly useful if they don't stick. As with all of microfabrication processes, ensuring cleanliness of the surfaces will help here. Increasing surface roughness is also a useful technique, as it increases the effective surface area onto which the film can be deposited. And when creating a thin film of a metal on an oxide, including an oxide-forming element between the layers can also ensure good adhesion.

5.3.2 Stress in thin films

The sign convention used in this text for normal stress is that tensile stresses are positive, whereas compressive stresses are negative. This is shown in Fig. 5.6. This is also consistent with the directions in Fig. 5.1.

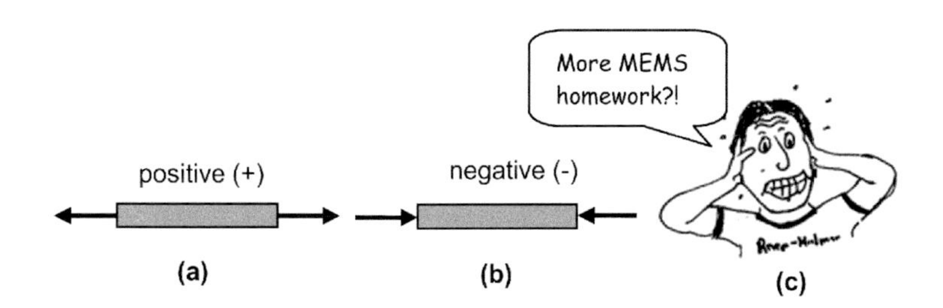

Fig. 5.6. (a) Tensile stress is positive; (b) Compressive stress is negative; (c) Stress in college life is inevitable.

Intrinsic stress

Stress in thin films can be characterized as either **intrinsic** or **extrinsic** stress, depending on the mechanism by which it develops. Intrinsic stresses are also known as **growth stresses**, as they result directly from the film nucleation process itself. Extrinsic stresses result from some externally imposed factor. Thermal mismatch stress is a good example of an extrinsic stress.

The processes of doping and sputtering explored in Chapter 2 can both result in stress within a thin film. In doping, the dopant material will almost certainly be of a different physical size than the substitutional site which it fills. If the dopant size is greater than the site size, the film is compressed, whereas a smaller dopant size than substitutional site leads to a film in tension. The constant bombardment of atoms during sputtering tends to densify thin films, leading to a compressive intrinsic stress. Other fabrication processes resulting in intrinsic stress include the appearance of **microvoids**, **gas entrapment** and **polymer shrinkage**. Theses stresses are summarized in Table 5.1.

Table 5.1. Common causes of intrinsic stress

Cause	Description	Resulting stress (+) tensile or (-) compressive
Doping	Adding impurities to a film can cause stress. The relative size of the added atoms/ions compared to the size of the substitutional site they fill affects the resulting stress.	Dopant size > site compressive (-) Dopant size < site tensile (+)
Sputtering	Bombardment of atoms densifies the film	compressive (-)
Microvoids	During deposition byproducts escape as gases, leaving behind gaps in the film.	tensile (+)
Gas entrapment	Gas gets trapped in the thin film.	compressive (-)
Polymer shrinkage	Curing causes polymers to shrink.	tensile (+)

Extrinsic stress: thermal mismatch stress

Consider a thin film deposited on a substrate at a deposition temperature, T_d, as shown in Fig. 5.7. (Both the film and the substrate are initially at T_d.) Initially the film is in a stress and strain free state. The film and substrate are then allowed to cool to room temperature, T_r. If the coefficients of thermal expansion of the two materials are different, and they almost always are, it will lead to an extrinsic stress known as **thermal mismatch stress**. Let us take a closer look at this stress.

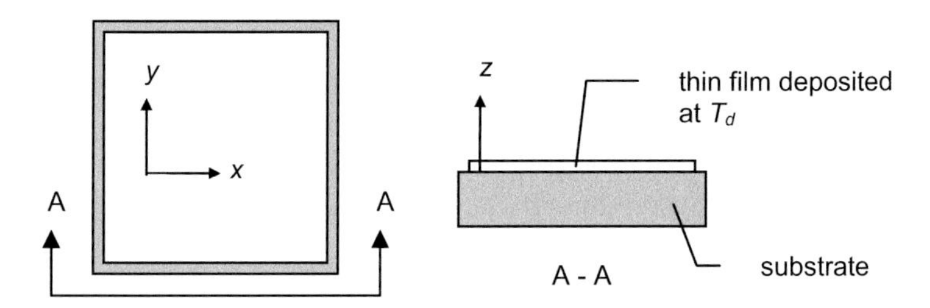

Fig. 5.7. Thermally deposited thin films can lead to thermal mismatch stress

Since the thin film is physically attached to the substrate, both materials must experience the same deformation, and therefore the same strain as they cool. As the substrate is much thicker and larger than the film, the strain experienced by both materials is the same as the thermal strain of the substrate alone. Assuming a constant thermal expansion coefficient for the substrate, $\alpha_{T,s}$,

$$\varepsilon = \varepsilon_{substrate} = \varepsilon_{substrate}(T_d) + \alpha_{T,s}(T_r - T_d) = \alpha_{T,s}(T_r - T_d). \tag{5.29}$$

Now the thin film would like to expand or contract by exactly the amount dictated by its thermal expansion coefficient, $\alpha_{T,f}$. However, it actually experiences two pieces of strain, one due to its own thermal expansion and an extra piece due to the fact that it is being dragged along by the the substrate

$$\varepsilon = \varepsilon_{film} = \varepsilon_{film}(T_d) + \alpha_{T,f}(T_r - T_d) + \varepsilon_{extra} = \alpha_{T,f}(T_r - T_d) + \varepsilon_{extra}. \tag{5.30}$$

This extra piece is the thermal mismatch strain, which is what causes the thermal mismatch stress.

Thermal mismatch strain is found by equating Eqs. (5.29) and (5.30) (remember, both materials must experience the same strain) and solving for ε_{extra}:

$$\varepsilon_{mismatch} = \varepsilon_{extra} = (\alpha_{T,f} - \alpha_{T,s})(T_d - T_r). \tag{5.31}$$

From Eq. (5.31) we see that the extra expansion or contraction of the thin film as it cools from T_d to T_r is directly proportional to the difference between its thermal expansion coefficient and that of the substrate, hence the term “thermal mismatch”.

And so what is the resulting stress from this strain? This is easily derived by making a few assumptions. Since the thickness of the thin film is much smaller than the in-plane dimensions, we have

$$\sigma_z = \tau_{xz} = \tau_{yz} = 0 . \tag{5.32}$$

At most, then, the remaining stresses are functions only of the in-plane directions. Assuming the thin film material to be isotropic, Eqs.

$$\varepsilon_x = \frac{1}{E}\left[\sigma_x - \nu(\sigma_y + \sigma_z)\right], \tag{5.14}$$

$$\varepsilon_y = \frac{1}{E}\left[\sigma_y - \nu(\sigma_x + \sigma_z)\right] \tag{5.15}$$

and

$$\tau_{xy} = G\gamma_{xy} \tag{5.17}$$

become

$$\varepsilon_x = \frac{1}{E}\left[\sigma_x - \nu\sigma_y\right] \tag{5.33}$$

$$\varepsilon_y = \frac{1}{E}\left[\sigma_y - \nu\sigma_x\right] \tag{5.34}$$

$$\tau_{xy} = G\gamma_{xy} \tag{5.35}$$

It can be shown that for any in-plane stress, we can always rotate the x-y axes so that $\tau_{xy} = 0$ and only σ_x and σ_y exist for the new directions. These axes are known as the **principal axes**. (Fig. 5.8.)

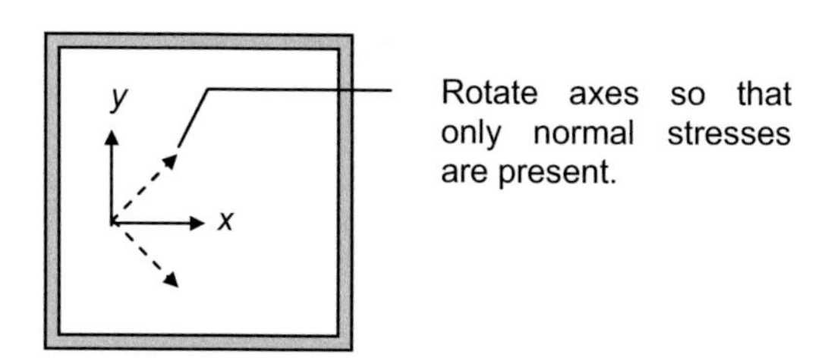

Fig. 5.8. Principal axes

This leaves only the x and y direction normal stress and strain.

We have already seen that thermal expansion tends to occur at the same rate in all directions. Hence,

$$\varepsilon_x = \varepsilon_y = \varepsilon_{mismatch}, \tag{5.36}$$

and Eqs. (5.33) and (5.34) must be equal. The only nontrivial solution is for the normal stresses to be equal as well, so that both Eq. (5.33 and (5.34 reduce to

$$\sigma = \frac{E}{1-\nu} \cdot \varepsilon, \tag{5.37}$$

which we recognize as being identical to Eq. (5.23) , the stress/strain relation for biaxial stress.

All that remains is to substitute the expression for thermal mismatch strain (Eq. (5.31)) into Eq. (5.37), resulting in

$$\sigma_{mismatch} = \frac{E}{1-\nu} \cdot (\alpha_{T,f} - \alpha_{T,s})(T_d - T_r). \tag{5.38}$$

In regards to Eq. (5.38) we see that if the deposited film has a larger thermal expansion coefficient than the substrate ($\alpha_{T,f} > \alpha_{T,s}$) the resulting stress is positive, and the film is in tension. If the converse is true and the film has a smaller thermal expansion coefficient than the substrate ($\alpha_{T,f} < \alpha_{T,s}$) the resulting stress is negative, and the film is in compression.

Here's a case where thermal mismatch stress can cause trouble. Consider a MEMS cantililever that we have miraculously fabricated on top of a sacrificial layer in a stress-free state. A thin film is then deposited on the cantilever at an elevated temperature, T_d and the whole assembly is allowed to cool to room temperature, T_r. (Fig. 5.9 (a).)

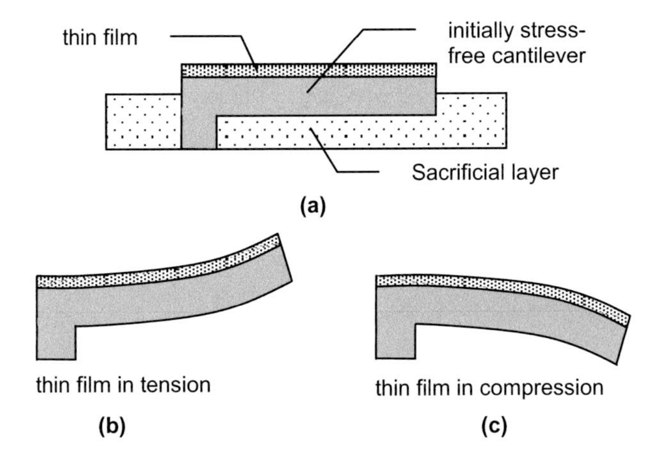

Fig. 5.9. (a) Cantilever before release; (b) Thin film in tension causing cantilever to bend up; (c) Thin film in compression causing cantilever to bend down

Depending on the thermal expansion coefficients of the film and substrate, there resulting thermal mismatch stress can be either positive (film in tension) or negative (film in compression). When the sacrificial layer is removed and the cantilever released, the stress in the film will cause the cantilever to bend. If the film is in tension, the cantilever will bend up, whereas a compressive film results in a downward bending cantilever, as shown in Fig. 5.9 (b) and (c). Fig. 5.10 shows a scanning electron microscope image of a layer of TiNi deposited on SiO_2. The Ti-Ni layer is in tension, causing the structure to bend upward.

Fig. 5.10. A layer of TiNi in tension causes upward bending of SiO_2. (From Wang, 2004)

Measuring thin fim stress

One of the most common methods for measuring stress in thin films is known as the **disk method**. In the disk method, the center of a substrate to which a thin film is added is subject to a deflection measurement test before and after the deposition process. Any change in deflection is directly attributable to stress within the thin film, thereby allowing the stress to be determined.

Fig. 5.11 shows a wafer before and after it has bowed due to the stress of a thin film. (The bow has been greatly exaggerated.) In the disk method, the stress is not measured directly, but rather calculated from indirect measurements of the radius of curvature of the wafer, *R*.

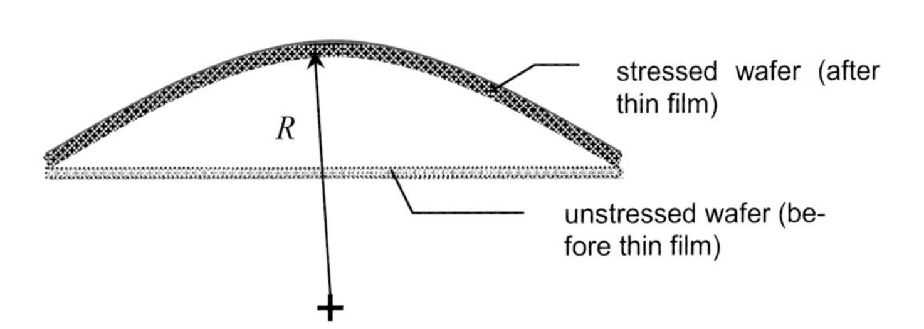

Fig. 5.11. Bowed wafer due to thin film stress

The disk method makes a number of assumptions in order to correlate the radius of curvature to the stress within the thin film. These include the following.

- The film thickness is uniform and small compared to the wafer thickness.
- The stress in the thin film is biaxial and uniform across its thickness.
- The stress in the wafer is equi-biaxial (biaxial at any location in the thickness).
- The wafer is unbowed before the addition of the thin film.
- Wafer properties are isotropic in the direction normal to the film.
- The wafer isn't rigidly attached to anything when the deflection measurement is made.

Given these assumtions, the stress is given by

$$\sigma = \frac{E}{1-\nu} \cdot \frac{T^2}{6Rt} \tag{5.39}$$

where T is the thickness of the substrate, t is the thickness of the thin film and $E/(1\text{-}\nu)$ is the biaxial modulus *of the substrate*, not the thin film. (The third assumption of equi-biaxial stress, the relative thickness of the substrate to the thin film and the continuity of stress are responsible for using the substrate's modulus and not that of the film.)

Though Eq. (5.39) is given here without derivation, it should have a familiar flavor to you. Namely, you should recognize that it relates some stress to some strain via an appropriate modulus of elasticity.

5.3.3 Peel forces

This idea of biaxial stress works very well for most thin films provided that you are looking far enough from the edges of the film. When you get close to the edge of the film, things are a bit different. At the very edge of the film, there is nothing physically touching the film in the plane direction; hence the edge itself is a stress-free surface. In this region, complicated stress gradients form, often resulting in a net force that tends to peel the film away from the surface. Such forces are known as **peel forces**.

Fig. 5.12 shows a thin film in tension on a substrate subject to peel forces at the edge. Note that in this case the film is a bit thicker at the edge. This is part of the complicated peel force phenomenon. Intuitively, however, the situation is similar to the rubber band of Fig. 5.4. In Fig. 5.4, the sections of rubber band to the right and the left of the person's grip are thicker than the rest of the band. Analogously in Fig. 5.12, the thin film is thicker near where the substrate "grips" the film.

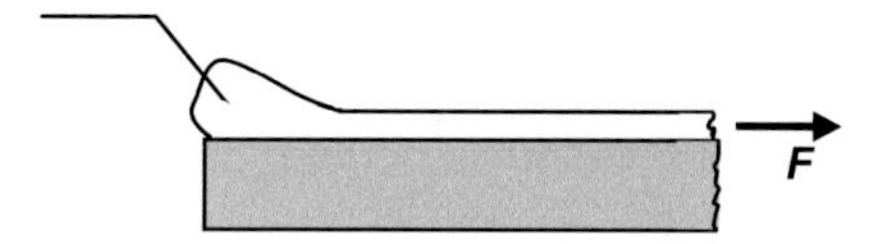

Fig. 5.12. Peel forces

References and suggested reading

Craig R (1999) Mechanics of Materials 2nd edn. John Wiley and Sons, Canada

Fung YC (1965) Foundations of Solid Mechanics. Prentice-Hall, Englewood Cliffs, NJ

Gere JM, Timoshenko SP (1997) Mechanics of Material, 4th edn. PWS Publishing, Boston

Malvern LE (1969) Introduction to the Mechanics of a Continuous Medium. Prentice-Hall, Englewood Cliffs, NJ

Madou MJ (2002) Fundamentals of Microfabrication, The Art and Science of Miniaturization, 2nd edn. CRC Press, New York

Maluf M, Williams K (2004) An Introduction to Microelectromechanical Systems Engineering, 2nd edn. Artech House, Norwood, MA

Senturia S (2001) Microsystem Design. Kluwer Academic Publishers, Boston

Wang Z (2004) Design, Modeling and Fabrication of a TiNi MEMS Heat Engine. Master's thesis, Rose-Hulman Institute of Technology

Questions and problems

5.1 Explain the concept of stress. Explain the concept of strain.

5.2 What two types of physical quantities does elastic theory linearly relate to each other? What is the generic constant of proportionality between the two quantities called?

5.3 What is Poisson contraction? What does the upper limit of $\nu = 0.5$ correspond to?

5.4 There are many moduli in solid mechanics: Young's modulus, the biaxial modulus, the shear modulus, etc. What common function do they serve?

5.5 What are the dimensions of stress? What is a typical set of units?

5.6 Crystalline Si is a cubic material. How does this complicate the stress/strain relationship?

5.7 In the context of thin films, what is the difference between intrinsic (growth) stress and extrinsic stress? Give an example of each.

5.8 In Eq. (5.39), what do you think the ratio $T^2/6Rt$ represents?

5.9 What are peel forces? Where do they occur and why?

5.10 Consider a *uniformly distributed* axial force, F, applied to a rectangular beam as shown in the figure.

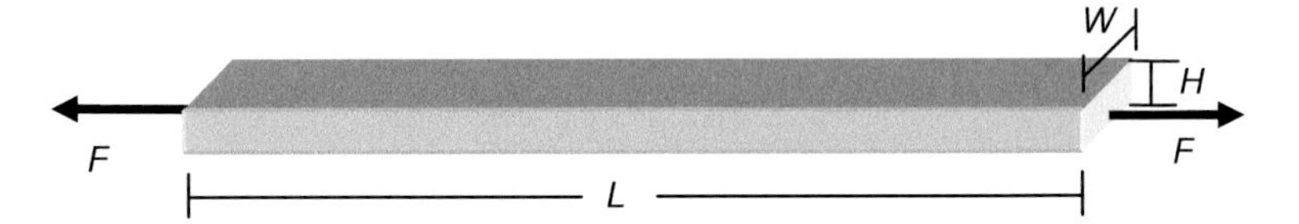

a. Find an expression for the stress, σ in terms of F, L, H, W and/or E.

b. In the beam is an elastic material, find an expression for the strain, ε in terms of F, L, H, W and/or E.

c. Find an expression for the change in length of the beam, ΔL resulting from the applied force F in terms of F, L, H, W and/or E.

d. If the beam is made of silicon with dimensions $L = 500$ µm, $H = 5$ µm and $W = 20$ µm, calculate the required force F to change the length of the beam by 1 µm. Assume for this problem that silicon is isotropic with $E = 160$ GPa.

5.11 A thin rod suspended between two fixed supports is initially in a stress free state. The rod is then uniformly heated resulting in a temperature change of the rod of ΔT. Because of the heating, the rod wants to expand. However, the fixed supports prevent this from happening resulting in a compressive stress in the rod.

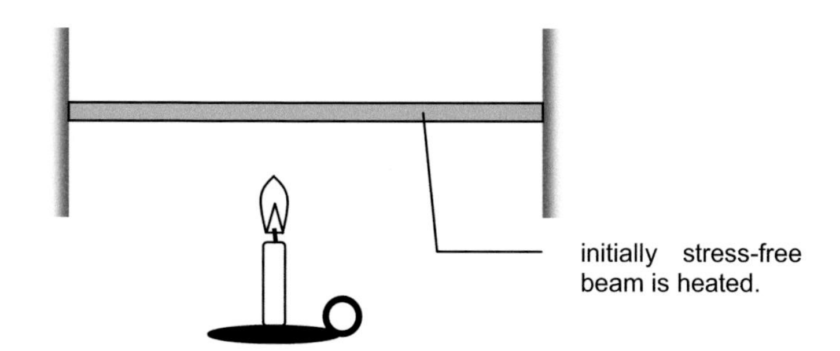

a. Find an expression for the resulting stress in the rod in terms of Young's modulus E, the thermal expansion coefficient α_T and the temperature change ΔT. Assume that the thermal expansion coefficient is constant.
b. If the rod is made of SiO_2 with $E = 69$ GPa and $\alpha_T = 0.55 \times 10^{-6}$ /°C, what stress will a 10°C temperature change produce?

5.12 A 1 μm aluminum film evaporated onto an initially flat silicon wafer causes it to bow. Optical measurements show that the deflection in the center of the wafer is 15 μm. If the wafer is 75 mm in diameter (a 3 inch wafer) with a thickness of 500 μm,

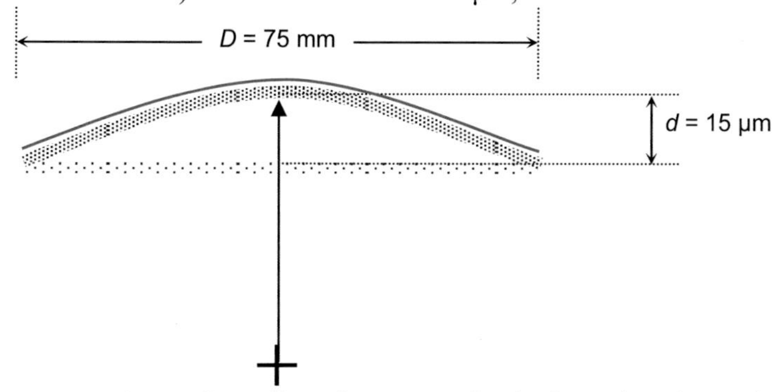

a. estimate the radius of curvature for the bowed wafer, and
b. estimate the stress in the aluminum film.

5.13 A thin layer of silicon dioxide is formed on a silicon substrate at a temperature of 950°C and then cooled to room temperature.

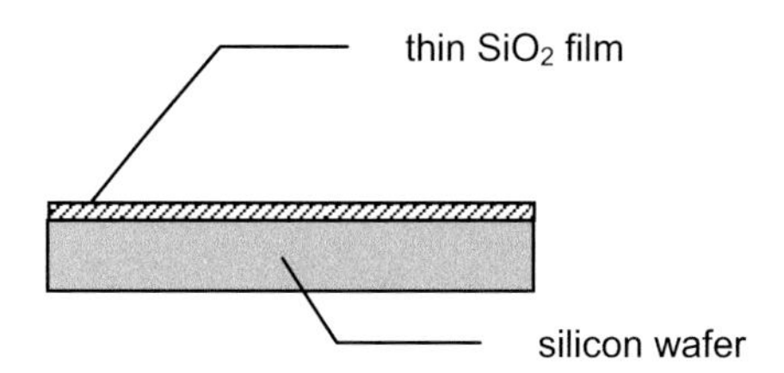

a. Estimate the resulting stress in the oxide layer after it has cooled.
b. Would you expect the wafer to bow up or down, or not at all? Why?

5.14 When a beam that is fixed on both ends is heated sufficiently, the resulting stress can cause it to bend out of plane in a phenomenon known as *buckling*. MEMS researchers are looking at using beam buckling as an infrared imaging method. Each beam in an array of thin beams acts as a pixel. At normal temperatures, the beams are flat. When heated sufficiently by infrared radiation, the beams buckle, thus reflecting light differently and creating an image. Such a beam/pixel is shown below.

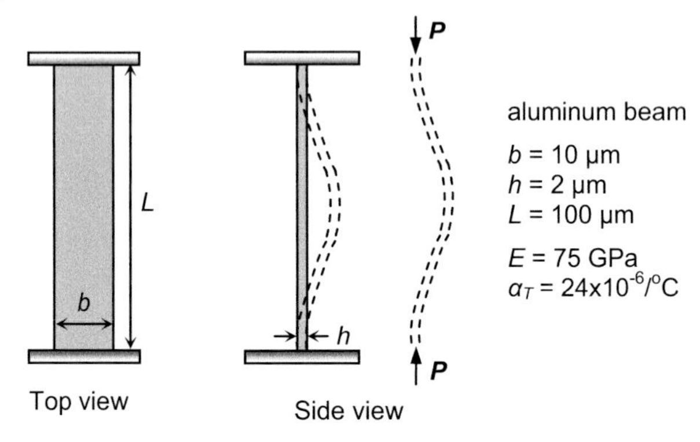

It can be shown that the critical axial force needed to buckle a rectangular cross sectioned beam is given by

$$P_{cr} = \frac{\pi^2 b h^3 E}{3L^2} \qquad ()$$

a. Find the normal stress, σ, corresponding to this critical force. Is the beam in tension or compression?
b. What change in temperature is needed to produce buckling? Assume that the unbuckled beam is stress/strain free. (Hint, consider the stress/strain to be uniaxial.)

5.15 A thin cylindrical polysilicon rod with the dimensions shown is anchored to a wall. A force $F = 0.0065$ N is applied to the free end of the rod, causing it to deform.

a. Calculate the normal axial stress in the rod, the associated strain and the change in length of the rod.

b. What increase in temperature ΔT would produce the same change in length?

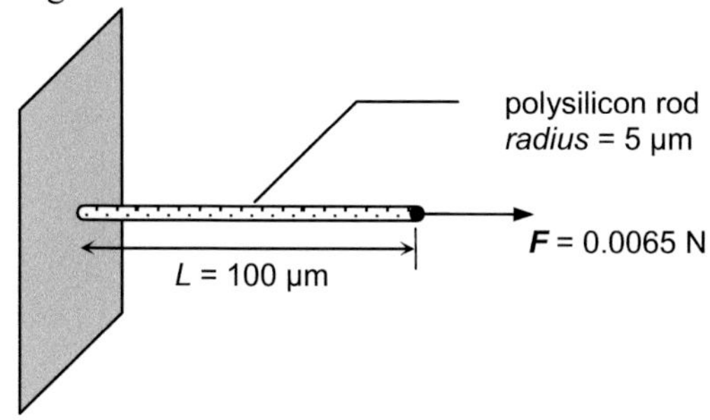

Part II—Applications

Chapter 6. Thinking about modeling

6.1 What is modeling?

A model is a system for interpreting, explaining, describing, thinking about and designing…another system. The models we will use are mathematical, generally involving geometry, trigonometry, calculus, statistics, and sometimes numerical and computer methods. The systems we model are (big surprise) MEMS!

Creating a mathematical model can be considered a design activity. A modeling practitioner works under conditions familiar to the student or practitioner of design: clients, goals, constraints, multiple feasible solutions, open-endedness, and iterative procedures that include testing and revision. With this in mind, here is a short outline of the process of designing mathematical models:

1. Establish the context of the system to be modeled.
2. Design, test, and revise the model.
3. Evaluate the design.
4. Document the design.

In the first step we define the basic problem and establish the goals and constraints of the modeling project. We revisit these definitions several times during the course of the modeling project, editing as necessary as we learn more about the system. In the second step, we design, test, and revise the model iteratively, refining the model until the results appear to be meeting the needs of the client. In the third step we perform a "reality check", comparing our results to other benchmark data, published results of related work, or auxiliary experiments involving the real physical system itself. If the evaluation indicates a shortcoming in the results, we may revisit earlier steps. In the fourth step, the documentation that has been assembled during the course of the project is put into final form.

T.M. Adams, R.A. Layton, *Introductory MEMS: Fabrication and Applications*,
DOI 10.1007/978-0-387-09511-0_6,

6.2 Units

Though we have done some calculations in the previous chapters of the text and have therefore already dealt with various physical quantities and their associated units, in modeling physical systems, meticulous care must be taken. Otherwise we cannot ensure that the modeling results are meaningful.

The virtuous complaints of students notwithstanding, modeling practitioners must be well-versed in both systems of units in common usage today: the Système International (SI) d'Unités, the modern form of the "metric" system, and the US customary system (USCS), also known as the foot-pound-second (FPS) system or the British Engineering system. Table 6.1 lists the names, units, and abbreviations for commonly encountered physical quantities such as time, length, force, and so forth in both systems of units.

Table 6.1. Units for common physical quantities.

Quantity	SI Unit	USCS Unit
Time	second (s)	second (sec)
Length	meter (m)	foot (ft)
Force	Newton (N)	pound (lb)
Mass	kilogram (kg)	slug
Energy	Joule (J)	foot-pound (ft-lb)
Power	Watt (W)	ft-lb/sec
Electric current	Ampere (A)	Ampere (A)
Temperature	degrees Celsius (°C)	degrees Fahrenheit (°F)
	Kelvin (K)	Rankine (R)

Prefixes are added to a units to create power-of-ten multiples of the basic units. The short name and symbol associated with commonly encountered prefixes are shown in Table 6.2. While these prefixes are defined only in terms of SI units, they have been adopted by some users of the US customary system as well—one might encounter the units of micro-inches (μin), for example.

Table 6.2. SI Prefixes

Power of 10	Short name	Symbol
10^3	kilo	k
10^{-3}	milli	m
10^{-6}	micro	μ
10^{-9}	nano	n
10^{-12}	pico	p

6.3 The input-output concept

When we model a system, we select some physical components to include in the model and we exclude others. To represent this separation of the system from its surroundings, we typically define a system *boundary* such that everything within the boundary is our *system*, everything outside the boundary is the system's *surroundings*, and interactions between the system and its surroundings can be represented as either *inputs* to the system or *outputs* to the surroundings. This input-output concept, and the *block-diagrams* we use to represent it, are useful parts of our vocabulary for communicating ideas about system models.

To illustrate the input-output concept, consider the automotive airbag system illustrated in Fig. 6.1. The basic principle of operation is that a sensor responds to a sudden vehicle deceleration by generating a signal that results in the rapid inflation of the airbag to cushion the impact of the driver being thrown forward towards the steering wheel. The figure illustrates the approximate physical layout of the system components.

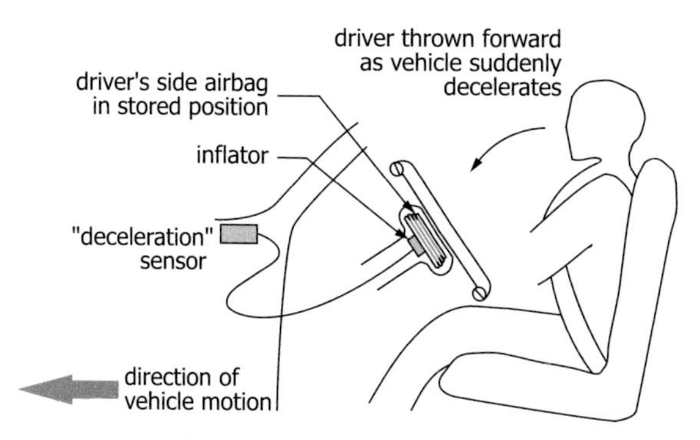

Fig. 6.1. Configuration of an automotive airbag system

If we imagine a boundary around the system that begins with the sensor and ends with the air bag, then the inputs to the system are the sudden deceleration of the vehicle and the driver's body being thrown forward to be cushioned by the bag. The system output is the bag behavior—sudden inflation to cushion the impact followed by sudden deflation immediately after impact. We sketch this input-output system in Fig. 6.2.

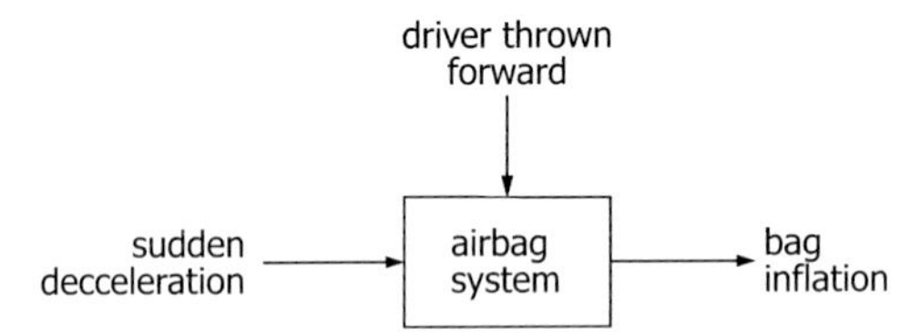

Fig. 6.2. Basic block diagram of an airbag system

The box, or *block*, in Fig. 6.2 represents the system. The outline of the block represents the system boundary, and the inputs and outputs are represented by arrows. This *block-diagram* is our simplest abstract representation of the airbag system. Compared to Fig. 6.1, the block diagram sacrifices information about spatial relationships and physical appearance to focus our attention on how we define the system, the boundary, the surroundings, and the inputs and outputs: essential information in communicating about the model and its design.

Looking inside the block, we can represent the system by a more detailed block diagram, as shown in Fig. 6.3. Here the major components of the system—the sensor, inflator, and airbag—are represented with their own blocks, inputs, and outputs. The symbol "*e*" represents a voltage signal. The boundary of the airbag system is indicated by the dashed line surrounding all the components.

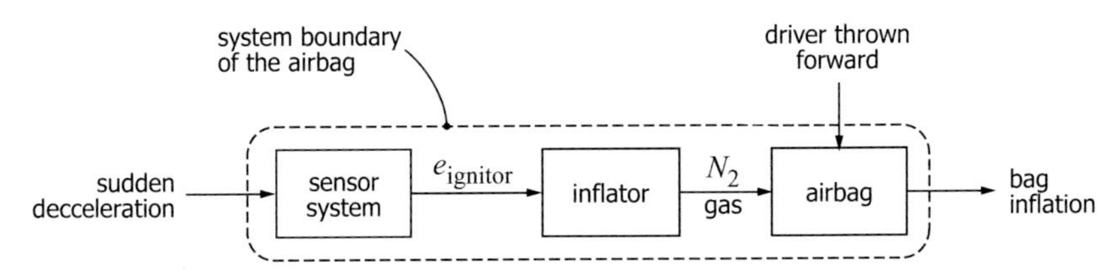

Fig. 6.3. A more detailed block diagram of an airbag system

Any or all of these blocks inside the system can be represented in greater detail. We'll focus on the sensor system because it commonly incorporates a MEMS accelerometer package. A typical configuration is shown in Fig. 6.4. The MEMS device interacts with the signal conditioning integrated circuits to produce the amplified signal that fires the ignitor if the deceleration is high enough.

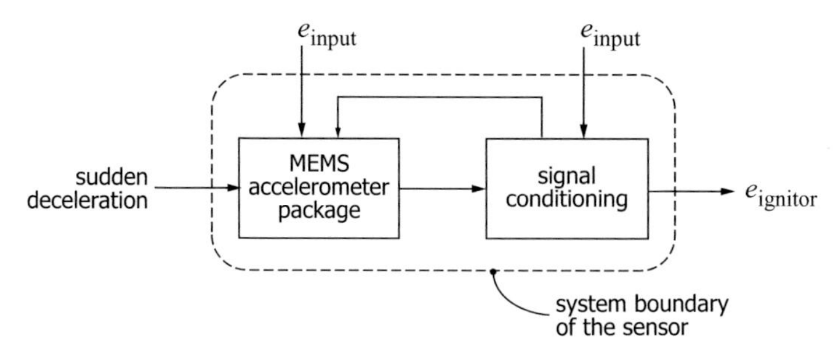

Fig. 6.4. The accelerometer sensor system

To conclude our example at one more level of detail, we'll consider a particular MEMS accelerometer package, the Analog Devices ADXL50, illustrated by the manufacturer's combined block diagram and circuit diagram shown in Fig. 6.5. The block labeled "sensor" is the MEMS device—a micromechanical element that detects sudden deceleration and creates an electrical output signal that is detected by the circuitry shown. This circuit, part of the Analog Device package, sends two output signals e_{ref} and e_{out} to a signal conditioning circuit that generates the signal that ignites the inflator. The resistors R_i, capacitors C_i, and connections to ground outside the system boundary have to be supplied and connected by the user.

In summary, we've illustrated how the input-output concept is expressed using block diagrams at various levels of detail and how, when selecting a particular system for study, we explicitly include some physical components and exclude others in defining the system we plan to model. The next step is to apply basic physical principles, such as Newton's second law or the concept of conservation of energy, and techniques such as circuit analysis to design our mathematical models.

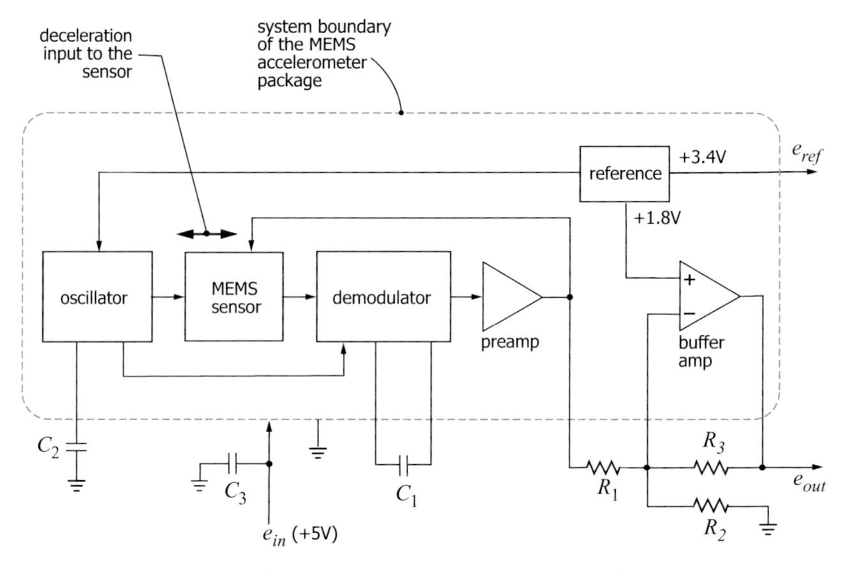

Fig. 6.5. Inside the Analog Device ADXL50 accelerometer

6.4 Physical variables and notation

In working with MEMS devices we find ourselves crossing the boundaries of traditional engineering disciplines and working with mechanical, electrical, fluid, and thermal quantities (among others). Because historically the engineering disciplines developed quite separately, each discipline developed its own nomenclature for its "own" physical quantities. Consequently different disciplines established the same notation to represent different physical quantities. For example, the letter V represents voltage in electrical nomenclature, volume in fluid mechanics nomenclature, and velocity in mechanics nomenclature. Today a modeler is very likely to encounter a system in which all three variables occur and the use of traditional nomenclature can lead to nomenclature confusion. Indeed you may have already encountered some of this confusion in the first half of this text!

Practitioners of the contemporary discipline of "system dynamics" have developed several methods for unifying both how one thinks about a multidisciplinary system and how one represents the physical variables in notation. For example, using the "force is like voltage" analogy, one might use the symbol "e_i" (*effort*) to represent forces, voltages, and pressures, and the symbol "f_i" (*flow*) to represent velocities, currents, and flow rates. In our opinion such a unified notation imposes a different sort of nomenclature confusion on our audience. Consequently, we have decided to use conventional disciplinary notation as much as possible but make substitutions where we think clarity is served.

For example, we use a lowercase q for charge and an uppercase Q for heat. We use v for velocity but e for voltage. We use E for an electric field, but E also represents total system energy as well as Young's modulus, a material property. And we use "dot notation" to indicate a transfer rate of several physical quantities; for example, the transfer rate of energy associated with work W per unit time is represented by $\dot{W}$.

Our primary goals are clarity and building on a reader's prior knowledge. Where notation confusion might arise, we define our terms where they are used to ensure that the mathematical expressions are placed in context of the model under discussion. For the sake of completeness, an exhaustive list of notation used in the entire text is included in Appendix A.

6.5 Preface to the modeling chapters

In the chapters that follow we will develop models for many different MEMS devices. In Chapter 7 general principles applicable to all MEMS transducers are discussed, and mainly qualitative descriptions of the principles of operation of each major class of MEMS device are given. The mathematics is kept to a minimum, reserving detailed models for individual devices for Chapters 8-11. In those subsequent chapters, four different types of MEMS transducers are treated separately. Near the beginning of each of those chapters a simple block diagram for a transducer is given, with much of the remaining sections devoted to developing a detailed model for that block.

Due to the multiple energy domains encountered in MEMS devices and their subsequent coupling, many of the physical concepts needed to model them usually fall outside the typical undergraduate's skill set. Therefore, Chapters 8-11 are presented in order of complexity and/or the sophistication of the tools needed to develop the given models based on that assumed skill set. A modeling case study for a specific device concludes each chapter.

The case studies in Chapters 8 and 9 are both sensors, whereas the case studies in Chapters 10 and 11 are actuators. As we will see, the output of a sensor is an electrical signal. Therefore, Chapters 8 and 9 are more laden with electrical systems concepts than the following two chapters. Conversely, Chapters 10 and 11 are denser in regards to mechanical concepts, since the desired output of an actuator is mechanical.

The various modeling schemes presented are not intended to be authoritative. Indeed, many other models for these devices exist. Nor are the presented models necessarily intended to give the reader a set of tools that are directly transferable to modeling another device. Rather, the reader will gain the *experience* of seeing the modeling process in action, as well as having gained a deeper understanding of the physics behind the functionality of MEMS devices.

References and suggested reading

Doebelin EO (1998) System Dynamics: Modeling, Analysis, Simulation, Design, Marcel Dekker, NewYork

Layton RA (1998) Principles of Analytical System Dynamics, Springer, Boston

Palm WJ III (2005) System Dynamics. McGraw-Hill, New York

"Analog Devices ADXL50 specification sheet" *DatasheetCatalog*. (2008) http://www.datasheetcatalog.com

Questions and problems

6.1 A common unit of mass seen in engineering practice is the *pound-mass*, often given the symbol lbm to distinguish it from the unit of force. On the surface of the earth, an object that *has a mass* of 1 lbm *weighs* one pound, which gives the unit its utility.
   a. How many lbm-ft/s^2 are in a pound, then?
   b. Gravitiational accleration on the moon is 1.63 m/s^2. How much does an object having a mass of 1 lbm weigh on the moon? What is the mass of the object on the moon?

6.2 Consider a thermal mass with an initial temperature of 13°C. The mass is heated until it reached a temperuatre of 56°C.
   a. What are the initial and final temperatures in K?
   b. What is the *change* in temperature in °C? in K?

6.3 Find an example of a MEMS sensor (i.e., a MEMS device that measures a physical quantity) via a jounanl publication or the Internet.
   a. Explain the operating principle for the sensor; that is, how does it measure the quantity of interest?

b. Create a block diagram showing the inputs(s) and output(s) of the device. Is there more than one input besides the physically measured quantity? If so, why?

6.4 Find an example of a MEMS actuator (i.e., a MEMS device that results in mechanical motion upon some stimulus) via a jounanl publication or the Internet.

a. Explain the operating priniciple of the actuator; that is, how does the stimulus result in motion?

b. Create a block diagram showing the inputs(s) and output(s) of the device. Is there more than one output besides mechanical motion? If so, why?

Chapter 7. MEMS transducers—An overview of how they work

7.1 What is a transducer?

A **transducer** is a device that converts power from one form to another for purposes of measurement or control. Historically, the word "transducer" referred only to devices that converted mechanical stimuli such as force or pressure into an electrical signal, usually voltage. Today the definition has been broadened to include all forms of input stimuli—mechanical, electrical, fluidic, or thermal—and output signals other than electrical.

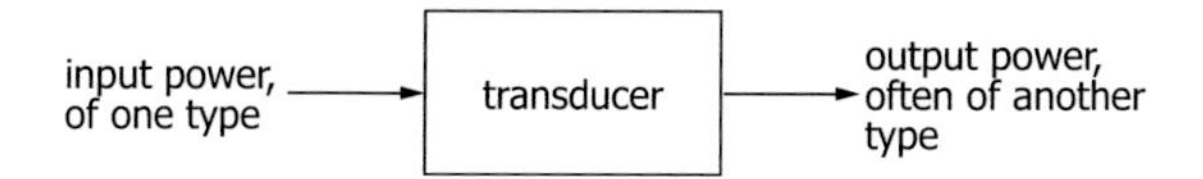

Fig. 7.1. Transducers are designed for either measurement or control.

Most commercialized MEMS transducers are electro-mechanical. That is, a mechanical motion, displacement, or stress accompanies the input power, whereas some type of electrical signal accompanies the output power, or vice versa. Other MEMS transducers on the market are thermo-electric and thermo-mechanical, converting heat energy into either an electrical signal or a mechanical motion. Chemical, optical, and biological MEMS transducers are in various stages of research and development.

In this chapter we discuss the basic principles of operation of MEMS transducers. We intend these overviews to be brief, complete, and at a level appropriate for an undergraduate audience. For brevity, we postpone most mathematical modeling to later chapters in which we discuss specific transducers in detail. For completeness, we touch on the important features of the transducers, even though some of these features are covered again in greater detail in later chapters. And to make the presentation accessible to our audience, we try to keep the engineering jargon and mathematics to a

T.M. Adams, R.A. Layton, *Introductory MEMS: Fabrication and Applications*,
DOI 10.1007/978-0-387-09511-0_7, © Springer Science+Business Media, LLC 2010

minimum, reserving most—not all!—of the technical depth to later chapters.

7.2 Distinguishing between sensors and actuators

Transducers are categorized as either sensors or actuators. **Sensors** are transducers used to measure a physical quantity such as pressure or acceleration. **Actuators** are transducers used to move or control a system, such as controlling the position of a movable microstructure. Sensors *measure* something; actuators *move* something.

In addition to differing in what they *do*, sensors and actuators differ primarily in the amount of input power they require. Sensors are low-input-power devices; actuators are high-input-power devices.

We want sensors to draw an acceptably low amount of power from the system they are monitoring. Otherwise the act of measuring alters the thing we're trying to measure—an effect called *loading*, illustrated in Fig. 7.2.

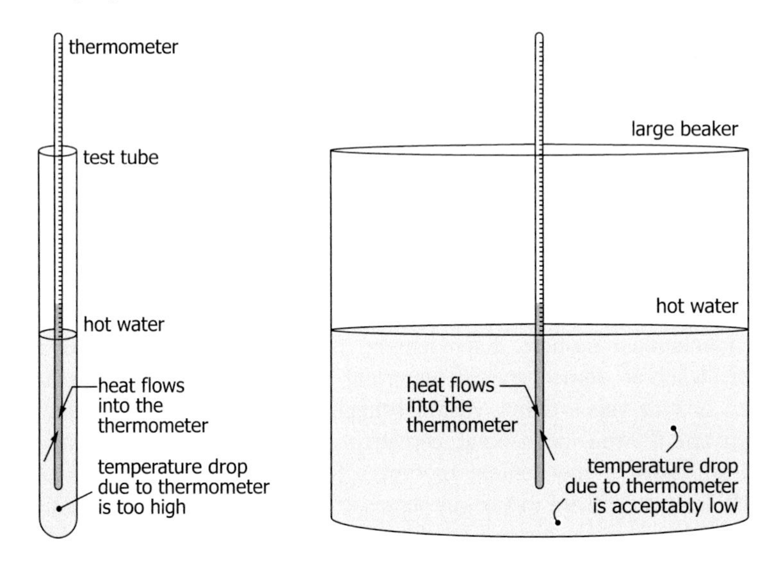

(a) sensor imposes a high load on the system (b) sensor imposes a low load on the system

Fig. 7.2. Illustration of sensor load

Suppose a mercury thermometer is used to measure the temperature of hot water in a narrow test tube, as shown in Fig. 7.2 (a). If the thermometer is at room temperature when inserted in the test tube, heat will transfer from the water to the thermometer, noticeably lowering the temperature of the water. By measuring the temperature, we've changed the temperature—the heat energy we extract is the sensor "load" on the system.

To reduce this effect, we might use thermometers only in cases where the volume of the liquid being measured is much larger than the volume of the thermometer, as in Fig. 7.2 (b). The sensor still draws heat energy from the system, but the volume of water is large enough that the temperature change of the system is negligible. The sensor load is acceptably small. Due to their small size, MEMS sensors often have an inherent advantage in this regard.

A generalized model of a sensor is shown in Fig. 7.3. The measured quantity is called the **measurand**—a physical quantity such as force, pressure, acceleration, and so forth. The measurand is the input to the sensor and the sensor is selected such that its loading effect on the measurand is negligible. In the thermometer example, the measurand is temperature. The sensor *output* is a number we obtain from a readout device or more generally a signal representing the measurand that is sent onward for processing. In the thermometer example, the output is the number corresponding to the height of the mercury column. Many sensors require a voltage supply as a second input to provide power for **signal conditioning**—circuits that help convert the output signal to a form useful for the application. (Signal conditioning is summarized Section 7.6).

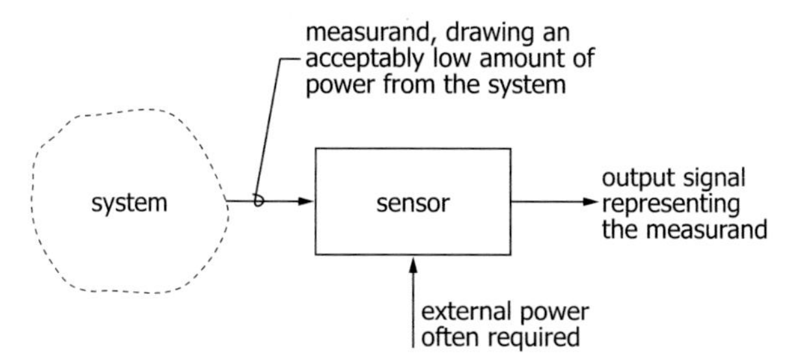

Fig. 7.3. Generalized model of a sensor: a relatively low-input-power device

Compared to sensors, actuators are relatively high-input-power devices. However much power is required to move the system, plus accounting for power losses, must be input or stored by the actuator before being delivered to the system—a consequence of the conservation of energy. The greater the power needed to move or control a system, the greater the power-input and power-conversion capabilities needed of the actuator.

For example, imagine a tennis player with a racket as an actuator for a tennis ball, as shown in Fig. 7.4 (a). The amount of power the human can deliver to the ball is generally sufficient to impart the desired motion to the tennis ball. If the tennis player tries to hit a basketball however, as in (b), it's unlikely that the human will have much of an effect on the ball. The actuator (the human plus the racket) has insufficient power to have the desired effect on the system (the basketball). In fact, the actuator may be damaged in attempting to control this system.

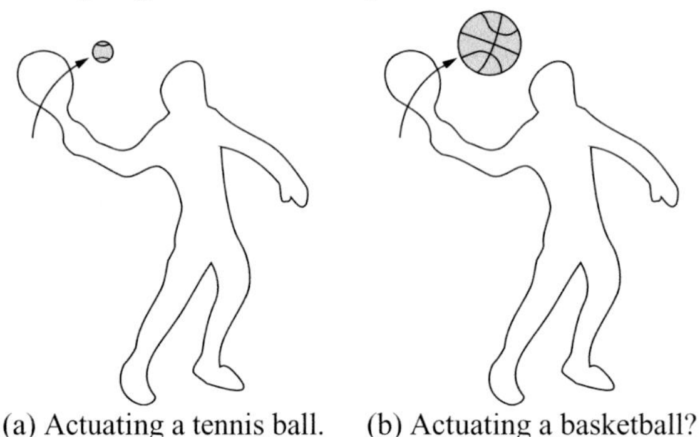

(a) Actuating a tennis ball. (b) Actuating a basketball?

Fig. 7.4. An actuator with (a) sufficient power and (b) insufficient power

A generalized model of an actuator is shown in Fig. 7.5. An actuator has two inputs: the *power* input that is converted into the actuation output; and a *control* input to tell the system what we want it to do. When we speak of an actuator being a high-input-power device, we refer to the power input. The magnitude of energy drawn by this input is often referred to as the actuator *drive energy*. The *actuation output* is the condition the actuator imposes on the system and the *system response* is what the system actually does. In the tennis ball example, the power input is the forcefulness of the stroke and the control input is the angle of the racket. The actuation is the impact and the system response is the ball's trajectory after impact. Ace! At the micro-scale of course the magnitude of input power to MEMS actuators is quite small.

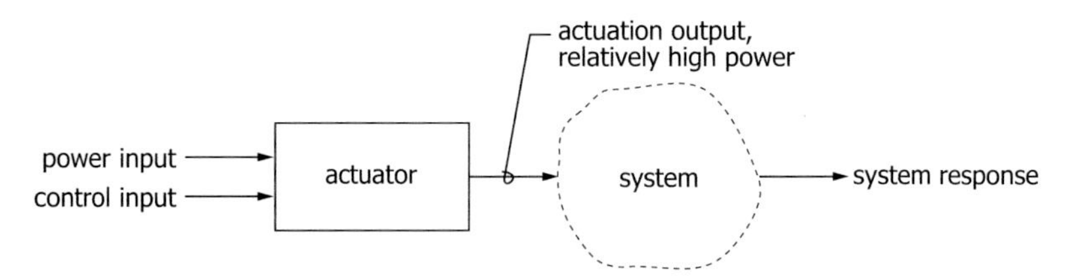

Fig. 7.5. Generalized model of an actuator: a relatively high-input-power device

Some physical principles are used only in sensing, e.g., the piezoresistive effect, while some are used only in actuation, e.g., bimetallic thermal expansion. Some physical principles, such as the piezoelectric effect, can be used for both sensing and actuation. Thus some MEMS transducers of one type can, at least in principle, be considered as the other type running "backwards". For example, a piezoelectric sensor may be thought of as an piezoelectric actuator working in reverse. However, the same physical device is rarely, if ever, used both ways since sensors and actuators have such dramatically different power requirements.

Lastly we show in Fig. 7.6 how sensors and actuators work together to monitor and control a physical system or process. The arrows indicate the direction of the flow of power and information signals. We record or preserve the data representing the system response when we debug a sensing/actuation system or need to determine if the performance of the entire system complies with design requirements.

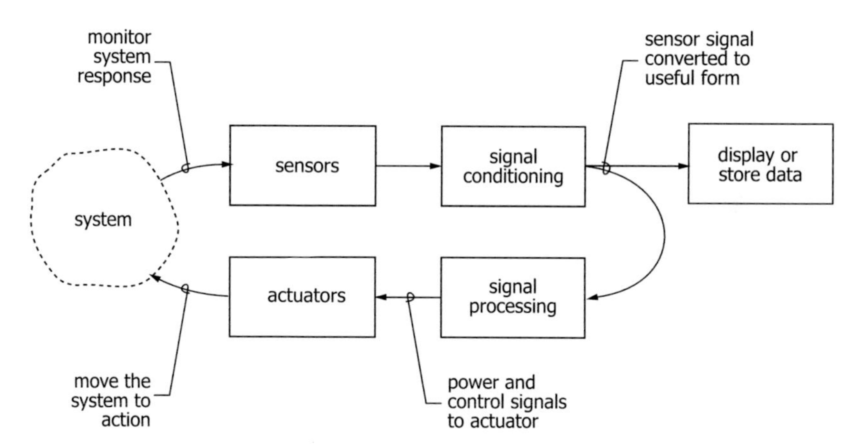

Fig. 7.6. Putting it all together: sensors and actuators working together in a measurement and control system

7.3 Response characteristics of transducers

MEMS sensors and actuators, like all transducers, have static and dynamic response characteristics useful for comparing the performance of different transducers. In this section we consider the basic response characteristics that we consider necessary for the student to understand information encountered on specification sheets provided by transducer manufacturers.

7.3.1 Static response characteristics

Static response characteristics are those generally associated with a **calibration** of the transducer, in which a steady (static) input is applied to the transducer and the resulting steady (static) output is measured. Basic static performance characteristics are illustrated in Fig. 7.7 for a general linear MEMS transducer with input variable x and output variable y. For a pressure sensor, for example, x would represent pressure input (Pa) and y would represent voltage output. For an electromechanical actuator x would represent voltage input and y would represent displacement output (μm).

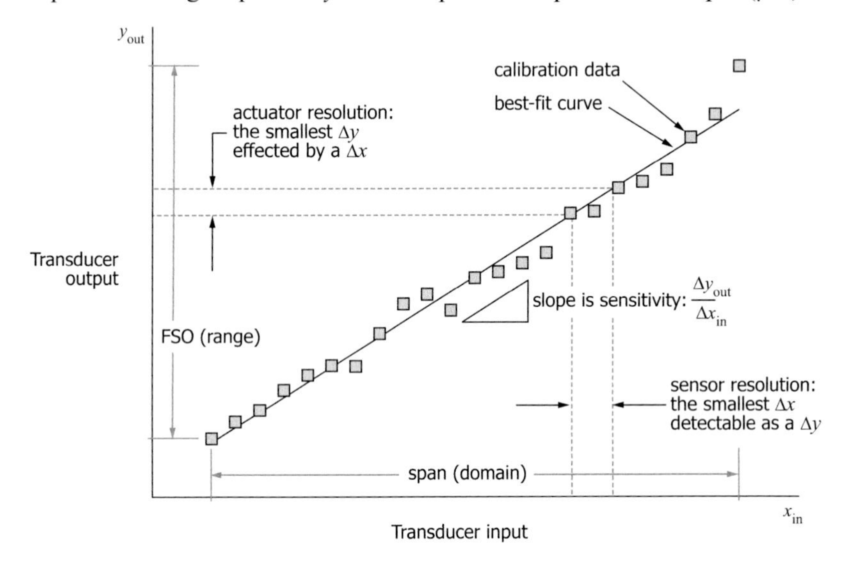

Fig. 7.7. Basic static response characteristics of a linear transducer.

Span (or full scale input) is the domain of input values over which a transducer produces output values with acceptable accuracy. The units of span are the units of the input. For example, a pressure sensor has span units of Pascals (Pa) or pounds per square inch (psi).

Full-scale output (FSO) is the range of output values corresponding to the span. The units of FSO are the units of the output. For example, the typical output units of a pressure sensor is volts (V) and the typical output units of an electromechanical actuator is micrometers (μm).

A transducer is linear if the ratio of output to input is constant. One way to visualize linearity is to plot the measured output values as a function of accurately known input values. If the best-fit curve is a straight line, like that shown in Fig. 7.7, then the sensor is linear. The degree to which the sensor output varies from this straight line is its *nonlinearity*.

Sensitivity is the constant of proportionality between output and input for a linear transducer. Graphically, sensitivity is the slope of the best-fit line. The units of sensitivity are the ratio of output units to input units. For example, for a pressure transducer with a full scale output of 15 mV and a span of 2 kPa, the sensitivity is 15 mV ÷ 2 kPa = 7.5 mV/kPa.

Accuracy is the degree to which a transducer output conforms to the true value of the measurand (sensor) or the desired output effect (actuator). Accuracy is determined by a calibration procedure and unless stated otherwise one can generally assume that accuracy is reported with a 95% *confidence interval*—the interval with a stated probability of containing the true value (sensor) or desired value (actuator). Consider, for example, a pressure transducer having a reported 2% accuracy. This means that if we take 100 measurements of pressure, in 95 cases the true value of the measurand (pressure) would lie on the interval from 2% less than the reading to 2% greater than the reading. Five times out of 100 the true pressure would lie outside this interval.

For a sensor, **resolution** is the smallest change in the measurand detectable in the readout. For example, suppose that a pressure sensor shows a steady reading. The pressure may be fluctuating (say, in a fractional Pa range) even though the readout is steady. These fluctuations are simply lower than the resolution of the sensor system. Sensor resolution is reported in units of the measurand, in this case pressure (Pa).

Actuator resolution is the smallest change in output that can effected by changing the input. For example, suppose an electromechanical actuator is holding a steady position. The input voltage to the actuator may be fluctuating (say, in a micro-volt range) even though the actuator output is steady. Such fluctuations are simply lower than the resolution of the actuator system. Actuator resolution is reported in units of the output effect, in this case displacement (μm).

7.3.2 Dynamic performance characteristics

Dynamic performance characteristics are those observed when a transducer responds to a "dynamic" input—one that changes over an interval of time rather than holding steady. We can cover most of the basics by considering just two input types, a step and a sinusoid, and the corresponding responses, the step response and the frequency response.

Step response

A **step input** is illustrated in Fig. 7.8 (a): at an instant the input to the transducer rises (or falls) from one constant value to a new constant value. The transducer output—its **step response**—follows suit, rising (or falling) from its previous constant value to a new constant value. Since no physical system can react instantaneously to a change in input, one characteristic of a step response is the **response time**—the time the output takes to reach its new steady state value. Long and short response time are illustrated in Fig. 7.8 (b); shorter is usually better.

Some systems "overreact" to the input, initially going beyond the desired value and oscillating before settling in to the steady output value corresponding to the input. *Overshoot* is the amount by which the initial response exceeds the desired value, as shown in Fig. 7.8(c). For example, suppose an actuator has a sensitivity of 3 nm/V and we apply a step input of 4V. We expect the actuator arm to move 12 nm (3 nm/V × 4 V = 12 nm). If the arm initially moves 17 nm before settling down to the desired 12 nm, then the overshoot is 5 nm. Small overshoot is usually desirable; in some cases a zero overshoot is required to prevent damage to the system.

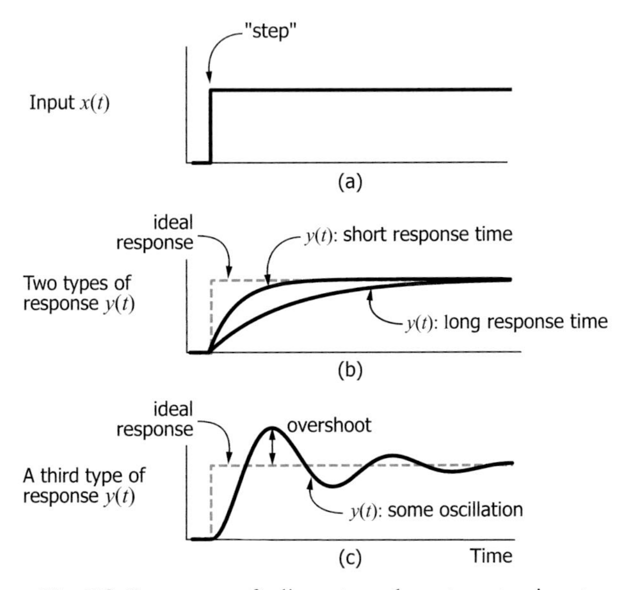

Fig. 7.8. Responses of a linear transducer to a step input.

Frequency response

A transducer responds to a sinusoidal input $x(t)$ by producing a sinusoidal output $y(t)$ at the same frequency, as shown in Fig. 7.9. *Amplitude* is the magnitude of the curve above (or below) its centerline. The *period* is the time interval of one cycle, usually reported in seconds; *frequency* f is the inverse of the period with units of cycles/s (Hz); *radial frequency* is given by $\omega = 2\pi f$ with units of rad/s. Thus we have two representations of the single concept "frequency"—and the student should be familiar with and able to use both the f and ω representations since both are used in practice.

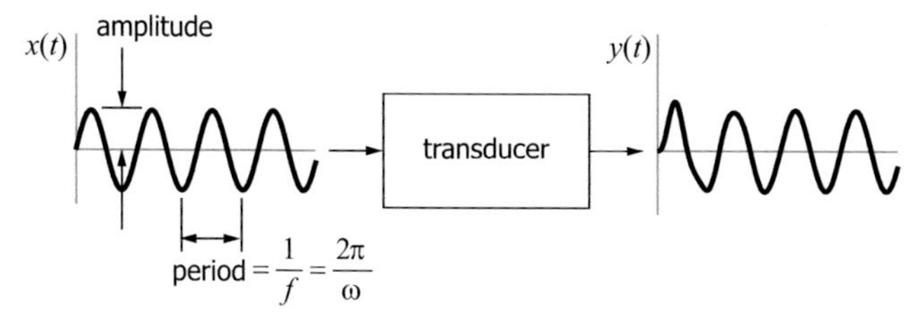

Fig. 7.9. Sinusoidal inputs and outputs have the same frequency, though not necessarily the same amplitude.

Though the input and output have the same frequency, their amplitudes and phase-alignment can be quite dissimilar. **Amplitude ratio** (or **gain**) is the ratio of the output amplitude to the input amplitude and is usually reported in units of *decibels* (dB), where dB = 20log(gain). A system's amplitude ratio can change dramatically as the input frequency changes: outputs can be magnified at some frequencies and attenuated (reduced) at others. **Resonance** is the tendency of some systems to exhibit large magnification at certain frequencies—the **resonance frequencies**—and not others. We illustrate these amplitude response characteristics in Fig. 7.10.

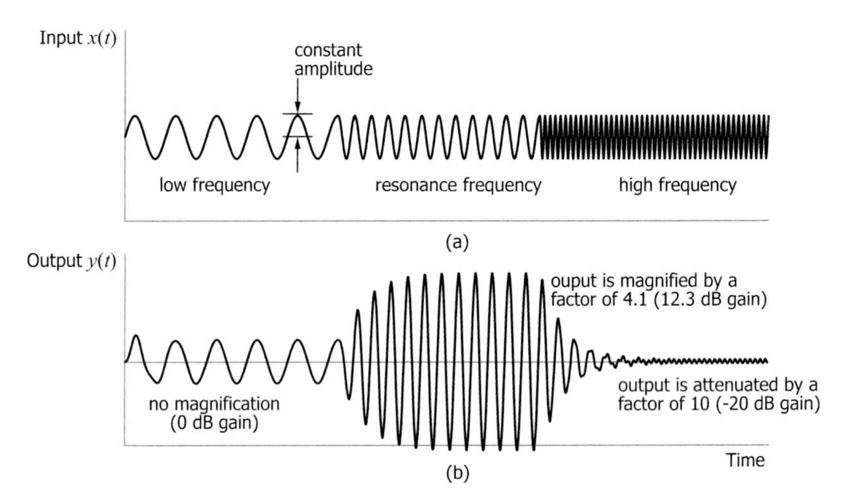

Fig. 7.10. Illustrating output amplitudes of a particular system as the input frequency varies

The input $x(t)$ in Fig. 7.10 (a) is a constant amplitude sine curve with a frequency that increases in three discrete stages over time. During the first time interval the input has a relatively low frequency; during the second interval the frequency is increased to the resonance frequency of the system used in this example; and during the third interval the frequency has increased yet again. The system response $y(t)$ is shown in (b). At the low frequency, the input and output amplitudes are equal, at the system's resonance frequency the output is magnified, and at the higher frequency the output is attenuated.

Phase is the amount of time by which an output sine wave is "misaligned" with the input sine wave, as shown in Fig. 7.11 (a). A phase shift of zero indicates that the output and input peaks and valleys are aligned, as in (b). Phase is often reported as the angle $\phi = \omega t$, where t is the time shift, ω is the radial frequency of the input in rad/s, and ϕ is the phase in radians. This phase angle is often reported in degrees, where deg = rad×180/π.

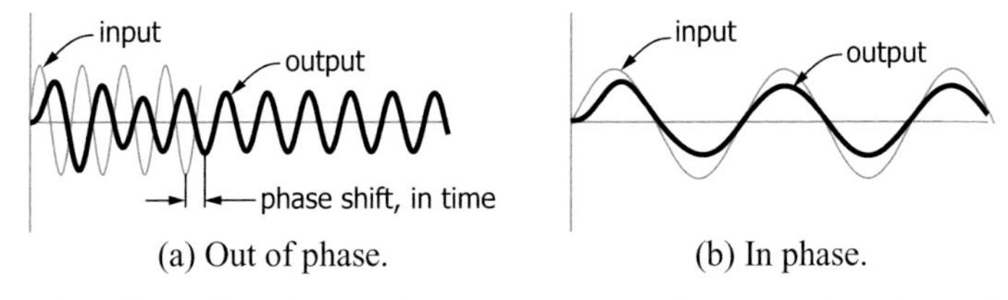

(a) Out of phase. (b) In phase.

Fig. 7.11. Illustrating phase shift between input and output as functions of time

So far we have illustrated the effect of input frequency on the amplitude ratio and phase at *specific* frequencies. If we compute the response over a *span* of frequencies, we can produce the *frequency response graph*, or *Bode plot*, shown in Fig. 7.12. The three frequencies labeled “low”, “resonance”, and “high” in this figure are the same frequencies used in the illustration of frequency response in Fig. 7.10. The differences between this graph (Fig. 7.12) and Fig. 7.10 are that this graph:

1. shows response as a function of *frequency* while the previous graph shows response a function of *time*;
2. shows the amplitude *ratio* while the previous graph shows only the *amplitudes*;
3. shows amplitude *and* phase while the previous graph shows *only* amplitude; and
4. shows the variability of amplitude and phase over a *span* of frequencies while the previous graph shows amplitudes for only *specific* frequencies.

For these reasons, Bode plots are useful for comparing the performance of transducers intended for periodic measurands as well as for understanding the operation of transducers based on resonance. One encounters these graphs regularly in practice—the student is encouraged to gain some familiarity with them.

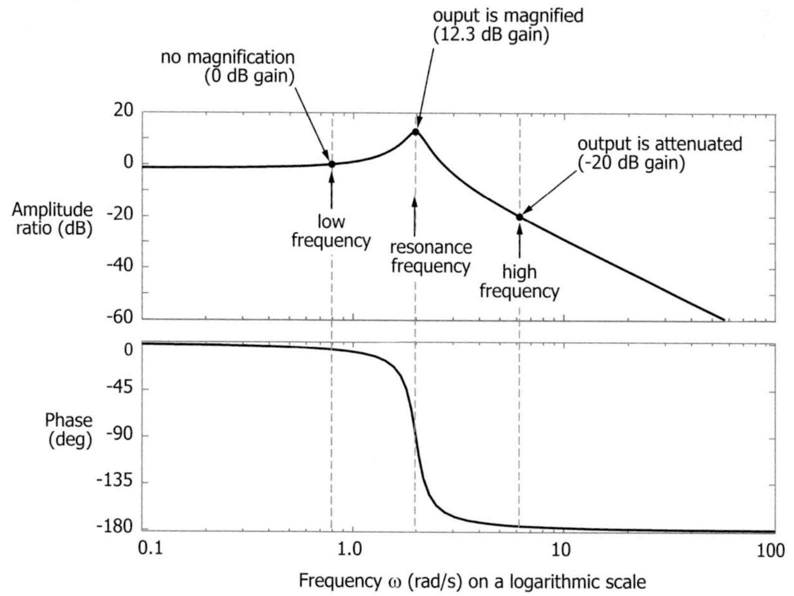

Fig. 7.12. An example of a frequency response plot (or “Bode plot”). Amplitude ratio and phase are continuous functions of input frequency.

7.4 MEMS sensors: principles of operation

In this section we discuss the principles of operation of MEMS sensors, aiming for brevity, completeness, and an introductory level of detail. Some of these technologies, commercialized more fully than others, receive a more detailed treatment in subsequent chapters: piezoresistivity in Chapter 9, capacitance in Chapter 10, piezoelectricity in Chapter 11, and thermal energy in Chapter 12.

7.4.1 Resistive sensing

In resistive sensing, the measurand causes a change to a material's electrical resistance. The change in resistance is usually detected using a circuit configuration called a bridge. We discuss two types of resistive sensing used in MEMS devices: piezoresistive sensing (commercialized) and magneto-resistive sensing (an emerging technology).

Piezoresistive sensing

The prefix "piezo-" means to squeeze or press. If by squeezing or pressing a material we cause its electrical resistance to change, the material is called **piezoresistive**. Most materials have this property, but the effect is particularly significant in some semiconductors.

The measurand of a piezoresistive sensor might be one of several mechanical quantities—acceleration, pressure, and force are the most common. In each case the sensor is designed so that the effect of the measurand is to "squeeze or press" a piezoresistive material inside the sensor, causing the material to deform. The deformation within a particular device is characterized by its geometry and the nature of the stress/strain relationship, many of which were covered in Chapter 5. Whatever the case, strain alters the material's **resistivity**, the property characterizing a material's ability to "resist" the flow of electrical current. It is this change in resistivity is that is sensed, albeit indirectly in most cases, and then related to the desired measurand.

Piezoresistance is illustrated in Fig. 7.13. An electrical resistor is fabricated from a piezoresistive material and installed in an electrical circuit. A force f causes strain in the material in one of several ways: bending, stretching, or twisting. These effects may also be applied in combination. The material's electrical characteristics are described by the familiar Ohm's law, $e = iR$, where e is voltage, i is current, and R is resistance—and in this case resistance varies depending on the amount of strain.

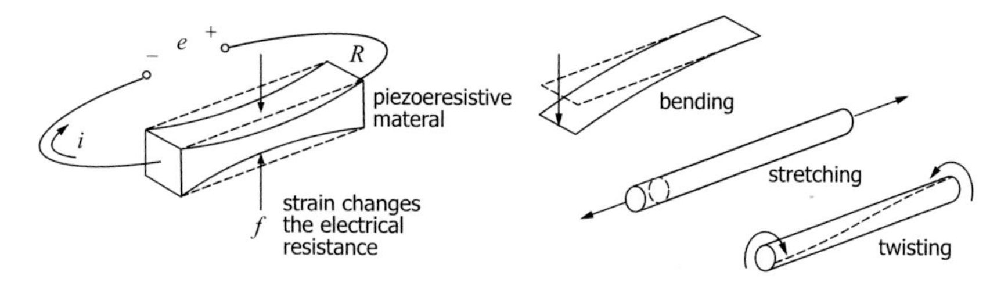

(a) The piezoresistive effect (b) Methods of creating strain

Fig. 7.13. Conceptual schematic of piezoresistance: electrical resistance varies with mechanical strain.

The resistance R of any conductor is determined by three physical parameters: resistivity, length, and cross-sectional area. Strain affects all three parameters. In metals the dominant effect is dimensional change, length and area, while in semiconductors the dominant effect is piezoresistivity. Since the sensing element in a MEMS piezoresistive sensor is usually doped single-crystal silicon, a semiconductor, we generally neglect the dimensional changes and consider only those changes associated with piezoresistivity. Additionally, we must consider the sensing element's ability to withstand repetitive small deformations without damage.

The typical changes in resistance associated with piezoresistive sensors are too small to be reliably measured directly. Consequently piezoresistors are usually wired into a configuration called a **Wheatstone bridge**—a circuit designed for detecting small changes in resistance. The bridge requires a constant input voltage and produces a small output voltage signal (usually in millivolts) that varies in response to the piezoresistive effect. Additional signal conditioning may be required to produce an output signal with the desired characteristics.

Magnetoresistive sensing

The prefix "magneto-" refers to processes carried out by magnetic means. If when subjected to a magnetic field a material's electrical resistance changes, the material is called **magnetoresistive**. Magnetoresistive sensors are one type in the class of sensors called *magnetometers*, the purpose of which is to detect magnetic fields. Thus the measurand for the magnetoresistive sensor is a magnetic field.

Magnetoresistance is illustrated in Fig. 7.14. An electrical resistor is fabricated from a magnetoresistive material and installed in an electrical circuit. A magnetic field B changes the resistance R of the material. If the resistor is installed in a bridge, similar to that used for piezoresistors, then the output voltage of the bridge is proportional to the strength of the magnetic field.

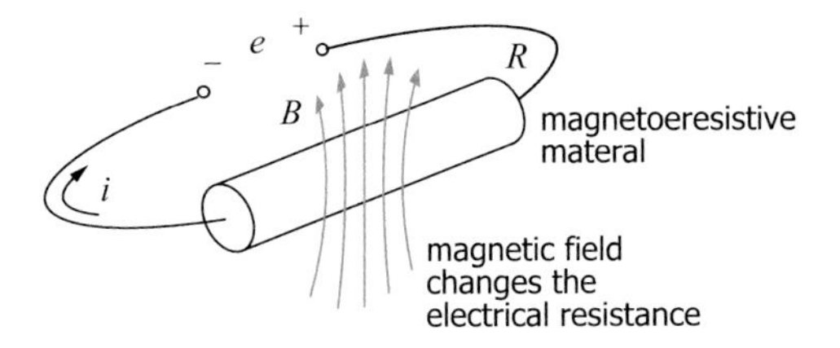

Fig. 7.14. Conceptual schematic of magnetoresistance: electrical resistance varies with the strength of the magnetic field

Thermo-resistive sensing

The prefix "thermo-" means heat. Most electrically resistive materials, when subjected to an environmental temperature change, undergo a resistance change as well. Often this is annoying, introducing error into sensor application. In such cases temperature compensation circuitry is included as part of the signal conditioning to remove the temperature effects so that the sensor reports only the phenomenon we're interested in such as stress, pressure, or a magnetic field.

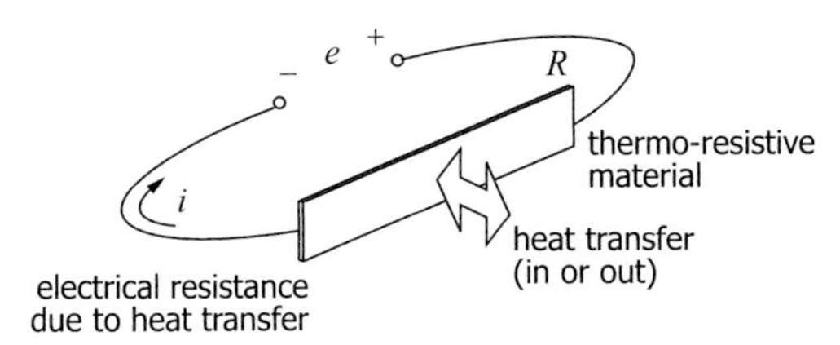

Fig. 7.15. Conceptual schematic of thermo-resistance: electrical resistance varies with the amount of heat transfer in or out

However, a heat-induced change in resistance can be exploited as a basic operating principle for a radiation or temperature sensor. In such cases, we use a material as the heat absorber that has an electrical resistance highly sensitive to heat, giving us a sensor with as wide a range and as high a sensitivity as possible. The concept is illustrated in Fig. 7.15.

7.4.2 Capacitive sensing

A voltage difference between two overlapping conductors separated by a nonconductive material causes charge to accumulate on the conductors. **Capacitance** is the characteristic that relates the charge to the voltage. In the simplest case, the relationship is described by the familiar linear relationship $q = Ce$, where q is charge, C is capacitance, and e is voltage. The basic concept is illustrated in Fig. 7.16.

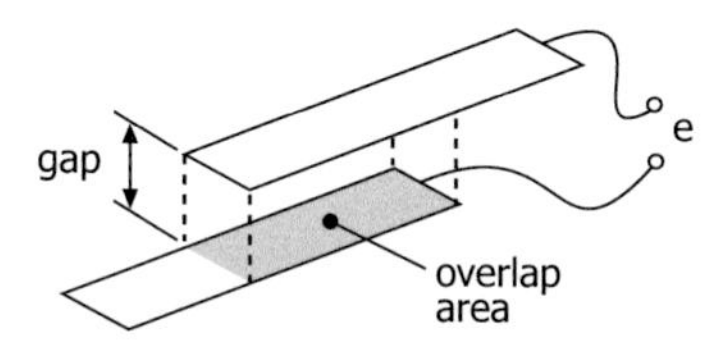

Fig. 7.16. Conceptual schematic of a capacitor: two conductors separated by a gap filled with a nonconductive material

Making one of the conductors movable is a common basis for capacitive sensing. Displacing one conductor with respect to the other produces a change in voltage. The effect is reversible—if we apply an input voltage to a capacitor with moveable conductors, the conductors move in response. Thus the capacitive effect can be used for both sensing (described here) and actuation (described in Section 7.5), though not at the same time.

The measurand for a capacitive sensor might be one of several mechanical quantities—in commercialized MEMS devices, acceleration and pressure are the most common. In each case the sensor is designed so that the measurand causes some portion of the capacitor assembly to move. The input motion alters the capacitance, changing the voltage difference between conductors.

The value of capacitance C is determined by three physical parameters: the overlap area, the gap size, and the permittivity of the material in the gap, where **permittivity** is a material property characterizing the material's ability to transmit or "permit" the electric field associated with the separation of charge. We can vary any of these parameters to create a variable capacitance, but the most common approach is to assume that the permittivity of the material in the gap (often just air) is constant and to design the input motion to vary either the overlap area, the gap size, or both.

We illustrate how the input motion might be used to vary the geometry of the capacitor in Fig. 7.17. In the first case, one plate is designed to move parallel to the other, changing the overlap area. In the second case, one plate moves either closer to or farther from the other, changing the gap distance. In either case, the geometry change causes a change in capacitance that causes a change in the voltage e across the plates. This voltage can be measured and correlated to the input motion.

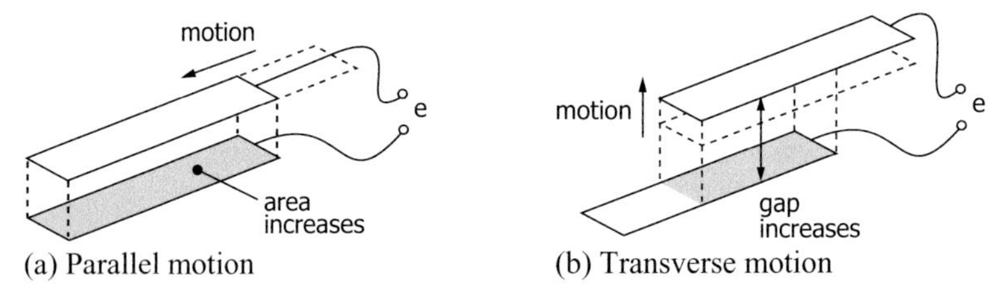

(a) Parallel motion (b) Transverse motion

Fig. 7.17. How the geometry of the capacitor can be changed by an input motion

A constant voltage or DC bias supplied to the sensor maintains a charge on the capacitor. Subtracting the DC bias from the total output voltage yields the voltage signal that represents the displacement input.

7.4.3 Piezoelectric sensing

We'll repeat here that the prefix "piezo-" means to squeeze or press. If we squeeze or press a **piezoelectric** material, the material produces an electric charge. The effect is reversible—if we apply a charge to a piezoelectric material, the material mechanically deforms in response. Thus the piezoelectric effect can be used for both sensing (described here) and actuation (described in Section 7.5), though not at the same time. Commonly used piezoelectric materials include naturally occurring crystals such as quartz, and man-made materials such as lead zirconium titanate (PZT).

The measurand of a piezoelectric sensor might be one of several mechanical quantities—acceleration and pressure are the most common. In each case the sensor is designed so that the effect of the measurand is to "squeeze or press" a piezoelectric material inside the sensor, causing the material to deform, as illustrated in Fig. 7.18. As the material deforms, positive ions become more concentrated on one side of the material, producing a positive charge on that side, and negative ions become more concentrated on the opposite side of the material, producing a negative charge on that side. Reversing the direction of the force, i.e., subjecting the material to tension rather than compression, reverses the polarity of the charge.

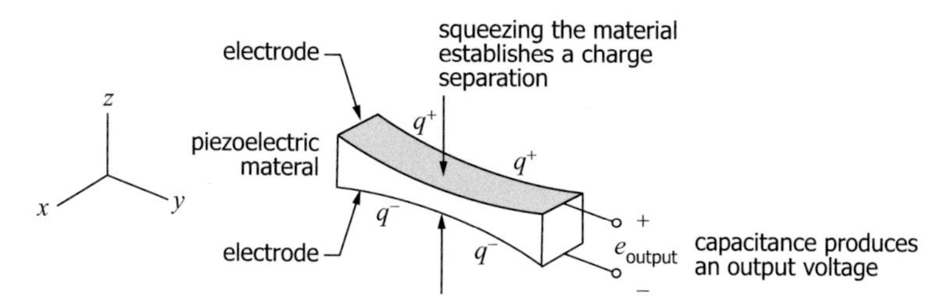

Fig. 7.18. Conceptual schematic of a piezoelectric sensor: mechanical deformation creates an electric charge

Electrodes are plated on the material to collect the charge. In the simplest one-dimensional case, a force f acting perpendicular to the electrodes (in the z-direction) gives rise to a charge given by $q = d_{33}f$, where q is charge and d_{33} is a material property called the **piezoelectric constant**. For many piezoelectric materials there is one value of the piezoelectric constant, d_{33}, for forces in the z-direction, perpendicular to the electrodes, and another value of the piezoelectric constant, d_{31}, for forces in the xy-plane, parallel to the electrodes. Other classes of piezoelectric materials have different piezoelectric constants arising from their crystalline structure.

Since the piezoelectric material is non-conductive, the assembly becomes a capacitor with all the characteristics associated with capacitors. The electrodes are the plates of the capacitor and the piezoelectric material resides in the gap between the plates. The area over which the electrodes overlap, the gap size, and the permittivity of the material in the gap all determine the value of capacitance. (Recall that *permittivity* characterizes the material's ability to transmit or "permit" an electric field.) Capacitance relates charge to voltage. In the simplest linear case, $e = q/C$, where C is capacitance, q is charge, and e is the voltage output of the sensor.

To measure this voltage, we might connect a voltmeter directly to the electrodes plated to the piezoelectric material. After all, voltmeters are designed to have a negligible loading effect on systems to which they are connected, that is, the current (and power) they draw from a system is quite small. For piezoelectric sensors, however, even a small current "loads" the sensor by drawing away charge. To prevent this effect, a special amplifier circuit is usually placed between the sensor and subsequent elements of the measurement system to buffer the sensor from the rest of the measurement system. Such amplifiers are classified generally as a *buffer amplifiers*, but those that are designed with features specialized for piezoelectric sensors are known as *charge amplifiers*.

7.4.4 Resonant sensing

You might already have guessed that resonant sensing is based on the phenomenon of resonance. Recall that **resonance** is the tendency of some systems to produce large amplitude outputs at certain frequencies—the **resonance frequencies**—and not others. Because resonance is a characteristic of frequency response, the reader may want to review relevant parts of Section 7.3 before proceeding.

In a resonant sensor, the physical component that oscillates at resonance frequencies is called (big surprise!) the **resonator**: often a thin mechanical structure such as a beam that vibrates with small displacements at high frequencies. A resonator has mass and stiffness (analogous to spring stiffness) and is lightly damped—*damping* is a frictional, energy-loss effect that limits the vibration amplitude. The resonance frequency is a function of the resonator mass, stiffness, and damping.

Two types of resonant sensor are discussed, distinguished by how the vibration of the resonator responds to the measurand. In the first sensor type the measurand affects the vibration *frequency*; in the second, the measurand affects the vibration *amplitude*.

Variable-frequency resonator: sensing strain

A resonator that responds to a mechanical input is shown in Fig. 7.19. The measurand for this resonant sensor may be one of several mechanical quantities—pressure and acceleration are the most common. In each case the sensor is designed so that the effect of the measurand is to impose a strain on the resonator, increasing its stiffness. If the resonator is vibrating, the increase in stiffness increases the resonance frequency—like tightening a vibrating guitar string and hearing the pitch rise.

The resonant sensor assembly includes both an actuator to "excite" the resonator and a frequency sensor to measure the vibration. The actuator causes the resonator to vibrate at or near its resonance frequency. Near resonance, the amplitude of the vibrations are large enough for the frequency sensor to measure. Using a feedback system, the frequency measurement is used to fine-tune the excitation frequency of the actuator, keeping the excitation frequency as nearly equal as possible to the measured resonance frequency.

As the measurand changes, causing the resonance frequency to change, the sensor "senses" the increased frequency and "tells" the actuator via the feedback loop to increase the actuation frequency to match. As the measurand fluctuates over its span, the internal sensor-actuator system keeps the resonator vibrating at resonance even as the resonance frequency changes.

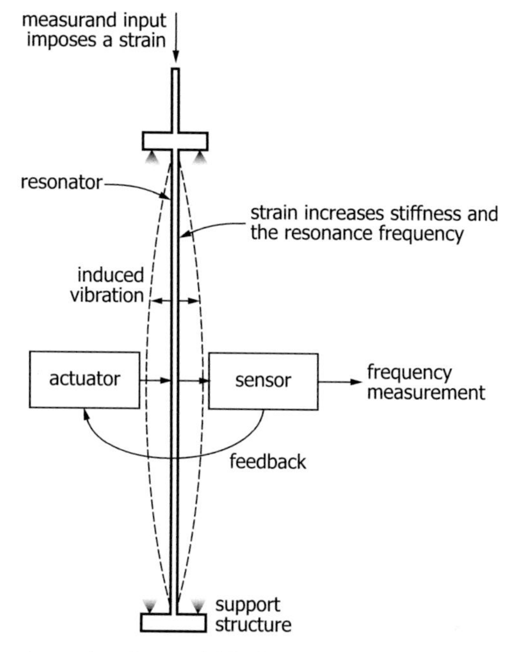

Fig. 7.19. Conceptual schematic of a variable-frequency resonant sensor. The measurand is a mechanical quantity that imposes a strain

Both the actuator and sensor subsystems are MEMS devices in their own right. The actuator might be a piezoelectric or magnetic system, for example, and the sensor might be a piezoelectric, piezoresistive, or optical system. Thus the total sensor system is actually a complete measurement and control system, like that illustrated in Fig. 7.6.

The output of the total sensor system is the frequency measurement, generally an AC voltage signal oscillating at the same frequency as the resonator. A resonant pressure sensor, for example, would have a sensitivity reported in units of pressure per unit frequency, such as kPa/MHz. Signal conditioning is used to convert a frequency signal to a DC voltage signal if desired, producing a sensitivity with units of "measurand per Volt", e.g., kPa/V.

Variable-amplitude resonator: sensing a magnetic field

A resonator that responds to a magnetic field input is shown in Fig. 7.20. In this system, the beam is designed to resonate always at the same constant frequency. The beam is an electrical conductor and carries an alternating current (AC) where the frequency of the current is set to the resonance frequency of the beam.

The alternating current alone cannot cause the beam to vibrate. In the presence of a magnetic field, however, the interaction between the current and the field produces a Lorentz force F that acts to bend the beam. The direction of the force (modeled as a vector cross product) is perpendicular to the directions of the current and field as shown in the small coordinate reference frame. The magnitude of the force is a function of current, beam length, and magnetic field strength.

The current alternates at the resonance frequency of the beam (that's why we use AC), causing the direction of the Lorentz force to alternate at the resonance frequency of the beam. The alternating-direction force causes the beam to vibrate at its resonance frequency. The greater the magnitude of the magnetic field B (the measurand), the greater the force and the greater the resulting amplitude of the vibration. An optical sensor is used to measure the vibration amplitude, giving us a measure of the magnetic field strength.

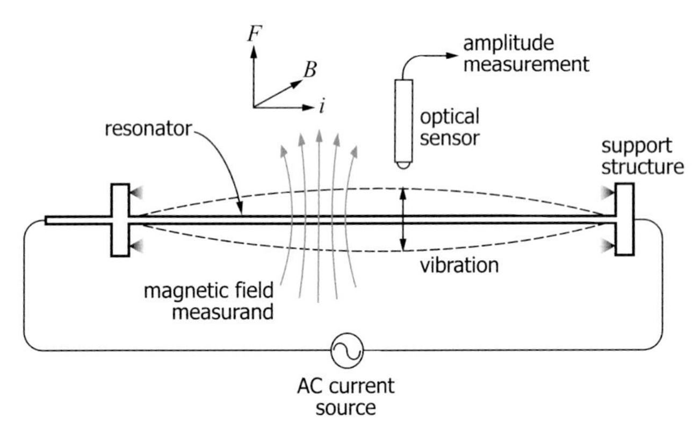

Fig. 7.20. Conceptual schematic of a variable-amplitude resonant sensor. The measurand is the magnetic field

7.4.5 Thermoelectric sensing

The prefix "thermo-" means heat. Heat is energy that flows from one point to another due to a temperature difference between the two points. In **thermoelectric** materials this flow of energy creates a voltage effect distributed between the two points. The effect is reversible—if we apply a voltage difference across a thermoelectric material a temperature difference is created. Thus the thermoelectric effect can be used for both sensing (described here) and cooling (described in Section 7.5).

The thermoelectric effect (or *Seebeck effect*) is illustrated in Fig. 7.21. Suppose we have a length of wire made of a thermoelectric material. We subject one end of the wire to a heat source and the other end to a *heat sink*—something to absorb the heat and dissipate it elsewhere. The heat flows from the source through the wire to the sink. Between any two points on the wire we find a voltage Δe that is a function of the temperature difference ΔT between those two points.

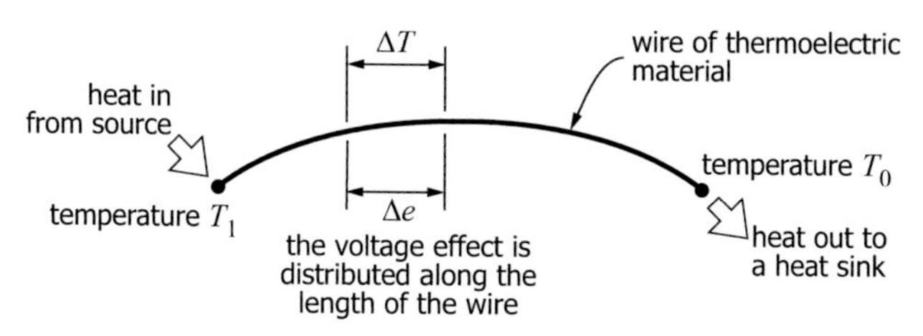

Fig. 7.21. The thermoelectric effect

The thermoelectric effect is the basis of thermocouple technology used for measuring temperature. A **thermocouple** consists of two wires of dissimilar thermoelectric materials, material A and material B, connected at their endpoints as shown in Fig. 7.22. One endpoint is exposed to a heat source and the other to a heat sink. The voltage output is a function of material selection and the temperature difference between the two endpoints.

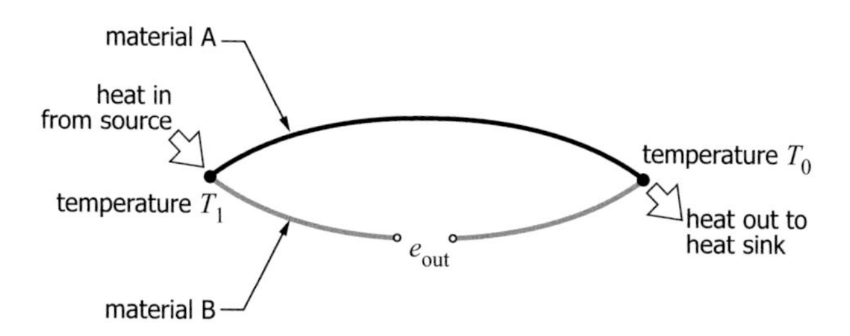

Fig. 7.22. The configuration of a thermocouple

To illustrate the thermocouple principle of operation, suppose that temperature T_1 is the unknown temperature we want to measure. The voltage output represents the temperature difference ΔT. If we use other means to determine T_0, we can compute T_1. Historically, T_0 was forced to be "zero" by placing the low-temperature junction in a container of ice water, creating a *reference junction* (or *cold junction*) at 0°C. Ice baths are somewhat impractical, and so today the temperature of the cold junction is usually measured directly using a thermistor, a resistance-temperature-detector (RTD), or an integrated-circuit (IC) sensor.

This raises the question of why, when another temperature sensor must be used to measure T_0, does one bother with the thermocouple in the first place? Why not just use the other temperature sensor to make the measurement? The answer: thermocouples have a much wider temperature span and are generally more rugged than thermistors, RTDs, and IC sensors.

If an increase in output voltage is needed, several thermocouple pairs can be assembled in series creating a *thermopile* like that shown in Fig. 7.23. One end of the thermopile, the "hot junction", is exposed to the heat source. The cold junction is exposed to a heat sink. The heat flow is evenly distributed so that every thermocouple pair experiences the same ΔT. The increase in output voltage is proportional to the number of thermocouple pairs in the thermopile.

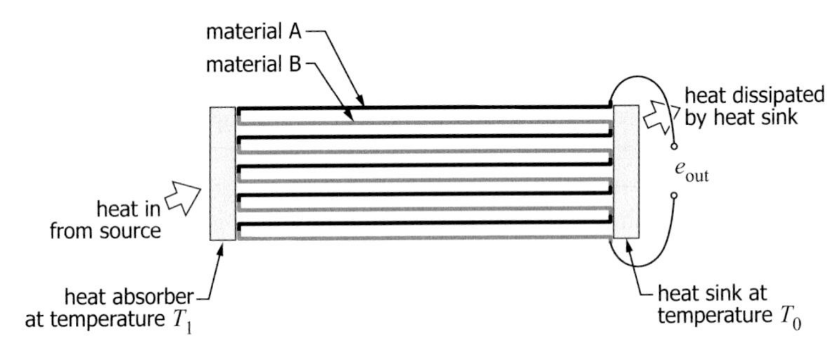

Fig. 7.23. The configuration of a thermopile

The primary applications of MEMS thermopiles are temperature measurement and infrared radiation measurement. A third application in which waste heat from other applications is converted to electrical energy, called energy "scavenging", is still in development.

For temperature measurement, the MEMS thermopile is similar to the macro-scale thermopile except in its scale and the materials used. The relationship between voltage and temperature is nonlinear, and can usually be expressed mathematically as a 5th- or higher-order polynomial. Signal conditioning is generally used to produce a direct temperature readout.

For infrared radiation (IR) measurement, the heat absorber is exposed to an IR source. Heat transfers to the thermopile via radiation, one of the three main mechanisms of heat transfer (discussed more fully in Ch. 12).

The thermopile responds to the heat transfer as usual, producing a voltage output. The instrument is calibrated to obtain the relationship between voltage output (V) and radiation power input in W/m^2.

Energy "scavenging" is an application still in development. The operating principle is the same—a temperature difference input produces a voltage output—but not for the purposes of sensing. Instead we use "waste heat" that would otherwise be lost to the environment to create a temperature differential across a thermopile, generating a voltage that can be used to provide energy to another MEMS device.

7.4.6 Magnetic sensing

Some transduction principles developed in previous sections have incorporated magnetic effects because a magnetic field was the measurand. For example, magnetically-induced resistance is discussed in the section on resistive sensing and magnetically-induced resonance is discussed in the section on resonant sensing. In this section we introduce sensors in which a magnetic or inductive effect underlies the principle of operation.

Magnetic transduction effects can be classified into two categories: *reluctance transduction*, based on changes in the energy stored in a magnetic field, and *inductive transduction*, based on charged particle interactions in a magnetic field. Either category can be used as a basis of operation for sensing.

Reluctance sensing: changes in the energy stored in a magnetic field

We can discuss the energy stored in a magnetic field by analogy to the energy stored in an electric field. Recall that the quantity that describes electrical field energy storage is capacitance. The magnetic analog of electrical capacitance is *permeance.* The inverse of permeance (*reluctance*) is the more commonly used term used in practice. Hence our term "reluctance sensing" to classify sensors based on changes in the energy stored in a magnetic field. The two most common types of reluctance sensing are those based on a variable gap (displacement) and eddy current sensing.

Recall that for capacitive energy storage, $q = Ce$, where q is charge, C is capacitance, and e is voltage (also called the *emf*, or "electromotive force"). The magnetic analog is given by $\phi = PM$, where ϕ is **magnetic flux** (analogous to charge), P is **permeance** (analogous to capacitance) referring to the *magnetic permeability* of a material, and M is **magnetomotance** (also called the *mmf*, or the "magnetomotive force", analogous to *emf*). Using these terms, one can model a "circuit" like that shown in Fig. 7.24 to illustrate reluctance-based energy storage.

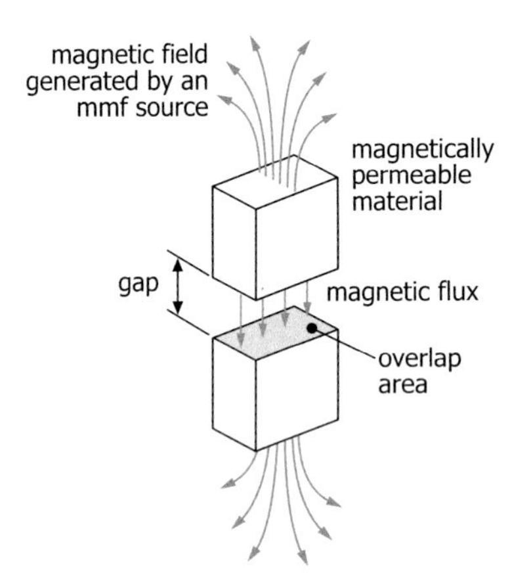

Fig. 7.24. Conceptual schematic of reluctance-based energy storage. Making the gap a variable distance is the basis for variable-reluctance displacement sensing.

The value of reluctance (and permeance) is determined by three physical parameters: the overlap area, the gap size, and material's **magnetic permeability**—a property characterizing the degree to which a magnetic field can "permeate" the material. We can vary any of these parameters to create a variable reluctance, but the most common approach is to assume that the permeability is constant and to design the input motion to vary either the overlap area, the gap size, or both.

Making the gap a variable length is a common basis for sensing a mechanical measurand. Typically one side of the gap is fixed and the other side of the gap attached to a movable arm or diaphragm. The sensor is designed so that the measurand (pressure or displacement, for example) causes some portion of the assembly to change the gap length. As the gap length changes, the mmf changes. We measure the mmf by the electrical current it induces in a coil.

A schematic of a general coil arrangement is shown in Fig. 7.25. On one side of the sensor, an AC voltage source creates a current through a wire wrapped or "coiled" around a magnetically permeable material. This generates a magnetic flux that "flows" across the air gaps and creates a magnetic field concentrated around the entire loop. The field alternates direction at the same frequency as the AC source. On the measurement side of the loop, the alternating field creates or "induces" a current in its coil. This current is our output variable. As the displacement input causes the gap length to change, the magnetic field strength changes, causing the output current to change. The key physical principle is the changing of the energy stored in the magnetic field.

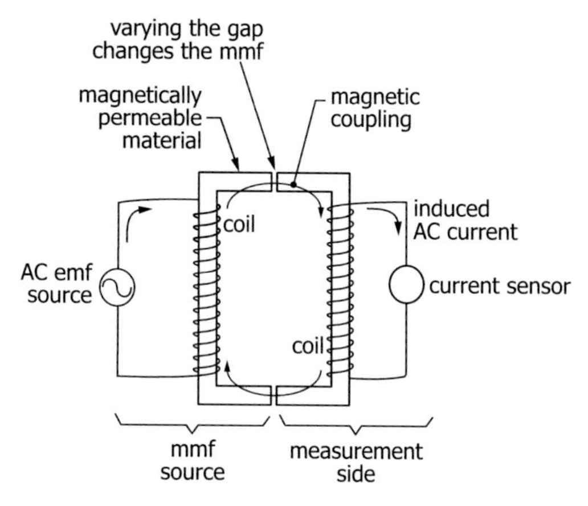

Fig. 7.25. A variable-reluctance displacement sensor: a displacement input varies the gap between coils. The measured output current varies with gap length.

Eddy current sensing is the second type of sensing based on changes in magnetic energy stored in a magnetic field. An *eddy current* is a current induced in a nonferromagnetic metal in the presence of a strong magnetic field. The eddy current in the "target" metal induces a magnetic field that acts to reduce the net strength of the original field. We sense this reduction in field strength using a coil like that shown in Fig. 7.25 to measure current. The most common application of eddy current sensors is to detect the position of a target, called "proximity" sensing.

Inductive sensing: charged particles moving in a magnetic field

Sensors of this type rely on the interactions between charged particles in the presence of a magnetic field. The "charged particles" are either in the form of a current in a conductor or a moving, conducting fluid. The magnetic field is usually provided by a permanent magnet or an electromagnet with a constant field strength. Moving coil sensors, Hall-effect sensors, and magnetic flowmeters are all of this type.

If a conductive coil is acted on by a force and is free to move in a magnetic field, a current is created or "induced" in the coil. Hence our term "inductive sensing" to classify sensors of this type. The direction and magnitude of the current are functions of the velocity of the coil motion, the magnitude and direction of the magnetic field, and the length of the conductor. The conductor, field, and force are often oriented along the three axes of a Cartesian coordinate system, as illustrated in Fig. 7.26. The current, flowing through a resistive circuit, creates a voltage we can measure to represent whatever input causes the force acting on the coil. The most common (non-MEMS) application of this technology is in dynamic microphones for sound reinforcement or recording.

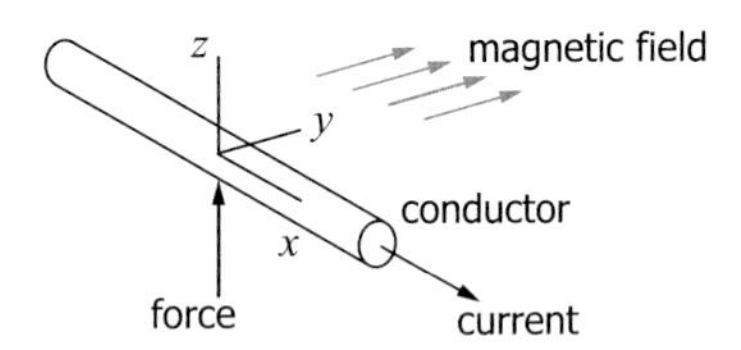

Fig. 7.26. Typical orientation of the conductor, magnetic field, and force in a moving coil inductive sensor

The forces arising from the motion of charged particles in a magnetic field also underlie the operation of Hall-effect sensing. Suppose a conductor is not free to move as in the moving coil example. The force that arises due to current flowing through a magnetic field acts on the charge carriers within the conductor, causing the current to "bend" within the conductor, as illustrated in Fig. 7.27. This nonuniform current *induces a voltage* across the conductor in the direction orthogonal to both the field and the current. The "appearance" of this induced voltage due to the nonuniform current density is the **Hall effect**.

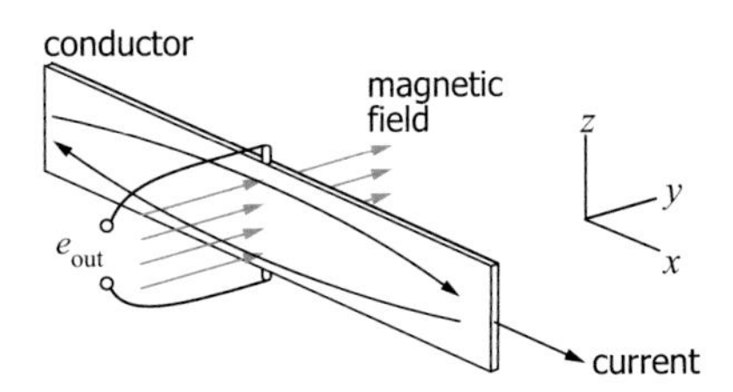

Fig. 7.27. The voltage generated by the Hall effect

Hall effect sensors are most commonly used as proximity sensors. We use a permanent magnet to produce the field and vary the magnet's distance from the conductor to vary the field strength. The closer the magnet comes to the current-carrying conductor, the greater the non-uniformity (the "bending") of the current in the conductor, and the greater the output Hall-effect voltage. Any mechanical measurand that creates this displacement can be the measurand for a Hall effect sensor.

In our final inductive sensor, a magnetic flow meter, a conducting fluid flows in a pipe. A magnetic field is oriented perpendicular to the flow, as shown in Fig. 7.28. The motion of the charged particles through the field induces a voltage that is perpendicular to both the fluid flow and the field. Placing electrodes in this orientation (z-direction), we measure an induced voltage that is proportional to the velocity of the fluid.

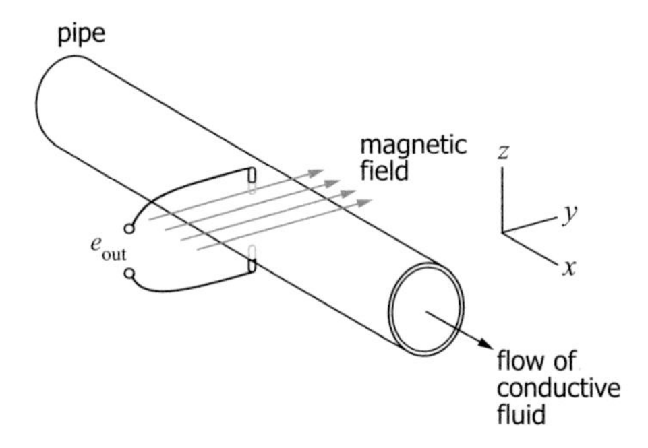

Fig. 7.28. A magnetic flow meter based on inductive sensing

7.5 MEMS actuators: principles of operation

In this section we discuss the principles of operation of MEMS actuators, again aiming for brevity, completeness, and an introductory level of detail. As in our discussion of sensors, some of these technologies are commercialized more fully than others and so receive a more detailed treatment in subsequent chapters: capacitance in Chapter 10, piezoelectricity in Chapter 11, and thermal actuation in Chapter 12.

7.5.1 Capacitive actuation

The capacitive effect is one of the physical principles that can be used for both sensing (described in Section 7.4) and actuation (described here), though not at the same time. In sensing, an input motion produces an output voltage. In actuation, an input voltage produces an output motion.

An example of a surface micromachined MEMS actuator is the linear **comb-drive** shown in Fig. 7.29. The actuator consists of rows of interlocking fingers; the center assembly is the moving part while the upper and lower figure structures (typically several hundreds of fingers each 10 microns long) are fixed to the substrate. Both assemblies are electrically insulated. Applying a voltage with same polarity to both moving and fixed structures will result the movable assembly to be repelled away from the fixed structures due to the electrostatic force. On the other hand, having opposite polarity the inter-digitated fingers are attracted toward each other. It is essential that one is careful that both assemblies are fabricated such that they are electrically insulated. The applied voltages can be either DC or AC. Due to the low inertial-mass these electrostatic comb-drive actuators can be moved at a very fast rate typically of the order of kHz.

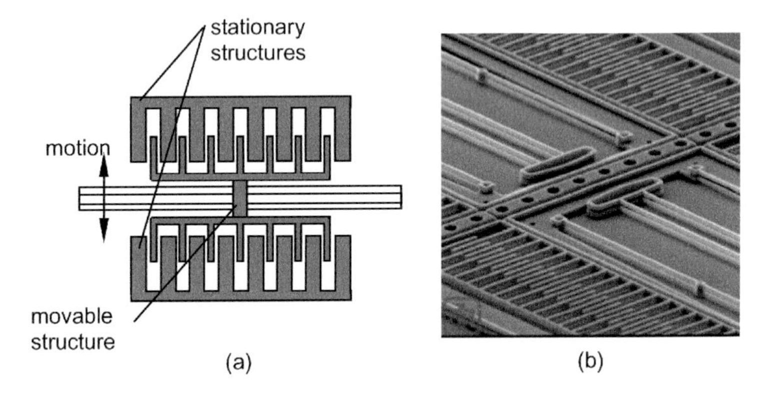

Fig. 7.29. Electrostatic comb-drives: (a) A schematic showing a typical layout. The center assembly is movable while the upper and lower structures are affixed to the substrate. (b) An SEM photo of an electrostatic comb-drive used to power a microengine. (Courtesy of Sandia National Laboratories, SUMMiT^{TM} Technologies, www.mems.sandia.gov)

7.5.2 Piezoelectric actuation

Like the capacitive effect, the piezoelectric effect is one of the physical principles that can be used for both sensing (described in Section 7.4) and actuation (described here), though not at the same time. We'll start with a reminder that the prefix "piezo-" means to squeeze or press. If we squeeze or press a **piezoelectric** material, the material produces an electric charge. The effect is reversible—if we apply a charge to a piezoelectric material, the material mechanically deforms in response. This second effect is the basis of piezoelectric actuation.

Piezoelectric actuation is illustrated in Fig. 7.30. In this simple case, a rectangular solid of piezoelectric material is sandwiched between two electrodes (top and bottom). We apply a voltage e across the electrodes. Since the piezoelectric material is non-conductive, the sandwich acts like a capacitor, establishing an electric field E. The field causes the piezoelectric material to undergo mechanical strain in three directions: contracting a small distance Δx along the x-axis, contracting a small distance Δy along the y-axis, and expanding a small distance Δz along the z-axis. Reversing polarity of the voltage reverses the direction of the deflections.

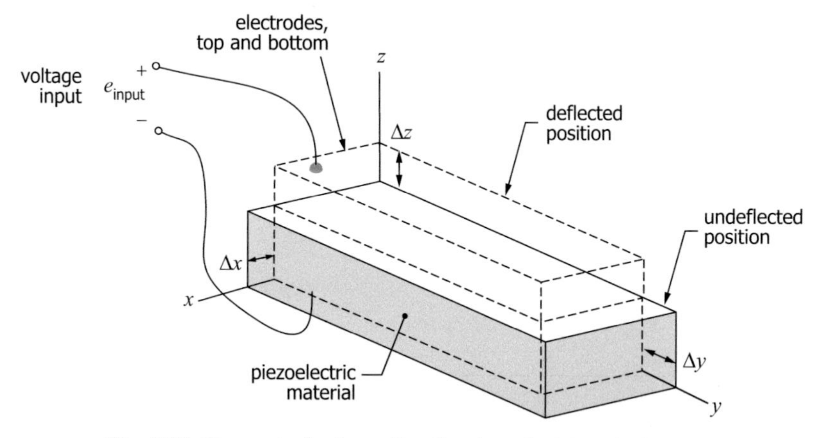

Fig. 7.30. Conceptual schematic of a piezoelectric actuator.

The magnitude of the small deflections depends on the strength of the electric field E, but the deflections are not the same in all directions. For the class of piezoelectric materials that is most commonly deposited as thin films in MEMS devices there is one value of the piezoelectric constant, represented by the symbol d_{31}, for deflection in the xy-plane, parallel to the electrodes, and another value of the piezoelectric constant, d_{33}, for deflection in the z-direction, perpendicular to the electrodes. These constants relate field strength to strain in the three coordinate directions; the relationships are given by $\varepsilon_x = \varepsilon_y = d_{31}E$ and $\varepsilon_z = d_{33}E$, where ε_x is strain in the x-direction, ε_y is strain in the y-direction, and ε_z is strain in the z-direction.

The deflections shown in the figure illustrate the basic physical phenomenon of piezoelectric actuation, but the deflections shown are much too small to be useful in MEMS devices. For example, applying an actuation voltage of several hundred volts across a bulk sample of PZT (lead zirconate titanate) produces a displacement of only a few microns. Thus a practical MEMS piezoelectric actuator is typically a stacked assembly including the piezoelectric material and its electrodes, a non-piezoelectric membrane or beam, and a support structure that combined have the effect of producing deflections that are both large enough to be useful and oriented in a particular direction.

Chapter 11 provides a detailed discussion of piezoelectric transducers, including mathematical modeling and a device case study.

7.5.3 Thermo-mechanical actuation

We saw back in Chapter 5 how thermal effects can lead to unintended stress in MEMS elements, causing them to bend or deform in unintended ways. However, we can take advantage of the fact that thermal effects can lead to mechanical forces. Specifically, we can purposely design a MEMS to include a piece that moves when heated. This is exactly what happens in a thermal actuator. Such a device, used in an optical switching application, is shown in Fig. 7.31.

Fig. 7.31. Thermal actuator produced by Southwest Research Institute. The actuator has been utilized in optical switching devices. (Courtesy Southwest Research Institute)

Bimetallic actuation

In Chapter 5 we explored the idea of thermal mismatch stress in thin films, which could result in the bending of a substrate. Essentially a bimetallic actuator actively controls thermal mismatch stress in order to bring about deformation at will, and hence, actuation.

A bimetallic actuator makes use of the differing coefficients of thermal expansion of different metals. In a bimetallic actuator, two thin layers of different metals are in direct contact with each other. When the actuator is heated, perhaps via the introduction of an electric current (Joule heating), the two different materials want to expand at different rates due to their differing coefficients of thermal expansion. As the two layers are physically hooked together, however, they are both forced to experience the same deformation, leading to the development of an internal stress. The stress in turn bends the actuator, causing motion.

Fig. 7.32 shows a bimetallic actuator consisting of a thin layer of nickel on a silicon membrane. The coefficient of thermal expansion of the nickel is greater than that of the silicon. When heated, this causes a compressive stress. As we learned in Chapter 5, this compressive stress will in turn bow the silicon membrane downward, causing actuation.

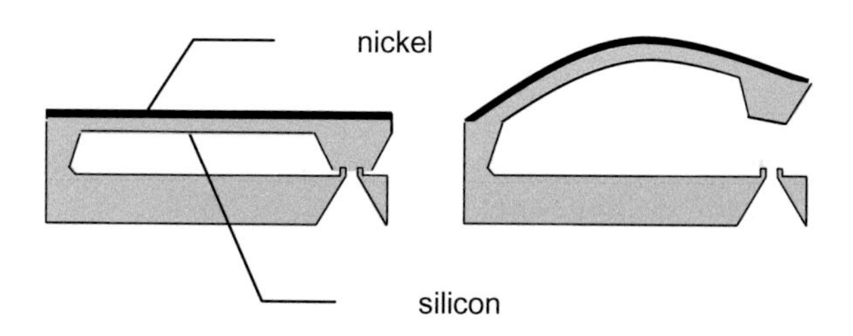

Fig. 7.32. A bimetallic thermal actuator. The differing coefficients of thermal expansion cause stress in the actuator when heated, and therefore deformation.

Thermopneumatic actuation

Thermopneumatic actuation relies on the large expansion of a fluid as it changes phase from liquid to gas. Fig. 7.33 illustrates the operation of such an actuator. When heated, a trapped fluid in the actuator begins to change phase from liquid to vapor. As the vapor has a much lower density than the liquid, the fluid expands and forces the walls of the space containing the fluid to expand, much like blowing up a balloon, causing the actuator motion.

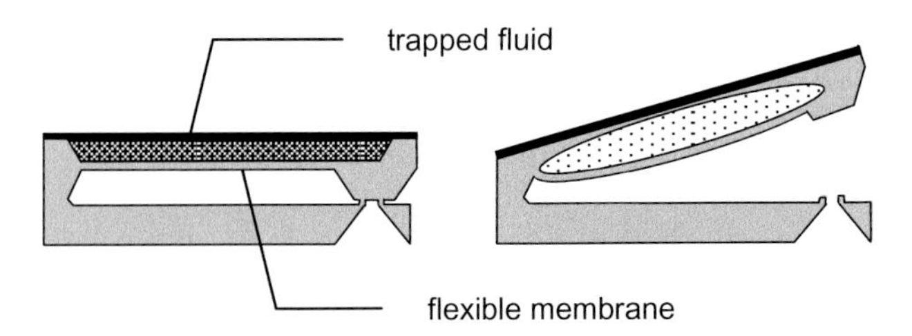

Fig. 7.33. A thermopneumatic actuator. The trapped fluid expands when heated, causing the flexible membrane to bow.

Thermopneumatic actuation is used in micropumps, microvalves, and inkjet print heads.

Shape memory alloy actuation

Several metal alloys exhibit the property called the *shape memory effect*, and are thus known as **shape memory alloys**, or **SMA**s. When within a certain temperature range, a shape memory alloy is in a solid phase called the austenite phase in which it behaves as a normal rigid solid. When cooled below a critical temperature, however, shape memory alloys take on a different solid phase called the martensite phase in which they become plastic and easily deformable. Upon reheating past this critical temperature, known as the martensite transition temperature, the shape memory alloy returns to the austenite phase and once again becomes a rigid, elastic solid. What's more, the shape memory alloy "remembers" its previous austenite phase shape. Fig. 7.34 illustrates the shape memory effect.

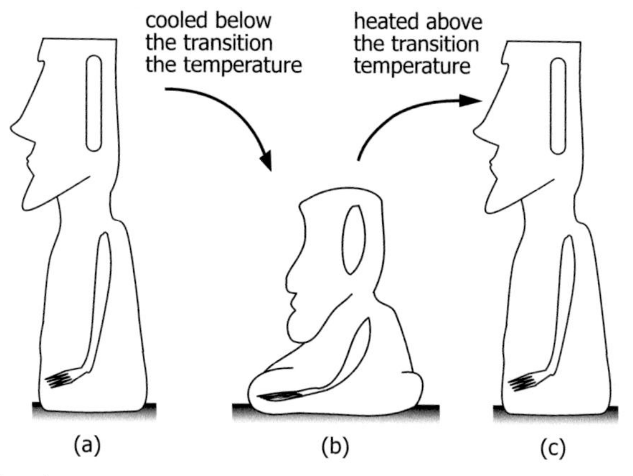

Fig. 7.34. The shape memory effect. (a) In the austenite phase, an SMA behaves as an elastic solid. (b) In the martensite phase, an SMA is plastic and easily deformable. (c) Returning to the austenite phase, an SMA recovers its original shape.

The MEMS designer can make great use of this shape memory effect as an actuation mechanism. Fig. 7.35 illustrates such an actuator, in this case, a valve. In (a), a shape-memory alloy (titanium nickel, or TiNi) is shown in its original austenite phase. This is the shape we want the alloy to "remember". In (b), cooled below the martensite transition temperature, the TiNi alloy is malleable enough to be easily deformed by a spring, keeping a valve closed. In (c), when the actuator is heated, the TiNi passes the martensite transition temperature and becomes a rigid solid in the austenite phase, overcoming the spring and pushing it away from the substrate, opening the valve.

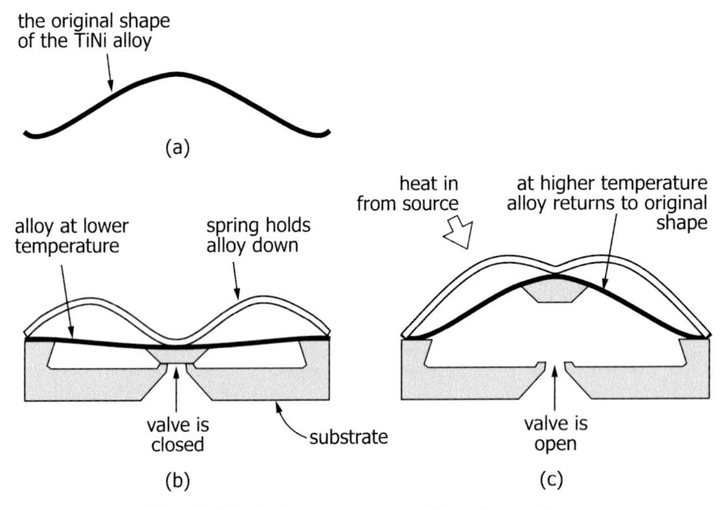

Fig. 7.35. A shape memory alloy thermal actuator

Hot arm actuation

Hot arm actuators, also known as Comtois actuators and heatuators, make use of the expansion of a material when heated. Unlike bimetallic actuators that rely on differing coefficients of thermal expansion of two different materials, a hot arm actuator need only consist of a single material. In a hot arm actuator, different regions of the actuator expand differently due to differences in geometry.

Fig. 7.36 shows a typical hot arm actuator. The actuator of Fig. 7.36 has two pads to the left where a voltage is applied. The long skinny region at the top of the figure is called the **hot arm**, and the fatter region to the bottom, the **cold arm**. The view of Fig. 7.36 is from above, looking down at the substrate.

During operation, a voltage applied across the pads creates an electric current through the actuator. As the current flows through the actuator, electrically resistive losses manifest themselves as thermal energy dissipated to the surroundings. The skinnier hot arm has a larger electrical resistance then does the fatter cold arm, and therefore dissipates more thermal energy, causing it to expand more than the cold arm. The two arms are anchored to the substrate at the left, and the different rates of expansion cause the actuator to bend as shown in Fig. 7.36 (b).

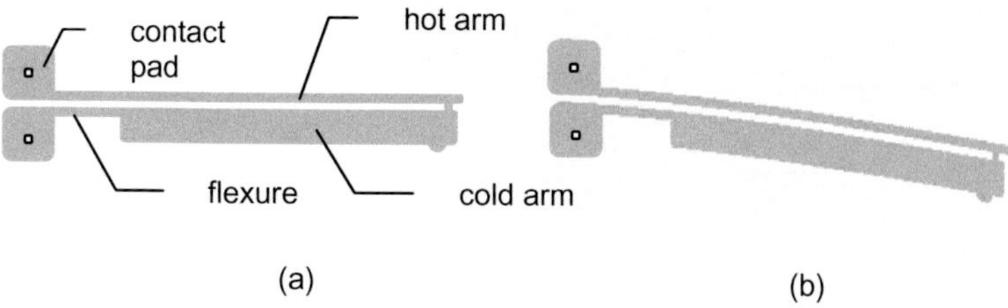

Fig. 7.36. A typical hot arm actuator: (a) Actuator in the undeflected position; (b) Actuator after a voltage is applied across the contact pads

7.5.4 Thermo-electric cooling

Recall that if two points in a material have different temperatures, heat is the energy that flows from the hotter point to the colder point. In **thermoelectric** materials, this movement of energy creates an electric field (described in Section Fig. 7.3). The effect is reversible—if we apply a voltage to a thermoelectric material, a flow of heat, and therefore a temperature difference, is created. This type of "actuation" can be used to draw heat away from a point, reducing its temperature—a process called *thermoelectric cooling*.

Thermoelectric cooling is illustrated in Fig. 7.37. Like a thermocouple, two dissimilar conductors are connected and one junction is exposed to a higher temperature location (the hot junction) and the other to a lower temperature location (cold junction). Unlike a thermocouple, a voltage source is used to create a current through the closed circuit. Current through a junction of dissimilar metals causes heat to be absorbed or rejected, depending on the direction of current, a phenomenon called the **Peltier effect**. The rate of heat transfer is proportional to the current. In (a) the current through the hot junction causes heat to be absorbed and transported to the cold junction. In (b) we reverse the direction of the current and the Peltier effect causes the direction of heat flow to change. Thus we can draw heat away from the cold junction, lowering its temperature further, and rejecting the heat through the hot junction.

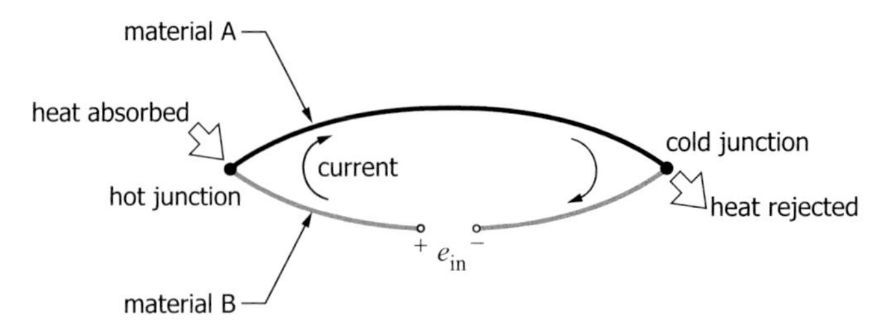

(a) drawing heat away from the hot junction

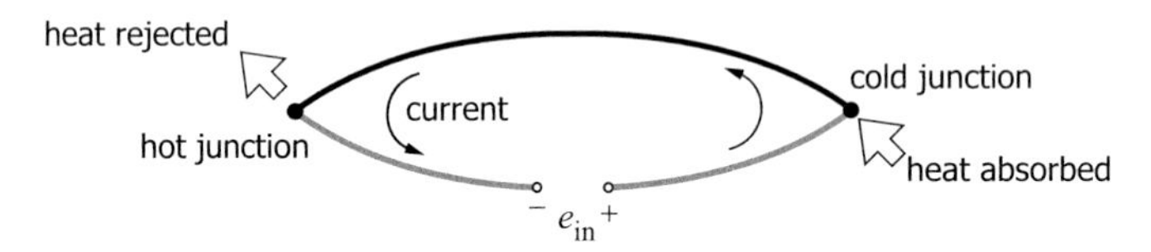

(b) drawing heat away from the cold junction

Fig. 7.37. Thermoelectric cooling

7.5.5 Magnetic actuation

Like the capacitive and piezoelectric effects, magnetic transduction is one of the physical principles that can be used for either sensing (described in Section 7.4) or actuation (described here), though not at the same time. Recall that magnetic transduction effects can be classified into two categories: *reluctance transduction*, based on changes in the energy stored in a magnetic field, and *inductive transduction*, based on charged particle interactions in a magnetic field. Unlike magnetic sensing, in which either approach can be used as a principle of operation, in magnetic actuation only reluctance transduction is used as a principle of operation. Variable reluctance relays, microvalves, solenoids, and magnetostrictive actuators are all of this type.

A schematic of a variable reluctance actuator is shown in Fig. 7.38. A current-carrying coil is wrapped around a magnetically permeable material. One arm of the material is free to move and is separated at one end by an air gap. The current in the coil creates a "magnetomotive force" or *mmf* (please recall that *mmf* and *emf* are not actually forces—these are just the traditional names for these electromagnetic phenomena). The *mmf* creates a flow of magnetic flux around the loop of material, completing what can be thought of as a magnetic "circuit". The magnitude of the flux depends on the magnitude of the *mmf*, the magnetic permeability μ of the material, the size of the gap, and the overlapping cross-sectional area of the material on either side of the gap. When the coil is turned "on", an electromagnetic force attracts the arm to the coil, attempting to close the gap. The magnitude of the force depends on the flux, the permeability, and the cross-sectional area. In the schematic, we suppose that there is a spring attached to the arm that acts to oppose the electromagnetic force such that if the magnetic coil is turned "off", the spring reopens the gap.

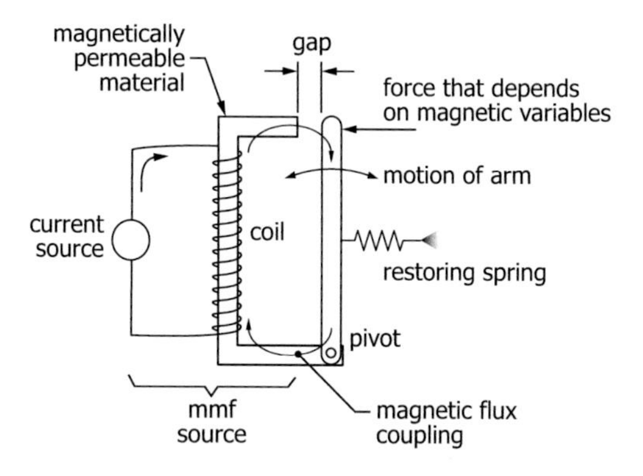

Fig. 7.38. Variable-reluctance actuator or relay

By varying the input current, we change the magnitude of the force acting on the arm, thereby changing the size of the gap. The movable arm is the actuator. We attach the arm to the external system component we want to move. If the current is applied in "on-off" fashion, that is, either no current (off) or maximum current (on), the actuator is called a *relay*, a device that holds either a fully open position or a fully closed position.

Due to the inherent three dimensionality of an electromagnet, magnetic actuators usually cannot be fabricated with the planar steps of typical micromachining as are capacitive (electrostatic) actuators. Though they tend to be more difficult to fabricate, magnetic actuators have the advantage of being able to produce larger forces and displacements over the modest values attainable in electrostatic actuation.

One may be tempted to believe that large forces can be obtained in magnetic actuators simply by employing large currents or a large number of coil turns. However, the greater the current or the number of coil turns, the greater the resistive heating that occurs. The rate at which this heat can be rejected to the surroundings imposes a limit on this particular design approach. Alternatives include selecting materials with a higher magnetic permeability and increasing the cross-sectional area of the material.

Magnetostrictive actuation

Ferromagnetic materials, in the presence of a magnetic field, expand along some axes and contract along others. The effect is called **magnetostriction** (you might think of the word as a combination of the words "magnetic" and "constriction"). The very small changes in dimension can be used as an actuator.

For example, suppose a sandwich is made of magnetostrictive materials with a piezoelectric material in the middle. In the presence of a magnetic field, and as the field strength changes, the magnetostrictive material dimensions change, applying strain to the piezoelectric material sandwiched between. The piezoelectric material generates a voltage output. The greater the magnitude of the magnetic field, the greater the voltage output generated by the piezoelectric material. We have a device that overall acts as a magnetic field sensor, a *magnetometer*, but that incorporates both an actuation technology (the magnetostrictive effect) and a sensing technology (the piezoelectric sensor).

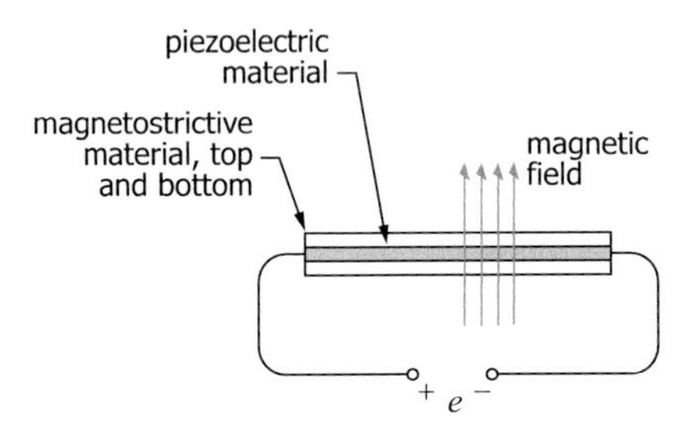

Fig. 7.39. Magnetostrictive actuation example

7.6 Signal conditioning

Signal conditioning encompasses all the methods used to convert sensor outputs to useful form: producing a signal with its desirable characteristics enhanced or with its undesirable characteristics attenuated for the next component in the measurement and control system. The process fits into the general scheme illustrated in Fig. 7.40 (this is Fig. 7.6, reproduced here for convenience). Some forms of signal conditioning have been introduced in previous sections where commonly associated with a particular transduction principle. For example, resistive sensors require a Wheatstone bridge to convert a small change in resistance to a reliably detectable voltage signal; capacitive transducers have historically required a charge amplifier to buffer the capacitor from loading effects. In this section we introduce three basic signal conditioning components commonly encountered in transducer applications: amplifiers, filters, and analog-to-digital converters.

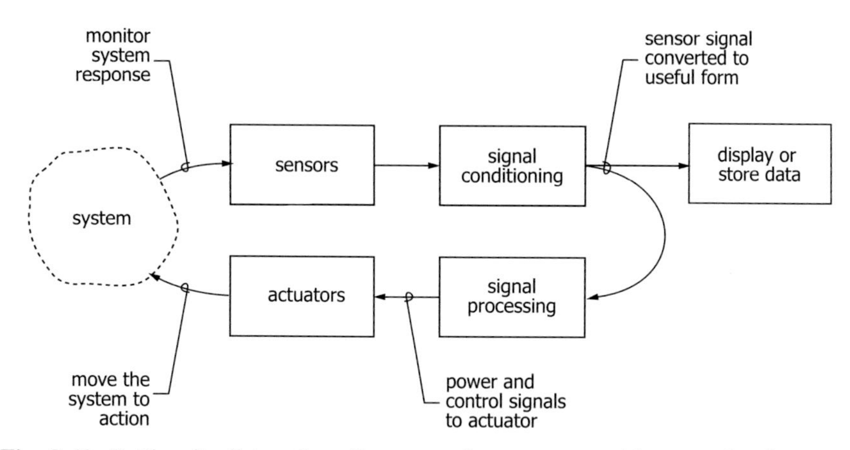

Fig. 7.40. Putting it all together: Sensors and actuators working together in a measurement and control system

A configuration for a simple signal conditioning system comprising an amplifier, filter, and **analog-to-digital converter (ADC)** is shown in Fig. 7.41. An **amplifier** is a device that increases (or decreases) the amplitude of a signal equally across all its frequency components to useful levels—useful in the sense that they are detectable and useable by the next component in the measurement and control system. Amplifiers are also used to buffer one system component from the next to address loading effects.

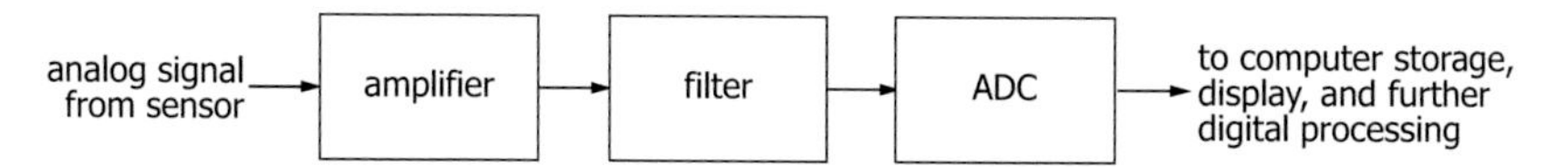

Fig. 7.41. A simple signal conditioning system: amplifier, filter, and analog-to-digital converter

A **filter** is a device that operates on a signal based on its frequency characteristics. First, any signal can be thought of (and mathematically modeled) as a linear combination of many periodic signals, each oscillating at a different frequency. A filter can be designed to reduce the amplitude of the signal at some select frequencies, increase the amplitude at other select frequencies, and let other frequencies "pass" through the filter with the amplitude unaffected. A filter is often used to reduce noise, improving the *signal-to-noise ratio*: the ratio of the amplitude of the desired signal to the amplitude of the noisy component of the signal. "Noise" is just a signal component we deem undesirable in the same way that a "weed" is just a plant we don't want in our garden. Filtering before the ADC lessens the likelihood of **aliasing**, an effect in which frequencies appear in the digitized signal that were not present in the original analog signal. An aliased signal lies to us!

An **analog-to-digital converter (ADC)** changes an analog signal to a digital signal. The basic differences between analog and digital signals are continuity and resolution. An **analog signal** is continuous in amplitude and over time; its resolution is effectively infinite (theoretically if not practically). A **digital signal** has discrete levels of amplitude that change only in discrete increments of time; its resolution is finite. Fig. 7.42 illustrates the difference.

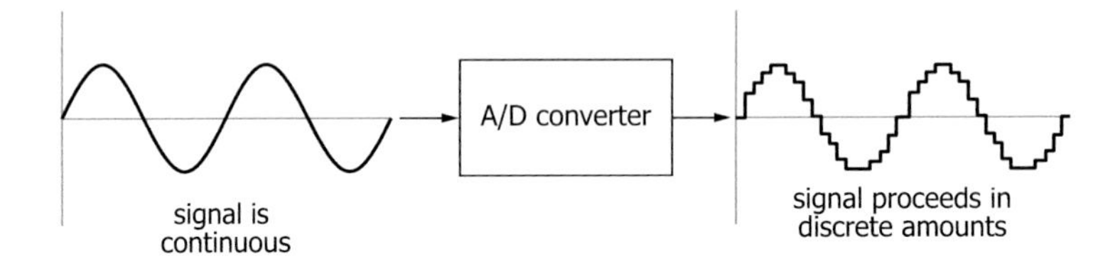

Fig. 7.42. Comparing analog and digital signals.

Analog-to-digital conversion is an important topic because of the necessary and widespread use of computers in measurement and control systems. The specifications of the conversion must be considered carefully by the designer—in going from an infinite resolution to a finite resolution, information is inevitably lost. The potential ill effects of the conversion are managed by careful selection of the conversion rate, number of bits used to represent the signal, and use of the full ADC input range.

7.7 A quick look at two applications

MEMS transducers, systems, and structures are found in industrial and automotive applications, imaging and fiber-optics communications, life-sciences applications, and radio-frequency devices. Some of our references and suggested readings go into these applications in great detail. Indeed, much of the second part of this text is devoted to a closer look at specific commercialized devices. In this section, we give a brief overview of two applications not covered in greater detail later, but that make use of the sensing and actuation technologies we've covered in this chapter.

7.7.1 RF applications

"RF" is the abbreviation for radio-frequency, referring to the range of frequencies of AC electrical signals used to produce or detect radio waves. In general, RF refers to frequencies from 30 kHz to 300 GHz. The primary application of RF MEMS devices is cell phones, operating in the ultra-high frequency (UHF) range from 300 MHz to 3000 MHz.

MEMS devices developed for RF applications include micromachined capacitors, inductors, resonators, filters and switches. All of these devices can be thought of as micro-scale circuit elements used in the design of micro-scale circuits that function in very much the same way as their macro-scale counterparts. As a very brief introduction to these components, we'll discuss RF switches.

Like any macro-scale switch, the purpose of a MEMS RF switch is to turn a circuit "on" or "off". An RF switch might be based on any of the actuation principles already discussed—the most common approach is electrostatics (capacitive). The two contacts of a switch are separated by a small gap. One contact is fixed to a substrate and the other to a movable membrane or cantilever beam. Applying a voltage to the electrostatic actuator creates a force that draws the capacitor plates towards each other, closing the switch. Remove the applied voltage and the contacts separate, opening the switch.

7.7.2 Optical applications

The two main applications of MEMS technology in optics are imaging and fiber-optics. "Imaging" includes applications such as detecting and displaying infrared radiation and image projection such as that used in digital video projectors. "Fiber-optics" MEMS involves the manipulation of light traveling in an optical fiber, primarily in communications applications.

In the case of an infrared imaging system, radiation strikes a thermo-resistive sensor element like that described in Section 7.4.1 Resistive sensing. Consider the voltage change due to the radiation that strikes one tiny resistive element as one pixel in a large array of such pixels. Readout electronics assemble the information from the array into an image. Ta-da!—an infrared night-vision display.

In image projection, a pixel-sized micro-mirror is attached to a electrostatically-actuated structure. Light is directed from a source onto the small mirror at an angle such that the angle of reflection directs the beam of light off-center, not striking the projection lens. When actuated, the micro-mirror aims the light beam at the projection lens, making that small pixel of light visible on the projection surface for the human eye to see. By controlling the speed of vibration of the mirror actuation, one can control the level of gray the viewer perceives in the projected image. Assemble as many pixels as desired to construct an array for conventional resolutions (like the 1440×900 pixel array of the computer screen I'm looking at now). To achieve full-color projection, one can employ three complete sets of mirrors, one set each for red, green, and blue.

In most fiber-optics applications, optical MEMS devices exist to manipulate tiny mirrors to direct and redirect light. While the applications are quite diverse, the primary function at the micro-scale is just using one of the actuation methods we've already discussed to manipulate a tiny mirror. Applications include tunable lasers (to change the frequency of a communications wave on demand), wavelength lockers (prevents a laser from drifting from its assigned wavelength), optical switches (an incoming optical signal can be "switched" or routed to one of several directions), and variable optical attenuators (adjusts the intensity of light in the optical fiber).

References and suggested reading

Beeby S, Ensell G, Kraft M, White N (2004) MEMS Mechanical Sensors, Artech House, Inc., Norwood, MA

Bryzek J, Roundy S, Bircumshaw B, Chung C, Castellino K, Stetter JR, Vestel M (2007) Marvelous MEMS, IEEE Circuits and Devices Magazine, Mar/Apr, pp 8-28

Comtois JH, Bright VM (1997) Applications for surface-micromachined polysilicon thermal actuators and arrays. Sensors and Actuators, 58.1:19-25

Busch-Vishniac I (1998) Electromechanical Sensors and Actuators, Springer, New York

Doebelin EO (2004) Measurement Systems: Application and Design 5th edn., McGraw Hill, New York

Fraden J (2004) Handbook of Modern Sensors 3rd edn., Springer, New York

Gad-el-Hak M (ed) (2002) The MEMS Handbook. CRC Press, New York

Madou MJ (2002) Fundamentals of Microfabrication 2nd edn., CRC Press, New York

"Microsystems Science, Technology and Components Dimensions." *Sandi National Laboratories.* 2005. http://www.mems.sandia.gov/
Maluf N, Williams K (2004) An Introduction to Microelectromechanical Systems Engineering, 2nd edn., Artech House, Inc., Norwood, MA
Read BC, Bright VM, Comtois JH (1995) Mechanical and optical characterization of thermal microactuators fabricated in a CMOS process. In: Proc. SPIE, vol. 2642 pp. 22-32

Questions and problems

7.1 Explain the difference between an actuator and sensor.
7.2 Give two examples of physical principles that can be used for both sensing and actuation.
7.3 Give two examples of physical principles that can be used only in MEMS sensors.
7.4 Gove two examples of physical principles that can be used only in MEMS actuators.
7.5 What is a measurand? What is sensitivity? Are these terms usually encountered with sensors or with transducers?
7.6 Compare and contrast the term "resolution" as used in photolithography and as used in measurement system.
7.7 What is the difference between static response and dynamic response of a transducer?
7.8 Map the physical components of the resonant sensor shown in Fig. 7.19 to the general measurement and control system shown in Fig. 7.6. To do this, sketch a diagram like Fig. 7.6 but replace the general terms in the diagram with specific components from the resonant sensor.
7.9 Explain the operating principles behind teach of the following sensing methods:
   a. Resistive sensing
   b. Capacitive sensing
   c. Piezoelectric sensing
   d. Resonant sensing
   e. Thermoelectric sensing
   f. Magnetic sensing
7.10 Explain the operating principles behind each of the following actuation methods:
   a. Capacitive actuation
   b. Piezoelectric actuation
   c. Thermo-mechanical actuation
   d. Magnetic actuation

7.11 What is signal processing? Why might it be needed?
7.12 In terms of quality, what is the difference between analog signal outputs and digital signal outputs?

Chapter 8. Piezoresistive transducers

8.1 Introduction

Piezoresistive transducers are based on the idea that a mechanical input (pressure, force, or acceleration for example) applied to a mechanical structure of some kind (a beam, a plate, or a diaphragm) will cause the structure to experience mechanical strain. Small piezoresistors attached to the structure undergo the same mechanical strain. The resulting deformation causes the piezoresistors' electrical resistance to change, allowing for the transducer to be used as a sensing device.

The electrical resistance of the piezoresistors is usually not sensed directly. Rather, the resistors are wired together in an electrical circuit configuration called a **Wheatstone bridge**. The bridge has a constant input voltage and produces a measurable output voltage that is proportional to the electrical resistance. These various inputs and outputs are illustrated in the block diagram in Fig. 8.1.

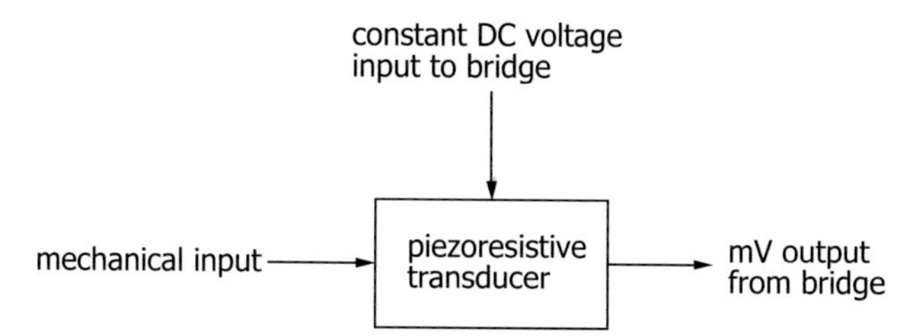

Fig. 8.1. System inputs and outputs for a piezoresistive transducer

If the transducer mechanical input is zero, then the mechanical structure sees zero strain, resulting in the resistors also experiencing zero strain. The bridge output voltage is therefore zero and the bridge is said to be balanced. When the transducer does have an applied input, however, then the mechanical structure and the resistors undergo strain, changing the electrical resistance of the piezoresistors. This in turn changes the currents in the

T.M. Adams, R.A. Layton, *Introductory MEMS: Fabrication and Applications*,
DOI 10.1007/978-0-387-09511-0_8,

bridge circuit such that the bridge now produces an output voltage. This bridge output voltage is proportional to the magnitude of the mechanical input.

The primary physical phenomenon that makes this possible is **piezoresistance**: the material property that the electrical resistance of the material changes when the material is subjected to mechanical deformation or strain. (Fig. 8.2.) An electrical resistor is fabricated from a piezoresistive material and installed in the bridge. A force f causes strain in the material in one of several ways: bending, stretching, or twisting. The piezoresistor is usually positioned such that its orientation makes it susceptible to strain in one primary direction. The material's electrical characteristics are described by Ohm's law, $e = iR$, where e is voltage, i is current, and R is resistance—and resistance is a function of strain.

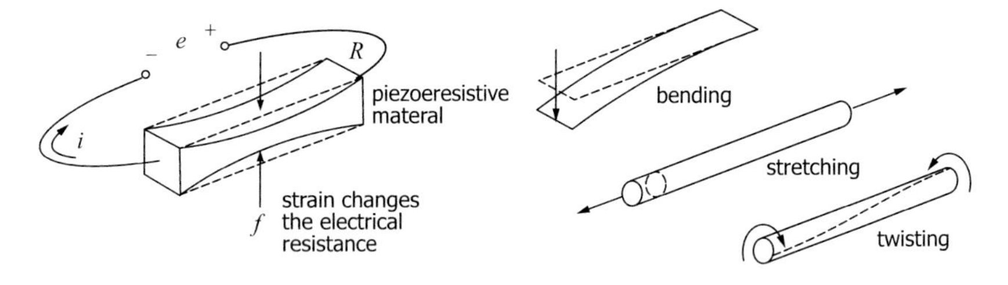

(a) The piezoresistive effect (b) Methods of creating strain

Fig. 8.2. Conceptual schematic of piezoresistance: electrical resistance varies with mechanical strain

Because piezoresistive transducers are based on electrical resistance, an energy-dissipating phenomenon, the effect is not thermodynamically reversible. Thus the piezoresistive effect can be used for sensing but not for actuation.

8.2 Modeling piezoresistive transducers

The block diagram of Fig. 8.1 illustrates that a piezoresistive sensor is an electromechanical system: it contains electrical and mechanical subsystems that respond to a mechanical input (force, pressure, or acceleration) to produce an electrical output (a voltage signal proportional to the mechanical input). In order to model the sensor we therefore begin by considering its primary electrical sub-system—the Wheatstone bridge.

8.2.1 Bridge analysis

A Wheatstone bridge is an electrical circuit that enables the detection of small changes in resistance. Of the several possible types of Wheatstone bridge configurations, we will examine just one: a voltage–sensitive, deflection–type circuit with a constant–voltage DC input, and ideal resistances (i.e., no impedance elements) in the arms of the bridge. Such a bridge is shown in Fig. 8.3.

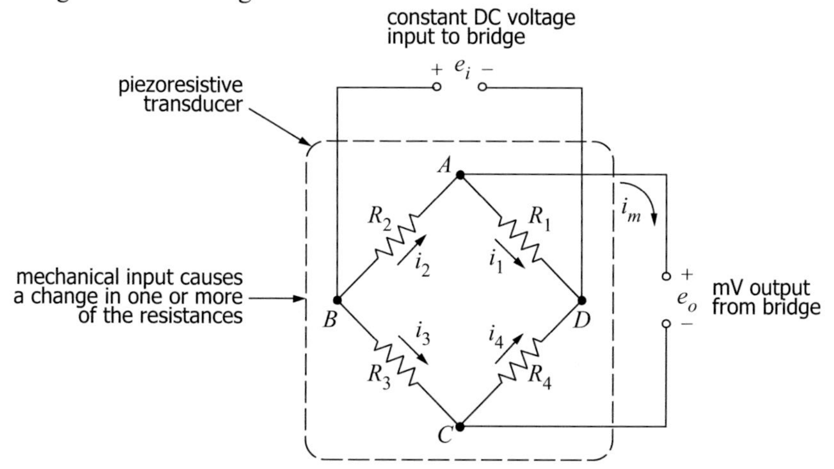

Fig. 8.3. Wheatstone bridge circuit inside a piezoresistive sensor. One or more of the resistors R_1 through R_4 may be piezoresistors affected by the mechanical input.

The input voltage e_i is supplied by a constant DC source. The four arms of the bridge each contain a resistor, R_1 through R_4 (at least one of which is made of a piezoresistive material). The output voltage e_o is the difference between the voltages at nodes A and C, that is,

$$e_o = e_A - e_C. \tag{8.1}$$

To obtain an expression for the output, therefore, we need expressions for the voltages at A and C. Using nodal analysis (applying Kirchhoff's current rule) at those two nodes,

$$\text{at } A: \; i_1 + i_m = i_2, \qquad \text{at } C: \; i_3 + i_m = i_4. \tag{8.2}$$

The output voltage is usually measured by a voltmeter with a high resistance, making the current i_m small enough to be negligible. Thus,

$$i_1 = i_2, \qquad i_3 = i_4. \tag{8.3}$$

The behavior of resistors is described using Ohm's law, $\Delta e = iR$, where Δe is the difference in voltage across the two ends of the resistor. Substituting for each current in Eq. (8.3) yields,

$$\frac{e_A - e_D}{R_1} = \frac{e_B - e_A}{R_2}, \qquad \frac{e_B - e_C}{R_3} = \frac{e_C - e_D}{R_4}. \tag{8.4}$$

The voltage at B is the input voltage and the voltage at D is the reference zero voltage. Making these substitutions yields

$$\frac{e_A}{R_1} = \frac{e_i - e_A}{R_2}, \qquad \frac{e_i - e_C}{R_3} = \frac{e_C}{R_4}. \tag{8.5}$$

Rearranging these two equations to solve for the voltages at A and at C and substituting into Eq. (8.1) yields the input-output relationship

$$e_o = \left(\frac{R_1}{R_1 + R_2} - \frac{R_4}{R_3 + R_4} \right) e_i. \tag{8.6}$$

The first conclusion we can draw from the analysis so far is the condition for *bridge balance*. If there is no mechanical input to the sensor, we would like to have zero voltage output. Zero output occurs if the parenthetical term in Eq. (8.6) is zero; that is,

$$\frac{R_1}{R_1 + R_2} - \frac{R_4}{R_3 + R_4} = 0. \tag{8.7}$$

We can rearrange this relationship to obtain

$$R_1 R_3 = R_2 R_4. \tag{8.8}$$

This condition can be met by the designer in more than one way. For instance, in what is called a *full-bridge*, all four resistors are identical piezoresistors with identical values of R. Or in a *half-bridge*, R_1 and R_2 could be identical piezoresistors with R_3 and R_4 being identical fixed resistors. In either case, selecting R_1, R_2, R_3, and R_4 such that Eq. (8.8) is true, we have $e_o = 0$ when there is no mechanical input to the transducer.

Continuing our bridge analysis, we know that the sensor is designed such that the resistors undergo a small change in resistance ΔR that produces a small change in output voltage Δe_o. Small changes like these are readily expressed mathematically using a Taylor-series expansion about the balanced condition. If we assume that all the resistances are subject to change due to applied strain, then the Taylor series has the form,

$$\Delta e_o = \frac{\partial e_o}{\partial R_1}\Delta R_1 + \frac{\partial e_o}{\partial R_2}\Delta R_2 + \frac{\partial e_o}{\partial R_3}\Delta R_3 + \frac{\partial e_o}{\partial R_4}\Delta R_4 + \text{higher order terms.} \tag{8.9}$$

We can usually assume that the higher-order terms are negligible. We obtain the partial derivative terms from Eq. (8.6), substitute them into Eq. (8.9), neglect the higher-order terms, and divide by e_i to obtain

$$\frac{\Delta e_o}{e_i} = \frac{R_2}{(R_1+R_2)^2}\Delta R_1 - \frac{R_1}{(R_1+R_2)^2}\Delta R_2 + \frac{R_4}{(R_3+R_4)^2}\Delta R_3 - \frac{R_3}{(R_3+R_4)^2}\Delta R_4 \cdot \tag{8.10}$$

This expression is the general form of the bridge model. To develop insight into the design of the sensor, however, it helps at this point to study a particular design—the full-bridge with four identical piezoresistors, that is, $R_1 = R_2 = R_3 = R_4 = R$. With these substitutions, Eq. (8.10) becomes

$$\frac{\Delta e_o}{e_i} = \frac{1}{4}\left(\frac{\Delta R_1}{R} - \frac{\Delta R_2}{R} + \frac{\Delta R_3}{R} - \frac{\Delta R_4}{R}\right). \tag{8.11}$$

Next we relate the resistance term $\Delta R/R$ to the mechanical properties of the piezoresistive material.

8.2.2 Relating electrical resistance to mechanical strain

Recall from Chapter 2 that a physical material of length L and constant cross-sectional area A has an electrical resistance R is given by

$$R = \frac{\rho L}{A}, \tag{9.12}$$

where ρ is the material's *resistivity*. The geometry of the resistor is illustrated in Fig. 8.4 for both a rectangular cross-section (as in a thin plate) and a circular cross-section (as in a thin wire).

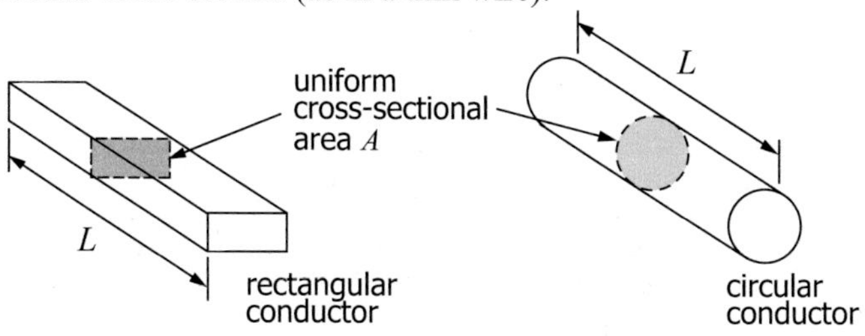

Fig. 8.4. Generalized resistor geometry

A change in resistance ΔR is produced by changes in any of the three quantities—resistivity, length, or area. Once again, small changes are modeled using a Taylor-series expansion. From Eq. (9.12) we obtain

$$\Delta R = \frac{\partial R}{\partial \rho}\Delta \rho + \frac{\partial R}{\partial L}\Delta L + \frac{\partial R}{\partial A}\Delta A + \text{higher order terms}\,. \tag{8.13}$$

We obtain the partial derivative terms from Eq. (9.12), neglect the higher-order terms of the series, and divide by R to obtain

$$\frac{\Delta R}{R} = \frac{\Delta \rho}{\rho} + \frac{\Delta L}{L} - \frac{\Delta A}{A}\,. \tag{8.14}$$

The first term on the right-hand side of Eq. (8.14) represents the piezoresistive property—a change in resistance due to a change in resistivity $\Delta\rho$ due to the application of mechanical stress. The second two terms represent changes in resistance due to changes in the geometry (length and area) of the resistor. In semiconductor materials, the first term dominates. In metals, the geometric terms dominate. In the next two sections we develop the model for both metals and semiconductors.

Gage factor for metal resistors

In metals, the relative change in resistance $\Delta R/R$ is due primarily to the changing geometry. In this case we can neglect the $\Delta\rho/\rho$ term in Eq. (8.14) and model the relative change in resistance using

$$\frac{\Delta R}{R} = \frac{\Delta L}{L} - \frac{\Delta A}{A}\,. \tag{8.15}$$

Piezoresistors made of metals are most commonly encountered in the form of *strain gages*. Strain gages are fabricated to be most responsive to uniaxial strain. Thus our analysis of the geometric effect in Eq. (8.15) is based on uniaxial strain applied to an isotropic material.

When a material is subjected to stress in one direction (we'll use a rectangular cross-section to illustrate), the length L of the material increases by a small amount ΔL. As we saw in Chapter 5, the height h and width w will both decrease by the amounts Δh and Δw due to Poisson contraction. (Fig. 8.5.) The strain ε_L in the direction of the applied stress is given by the ratio of the change in length to the original length, $\varepsilon_L = \Delta L/L$.

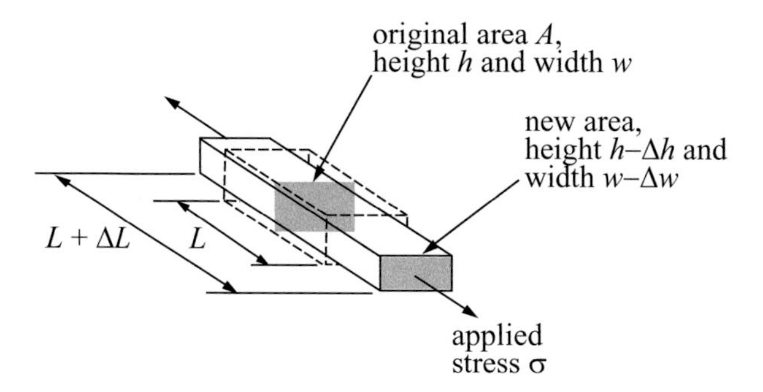

Fig. 8.5. Dimensional changes due to applied stress in one direction

The cross-sectional area of the unstressed specimen is $A_0 = hw$. When stress is applied, the cross-sectional area becomes $A_\sigma = (h-\Delta h)(w-\Delta w)$, where the subscript σ indicates stress has been applied. The change in area is given by $\Delta A = A_\sigma - A_0$,

$$\Delta A = (h - \Delta h)(w - \Delta w) - hw\,. \tag{9.16}$$

Dividing by A yields

$$\frac{\Delta A}{A} = \left(1 - \frac{\Delta h}{h}\right)\left(1 - \frac{\Delta w}{w}\right) - 1\,. \tag{8.17}$$

The transverse strain terms $\Delta h/h$ and $\Delta w/w$ for metals and cubic crystals can be expressed in terms of Poisson's ratio ν and the axial strain ε_L by

$$\frac{\Delta h}{h} = \nu\varepsilon_L \quad \text{and} \quad \frac{\Delta w}{w} = \nu\varepsilon_L\,. \tag{8.18}$$

Substituting into Eq. (8.17) and simplifying yields

$$\frac{\Delta A}{A} = -2\nu\varepsilon_L + \nu^2\varepsilon_L^2\,. \tag{8.19}$$

Because ν and ε_L are both small, we can consider the product of their squares negligible, yielding

$$\frac{\Delta A}{A} = -2\nu\varepsilon_L\,. \tag{8.20}$$

Returning to our change-of-resistance relationship Eq. (8.15), substituting Eq. (8.20) and $\varepsilon_L = \Delta L/L$, and combining terms yields

$$\frac{\Delta R}{R} = (1 + 2\nu)\varepsilon_L . \tag{8.21}$$

This relationship models the change in resistance in metals used in strain gages as a function of a material property, Poisson's ratio, and the applied mechanical input, uniaxial strain.

The sensitivity of the sensor depends on the ratio of $\Delta R/R$ to ε_L. This ratio is called the **gage factor**, F, a dimensionless number defined as

$$F = \frac{\Delta R/R}{\varepsilon_L} . \tag{8.22}$$

Applying this definition to Eq. (8.21), we obtain the expression for gage factor for metal strain gages given by

$$F = 1 + 2\nu . \tag{8.23}$$

Gage factor is a *figure of merit* used both to compare the predicted performance of candidate materials for use as piezoresistors and to guide us in positioning the piezoresistors on the mechanical structure that is subjected to the input strain. We discuss the placement and orientation of resistors following the discussion of gage factor for semiconductors.

Gage factor for semiconductor resistors

In semiconductors, the relative change in resistance $\Delta R/R$ is due primarily to changes in resistivity, not changes in geometry. Neglecting the geometric terms $\Delta L/L$ and $\Delta A/A$ in Eq. (8.14) yields

$$\frac{\Delta R}{R} = \frac{\Delta \rho}{\rho} . \tag{8.24}$$

For a piezoresistor subjected to longitudinal and transverse stresses, the resistivity change is

$$\frac{\Delta \rho}{\rho} = \pi_L \sigma_L + \pi_T \sigma_T , \tag{8.25}$$

where σ_L and σ_T are the longitudinal and transverse stresses, and π_L and π_T are the longitudinal and transverse **piezoresistance coefficients** of the material. In practice, longitudinal means "in the direction of current" and transverse means "perpendicular to the direction of current". Alternatively—and slightly more useful in our discussion of gage factor—resistivity can be expressed in terms of strain,

$$\frac{\Delta\rho}{\rho} = \gamma_L \varepsilon_L + \gamma_T \varepsilon_T , \tag{8.26}$$

where γ_L and γ_T are the longitudinal and transverse **elastoresistance coefficients** of the material.

The two models of Eqs. (8.25) and (8.26) are different representations of the same effect, and are related to each other via the relationships between stress and strain in the longitudinal and transverse directions. However, because the semiconductor materials tend to be anisotropic, the linear Young's modulus relationship for uniaxial stress/strain, $\sigma = E\varepsilon$, *does not apply* here. Rather, the more general relation given in Eq. (5.25) is necessary.

Substituting Eq. (8.26) into Eq. (8.24) we obtain

$$\frac{\Delta R}{R} = \gamma_L \varepsilon_L + \gamma_T \varepsilon_T . \tag{8.27}$$

This relationship models the change in resistance in semiconductors used in piezoresistive applications as a function of a material property, the elastoresistance coefficients, and the applied mechanical inputs, longitudinal and transverse strains.

Dividing by ε_L, we obtain an expression for the gage factor,

$$F = \gamma_L + \gamma_T \frac{\varepsilon_T}{\varepsilon_L} . \tag{8.28}$$

Both the elastoresistance coefficients and the strains depend on the orientation of the resistor.

Table 8.1 gives approximate ranges of gage factors for different materials.

Table 8.1 Approximate gage factors for different materials

Material	Gage factor, F
Metals	2-5
Cermets (Ceramic-metal mixtures)	5-50
Silicon and germanium	70-135

Physical placement and orientation of piezoresistors

Using the concept of gage factor, we can learn something about the placement of the piezoresistors on the mechanical structure of the sensor. From Eq. (8.22) we see that $\Delta R/R = F\varepsilon_L$ for both metal and semiconductor piezoresistors. Substituting for the $\Delta R/R$ terms in the full-bridge bridge relationship Eq. (8.11), we obtain

$$\frac{\Delta e_o}{e_i} = \frac{F}{4}\left(\varepsilon_1 - \varepsilon_2 + \varepsilon_3 - \varepsilon_4\right). \tag{8.29}$$

This model indicates that if the strains are equal in magnitude and have the same sign, then the bridge will produce *zero* output—not a useful outcome. To obtain a useful voltage output, we therefore position resistors 2 and 4 to undergo strain in the opposite sense to the strain of resistors 1 and 3.

For example, consider the piezoresistive accelerometer shown in Fig. 8.6. The accelerometer housing is secured to a body undergoing the acceleration we want to measure—the vibration of a machine, for example. In response to the acceleration of the body, the seismic mass accelerates causing the cantilever beam to bend. The bending applies strain to the piezoresistors. Physical placement and orientation of the resistors is influenced by three considerations: the bridge configuration (which resistor is wired into which leg of the bridge), the region of maximum stress, and the direction of the strain to which we want the resistor to respond.

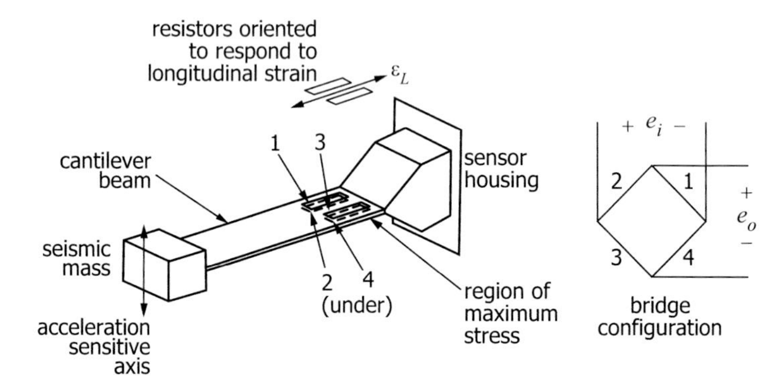

Fig. 8.6. Placement of resistors on a beam-type piezoresistive accelerometer sensitive to acceleration in one direction

The bridge configuration in this example is determined by the sign of the strain in each resistor location. If the beam bends downwards, the two resistors on top of the beam are in tension (a positive ε for p-type semiconductors) at the same time the resistors underneath the beam are in compression (a negative ε for p-type semiconductors). If the acceleration changes direction, the regions of tension and compression swap. For the strains to add up and not cancel each other out in Eq. (8.29) we can place resistors 1 and 3 side by side on one side of the beam (here both are shown on top of the beam) and resistors 2 and 4 on the other side of the beam. Then ε_1 and ε_3 have the same sign, opposite that of ε_2 and ε_4, and the bridge produces a total positive or a total negative output voltage Δe_o.

The four resistors are placed at the base of the beam because this is the region of maximum stress. Any other placement reduces the sensitivity of the sensor.

The resistors are oriented to be responsive to longitudinal strain, that is, the direction along the longitudinal axis of the cantilever beam. Transverse effects are negligible in this configuration.

These three factors—bridge configuration, location and orientation of maximum stress and strain, and the orientation of the resistors that respond to the strain—are used to determine the optimum placement and orientation of piezoresistors in a transducer. We revisit these concepts in the case study that follows: a piezoresistive pressure transducer.

8.3 Device case study: a piezoresistive pressure sensor

In this section we study a piezoresistive pressure sensor with operational parameters similar to those seen in automotive applications. The characteristics described in this case study are representative of a class of pressure sensors that are similar in size and use to the Omega PX409 pressure transducer.

The basic configuration of this piezoresistive sensor is shown in Fig. 8.7. The sensor housing is stainless steel, about 80 mm long, with a pressure fitting at one end and an electrical connection for input and output voltage at the other end. Sensors of this type can typically be ordered to measure pressure ranges from 0–6.9 kPa (1 psi) to 0–34.5 MPa (5000 psi). The DC excitation voltage is usually between 5–10 V. When selecting a particular sensor make and model, one usually has a choice of electrical output signals: 0–100 mV, 0–5 V or 4–20 mA.

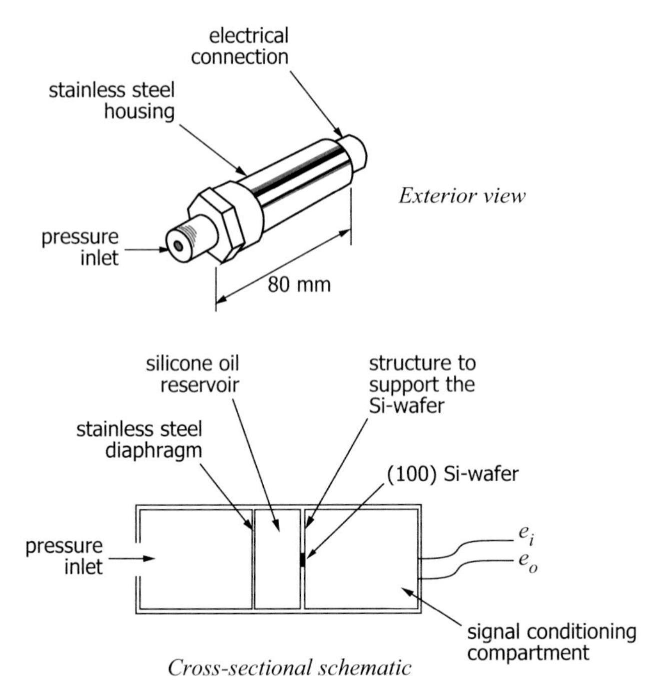

Fig. 8.7. A commercially available piezoresistive pressure sensor. (Adapted from the Omega PX409 pressure transducer.)

The cross-section shows that the working fluid whose pressure we want to measure enters the sensor through the pressure inlet fitting and imposes pressure on the stainless steel diaphragm. A small volume of silicone oil transfers the pressure from the stainless-steel diaphragm to the (100) Si-diaphragm. The induced stress and strain of the wafer is detected by piezoresistors on the wafer in a bridge configuration, producing an electrical output signal proportional to the inlet pressure.

In this example, we examine a pressure sensor with a 0–1 MPa (145 psi) full scale input, 0–100 mV full scale output, a 10 VDC excitation, and p-Si piezoresistors.

We begin by discussing the mechanical properties of the diaphragm. In this example, the (100) Si-diaphragm is a square, 1.2 mm on each side, 80μm thick, oriented with the <110> directions bisecting the square as shown in Fig. 8.8. When pressure is applied from below the largest upwards deflection of the diaphragm is at the center of the square. Consequently, the largest stress σ_C, located midway along each edge, is directed towards this *center*. Stress σ_B, at the same location, is directed parallel to the edge, along the *boundary* of the diaphragm. Based on published experimental results for a square diaphragm of this type and size, we estimate $\sigma_C = 45.0$ MPa and $\sigma_B = 22.5$ MPa. The strains in the same two directions are $\varepsilon_C = 152\ \mu\varepsilon$ and $\varepsilon_B = -17\ \mu\varepsilon$, (Recall that the symbol $\mu\varepsilon$ means *microstrain*, where 1 microstrain = 10^{-6} strain.) The maximum stress σ_C is lower than the fracture stress of the Si-diaphragm (360 MPa) by safety factor of 8.

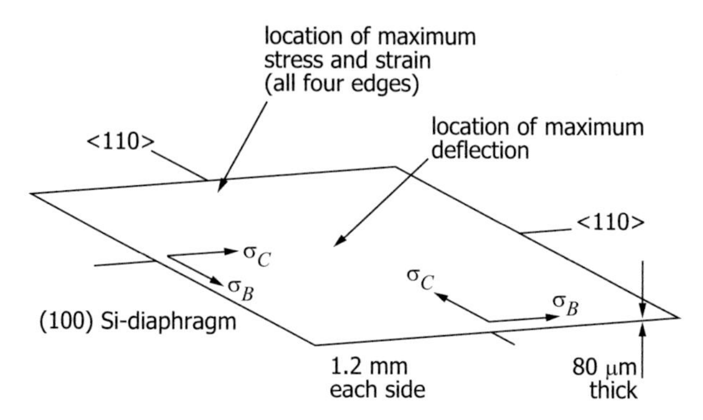

Fig. 8.8. Orientation of the Si-diaphragm and points of maximum stress and strain

The four p-type piezoresistors are placed at the four locations of maximum stress and strain. From the bridge model, we know we want resistors 1 and 3 to undergo a positive $\Delta R/R$ at the same time resistors 2 and 4 undergo a negative $\Delta R/R$. We accomplish this by orienting the resistors as shown in Fig. 8.9. Resistors 1 and 3 are oriented with their *longitudinal axes* in the direction of maximum stress (σ_C), creating a net positive $\Delta R/R$. Resistors 2 and 4 are oriented with their *transverse axes* in the direction of maximum stress (σ_C), creating a net negative $\Delta R/R$.

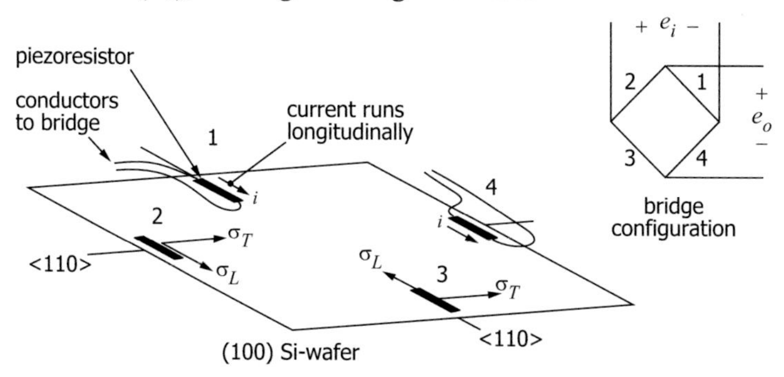

Fig. 8.9. Location and orientation of four piezoresistors on the square diaphragm

The elastoresistance coefficients of the resistors in the <110> direction can be found in published tables to be

- $\gamma_L = 120$ <110>
- $\gamma_T = -54$ <110>.

Recall, from (8.27), that the change in resistance $\Delta R/R$ is a function of the longitudinal and transverse elastoresistance coefficients and strain. We group the analysis below into two columns: the left column for resistors 1 and 3 and the right column for resistors 2 and 4. We write (8.27) for each case:

$$\frac{\Delta R_{1,3}}{R} = \gamma_L \varepsilon_L + \gamma_T \varepsilon_T , \qquad \frac{\Delta R_{2,4}}{R} = \gamma_L \varepsilon_L + \gamma_T \varepsilon_T . \tag{8.30}$$

Comparing Fig. 8.8 to Fig. 8.9, we see that the longitudinal strain of resistors 1 and 3 is ε_C (towards the center of the diaphragm) while the longitudinal strain of resistors 2 and 4 is ε_B (along the boundary of the diaphragm). Substituting for ε_L yields

$$\frac{\Delta R_{1,3}}{R} = \gamma_L (\varepsilon_C) + \gamma_T \varepsilon_T , \qquad \frac{\Delta R_{2,4}}{R} = \gamma_L (\varepsilon_B) + \gamma_T \varepsilon_T . \tag{8.31}$$

where parentheses have been used to highlight the substitutions.

The converse is true for the transverse strains. The transverse strain of resistors 1 and 3 is ε_B (along the boundary) while the transverse strain of resistors 2 and 4 is ε_C (towards the center). Substituting for ε_T yields

$$\frac{\Delta R_{1,3}}{R} = \gamma_L (\varepsilon_C) + \gamma_T (\varepsilon_B), \qquad \frac{\Delta R_{2,4}}{R} = \gamma_L (\varepsilon_B) + \gamma_T (\varepsilon_C). \tag{8.32}$$

We're ready to obtain a numerical value for the two $\Delta R/R$ terms:

$$\begin{aligned} \frac{\Delta R_{1,3}}{R} &= (120)(152 \times 10^{-6}) + (-54)(-17 \times 10^{-6}) \\ &= 19.1 \times 10^{-3} \\ \frac{\Delta R_{2,4}}{R} &= (120)(-17 \times 10^{-6}) + (-54)(152 \times 10^{-6}) \\ &= -10.2 \times 10^{-3} \end{aligned} \tag{8.33}$$

Thus, as we'd planned, resistors 1 and 3 see a positive change in resistance (of about 2%) and resistors 2 and 4 see a negative change in resistance (of about 1%). As a reality check on the analysis to this point, we compute the gage factor F for resistors 1 and 3, oriented longitudinally in the direction of maximum stress (σ_C). The gage factor is computed below, with a result that lies in the expected range ($70 \le F \le 135$) for silicon.

$$
\begin{aligned}
F &= \frac{\Delta R_1/R}{\varepsilon_C} \\
&= \frac{19.1\times 10^{-3}}{152\times 10^{-6}} \\
&= 126
\end{aligned}
\tag{8.34}
$$

Next, we determine the ratio of electrical output to input at the full pressure load of 1.0 MPa using the full-bridge model and the values of $\Delta R/R$,

$$
\begin{aligned}
\frac{\Delta e_o}{e_i} &= \frac{1}{4}\left(\frac{\Delta R_1}{R} - \frac{\Delta R_2}{R} + \frac{\Delta R_3}{R} - \frac{\Delta R_4}{R}\right) \\
&= \frac{1}{4}\left[19.1 - (-10.2) + 19.1 - (-10.2)\right]\times 10^{-3} \\
&= 14.7\ \text{mV/V}.
\end{aligned}
\tag{8.35}
$$

With a 10 VDC excitation voltage, the bridge full load output is 147 mV. Since the full load is 1.0 MPa, the sensitivity η is given by

$$
\begin{aligned}
\eta &= \frac{147\ \text{mV}}{1\ \text{MPa}} \\
&= 147\ \text{mV/MPa}\ (1.01\ \text{mV/psi})
\end{aligned}
\tag{8.36}
$$

To meet our goal of a 0–100 mV full scale output, we add an amplifier circuit to *attenuate* (reduce) the 147 mV bridge output by a factor of 100/147 = 0.682. This factor is called a *gain* K and is often reported in the units of decibels (dB) as follows,

$$
\begin{aligned}
K &= 20\log_{10}\left(\frac{100}{147}\right) \\
&= -3.32\ \text{dB}
\end{aligned}
\tag{8.37}
$$

The simplest op-amp circuit that provides positive attenuation comprises two inverting amplifiers in series, as shown in Fig. 8.10. We have to select the resistors R_f and R_o to obtain the desired gain of 0.0682 (or –3.32 dB).

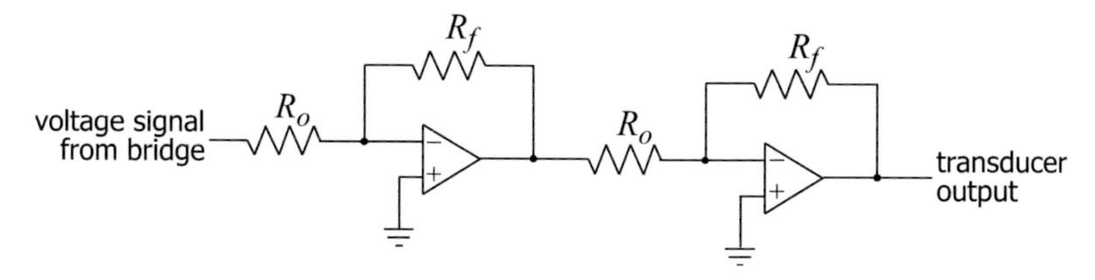

Fig. 8.10. Simple non-inverting attenuation circuit

The gain of an inverting amplifier is the ratio of R_f to R_o, that is,

$$K = -\frac{R_f}{R_o}, \tag{8.38}$$

and therefore the gain of the two inverting op-amps in series is given by

$$K = \left(-\frac{R_f}{R_o}\right)\left(-\frac{R_f}{R_o}\right) = +\left(\frac{R_f}{R_o}\right)^2. \tag{8.39}$$

We write a computer program to sort through the combinations of readily obtainable resistors looking for values of K close to 0.682. We find that selecting $R_f = 6.2\ \Omega$ and $R_o = 7.5\ \Omega$ produces a gain $K = 0.683$, which is within 0.2% of the desired value, close enough to be useful. This amplifier circuit is placed in the signal-conditioning compartment of the sensor. Our electrical output-to-input ratio now includes the amplifier gain,

$$\begin{aligned}\frac{\Delta e_o}{e_i} &= \frac{K}{4}\left(\frac{\Delta R_1}{R} - \frac{\Delta R_2}{R} + \frac{\Delta R_3}{R} - \frac{\Delta R_4}{R}\right) \\ &= \frac{0.683}{4}\left[19.1 - (-10.2) + 19.1 - (-10.2)\right] \times 10^{-3} \\ &= 10.0\ \text{mV/V}.\end{aligned} \tag{8.40}$$

With a 10 VDC excitation voltage, the bridge full load output is 100 mV and the sensitivity η is now, as desired,

$$\eta = 100\ \text{mV/MPa}\ (0.69\ \text{mV/psi}). \tag{8.41}$$

In summary, this example illustrates how each of the important factors—bridge configuration, location and orientation of maximum stress and strain, and the orientation of the resistors that respond to the strain—affect the design and sensitivity of a piezoresistive transducer.

References and suggested reading

Beeby S, Ensell G, Kraft M, White N (2004) MEMS Mechanical Sensors. Artech House, Inc., Norwood, MA

Busch-Vishniac I (1998) Electromechanical Sensors and Actuators. Springer, New York, NY

Clark, SK and Kensall DW (1979) Pressure sensitivity in anisotropically etched thin-diaphragm pressure sensors. IEEE Trans on Electron Devices, ED-26.12

Lin et al. (1999) Piezoresistive pressure sensors, J of Microelectromechanical Systems, 8.4

Madou MJ (2002) Fundamentals of Microfabrication 2nd edn. CRC Press, New York

Questions and problems

8.1 In Problem 4.21 you are asked to create a process flow to fabricate a simple MEMS cantilever whose deflection can be sensed using a piezoresistor created via boron doping. There is something impractical about this particular design. What is it? How would you modify the design?

8.2 What is a Wheatstone bridge? Considering a Wheatstone bridge by itself, what is the measurand? What is the output?

8.3 What is the major source of resistance change in metal materials upon deformation? in semiconductor materials?

8.4 What is meant by longitudinal and tranverse strain?

8.5 Why are piezoresistors gernally placed in the region(s) of maximum stress within a transducer?

8.6 Suppose we have a cantilever beam type piezoresistive accelerometer wired in a *half-bridge* configuration in which two resistors of the bridge actively respond to the input strain and the other two resistors are used for *temperature compensation*. The resistors used for temperture compensation are subject to the same temperature as the active resistors, but not subject to the input strains, and are configured in the bridge so that the temperature has no effect on bridge output. If all four resistors are at the same temperature and undergo the same temperature-induced strain ε_T, determine which resistor is wired into which leg of the bridge to provide temperature compensation.

8.7 Determine the gage factor for the following materials: aluminum, titanium, gold. (Hint: You will have to look up the appropriate material properties. What class of materials are these? What material property or properties will you require, then?)

8.8 Using the values of stress and strain given in the case study and assuming the same diaphragm geometry, determine the gage factor for each combination of the materials and orientation possbile from the elastoresistance coefficients given in the table below.

Elastoresistance coefficients for different orientations of p-type and n-type silicon and germanium

Material	γ_L <110>	γ_T <110>	γ_L <100>	γ_T <100>
n-Si	-81	47	-108	40
p-Si	120	-54	77	-51
n-Ge	-96	40	-60	39
p-Ge	63	-28	37	-27

8.9 Using the gage factors found in Problem 8.8, determine the respective sensitivities of a 1 MPa full load pressure transducer corresponding to each material/orientation combination. Express your answers in mV/V/MPa.

8.10 A configuration of p-Si resistors on a square diaphragm is suggested in which *two* resistors are used to sense the maximum stress σ_C. (I.e., two side-by-side resistors are placed at locations 1 and 3 in Fig. 8.9, but wired in series.) Use the following parameters (as in the case study):

- 0–1 MPa (145 psi) full scale input
- 10 VDC excitation,
- p-Si piezoresistors
- square (100) Si-diaphragm, 1.2 mm on each side, 80 μm thick, and the <110> directions bisect the square orthogonally
- $\sigma_C = 45.0$ Mpa, $\sigma_B = 22.5$ Mpa, $\varepsilon_C = 152$ με, and $\varepsilon_B = -17$ με

Determine:

a. the new bridge equation (i.e., e_o in terms of e_i and the various resistances)
b. the value of $\Delta R/R$ for each resistor
c. the value of $\Delta e_o/e_i$
d. the transducer sensitivity η (with no amplifier circuit)

8.11 One study has shown that the stress in a square (100) p-Si diaphragm can be estimated using

$$\bar{\sigma}_j = \frac{\sigma_j}{p}\left(\frac{h}{l}\right)^2.$$

where $\bar{\sigma}$ is a nondimensional parameter determined experimentally, σ_j is the stress in the j-direction, p is the applied pressure, h is the diaphragm thickness, and l is the length of the side of the diaphragm. For p-Si <110> resistors located close to the edge of the diaphragm, $\bar{\sigma}_C \approx 0.2$ and $\bar{\sigma}_B \approx 0.1$. Use the following parameters:

- 0–100 MPa full scale input

- 10 VDC excitation
- square diaphragm, 1 mm on each side, 50 μm thick with piezoresistors arranged as in Fig. 8.9
- (100) p-Si with <110> direction piezoresistors so that $\pi_L = 72\times10^{-11}$ Pa^{-1} and $\pi_T = -66\times10^{-11}$ Pa^{-1}

Determine:

a. σ_C, σ_B
b. the value of $\Delta R/R$ for each resistor
c. the value of $\Delta e_o/e_i$
d. the transducer sensitivity η (with no amplifier circuit)

8.12 Given that $\Delta e_o/e_i$ = 15 mV/V for a particualar piezoresistive sensor, use the following list of resistors to design an amplifier that gives a full-scale output of 100 mV for a 5V DC input.

Readily-obtained resistors for use in an op-amp circuit (Ohms)

1	8.2	22	56	150	390
1.5	9.1	24	62	160	430
2.7	10	27	68	180	470
4.3	11	30	75	200	510
4.7	12	33	82	220	560
5.1	13	36	91	240	620
5.6	15	39	100	270	680
6.2	16	43	110	300	750
6.8	18	47	120	330	820
7.5	20	51	130	360	910

8.13 Reconsider the piezoresistive accelerometer shown in Fig. 8.6. When a force P is applied to the seismic mass m, it can be shown from solid mechanics principles that the strains on the top and bottom of the beam are given by

$$\varepsilon = \frac{6PL}{Ebh^3},$$

where E is Young's modulus and the dimensions L, b and h are given in the figure below. (Placement of the piezoresistors, bridge configuration, etc. are the same as in Fig. 8.6.) Recall that the strains on the top and bottom are equal in magitude but opposite in sign.

a. What is the gage factor in terms of the piezoresistance coefficients for this configuration? Why?
b. Find $\Delta e_o/e_i$ in terms of material properties, geometry and the force P.
c. Find the sensitivity η in terms of material properties, geometry and the mass m. (Hints: Ignore the mass of the beam itself. What

is the measurand for an accelerometer and the associated inputs/outputs?)

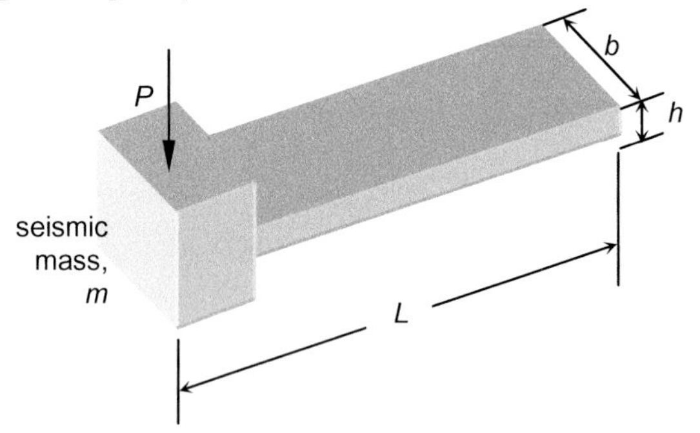

Chapter 9. Capacitive transducers

9.1 Introduction

Capacitive transducers can be thought of as a modification of a standard electrical capacitor that allows at least one electrode to move. In the case of sensing, a mechanical input (pressure, force, or acceleration for example) applied to the transducer changes its electrical characteristics, which constitute the sensed output. Capacitive actuation makes uses the same principle with the input and output switched; that is, an electrical input causes at least one electrode to move, thereby causing actuation.

You probably first encountered capacitors—along with inductors and resistors—in an introductory physics or electrical circuits course. You may recall that capacitors are energy-storage devices (and hopefully you were cautioned about mishandling them…capacitors in home appliances can cause serious injury if their stored energy is not safely discharged before making repairs). We use that stored energy in both sensors and actuators to relate particular inputs to particular outputs.

We begin by reviewing some of the basic characteristics of simple parallel plate capacitors and then develop the electromechanical relationships used to design capacitive actuators and sensors. We then move on to looking at the principles required to understand capacitive sensing in particular. In practice, capacitive sensing rarely involves simply measuring the resulting capacitance of a sensor while subject to some static mechanical input. Rather, an alternating current (AC) signal is input to the device along with the mechanical quantity of interest, with the information about the measurand being contained within the AC output signal. We therefore devote a significant portion of the chapter to understanding the nature of the electrical subsystems of capacitive sensors. We conclude with a brief case study of a capacitive accelerometer.

T.M. Adams, R.A. Layton, *Introductory MEMS: Fabrication and Applications*,
DOI 10.1007/978-0-387-09511-0_9, © Springer Science+Business Media, LLC 2010

9.2 Capacitor fundamentals

9.2.1 Fixed-capacitance capacitor

Conceptually, a capacitor consists of two electrodes separated by a gap filled with a non-conducting material, traditionally called a **dielectric**. The electrodes can have a variety of geometries. For our introductory model we will suppose that the electrodes are rectangular flat plates with area A, separated by a gap d, as shown in Fig. 9.1. This simple geometry is the basis for the symbol used for capacitors in electrical circuit diagrams.

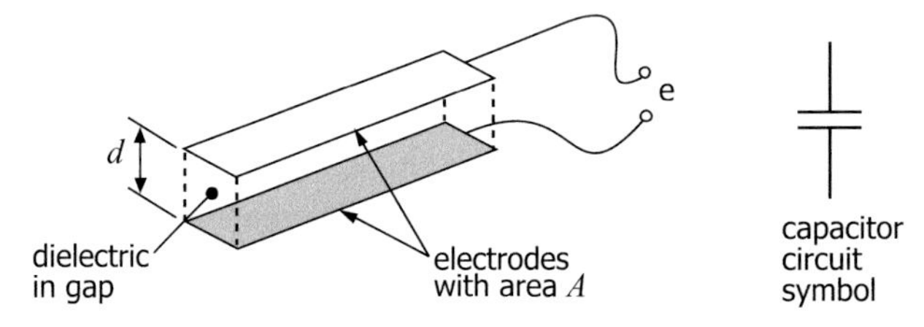

Fig. 9.1. Conceptual schematic of a capacitor: two conductors separated by a gap filled with a nonconductive material. The plates are a fixed distance apart.

The capacitance C of this ideal capacitor is given by

$$C = \frac{\varepsilon A}{d}. \tag{9.1}$$

where ε is the permittivity of the dielectric material. **Permittivity** is a material property characterizing a material's ability to transmit or "permit" the electric field associated with the separation of charge.

In practice, non-MEMS capacitors have external packaging like those shown in Fig. 9.2. Inside these packages, the geometry and materials of the electrodes and dielectrics may vary, but all can be modeled as if they were simple fixed-parallel-plate capacitors like that shown in Fig. 9.1.

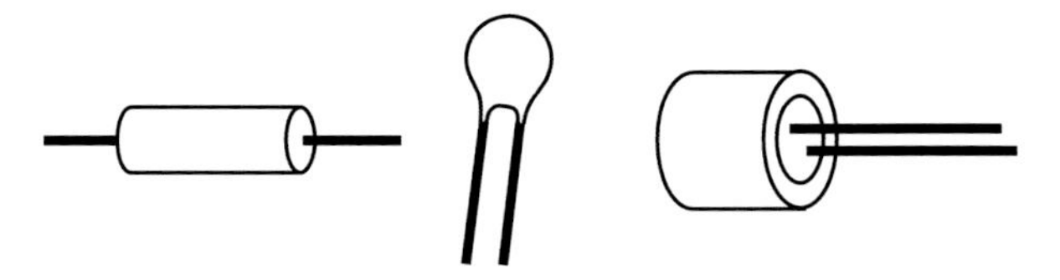

Fig. 9.2. Three examples of the packaging of conventional, non-MEMS capacitors

For these commonly encountered capacitors, the magnitude of the charge separation q is linearly related to the applied voltage e by

$$q = Ce, \tag{9.2}$$

where C is the **capacitance**—representing the *capacity* of the device to store energy. By "charge separation", we mean the collection of excess electrons on one plate and the corresponding deficiency of electrons on the other plate resulting from a power input to the capacitor from some external power supply. (Note that even though we model ideal voltage sources as if they provided no current at all, all real sources actually deliver *power*, the product of voltage and current.) We can't increase the energy stored in any system without an input of power, a direct result of the conservation of energy principle.

The stored energy E (a form of potential energy) is a function of the charge separation q given by

$$E = \int e(q)dq = \int \frac{q}{C} dq = \frac{q^2}{2C} \tag{9.3}$$

In this linear case, you might have previously seen an energy expression written in terms of the applied voltage, obtained by simply substituting Ce for q in the rightmost term of (9.3) to get

$$E = \tfrac{1}{2} Ce^2 . \tag{9.4}$$

This expression brings to our attention that in order to store more energy we can increase the capacitance C, the applied voltage e, or both (within the physical limitations of the device, that is). Equation (9.4) is equivalent to the rightmost term of Eq. (9.3) as long as the linear constitutive behavior Eq. (9.2) holds.

Because of charge separation, the two plates have opposite polarities, one positive and one negative. You may recall from introductory physics that a force of attraction exists between two particles of opposite polarity, and the same holds true for these two plates. With the two plates fixed, however, no relative motion between them is allowed. Thus this force—which is present in all capacitors when charged—can do no work. We get the force to do work by making one of the plates movable. A movable capacitive electrode is the basic principle of operation for most (though not all) capacitive transducers.

To summarize this brief review, by putting power into a capacitor we create a separation of charge that simultaneously creates a store of energy and a force of attraction between plates. Making one of the capacitive plates movable is the most common approach to exploiting these electro-mechanical interactions in the design of capacitive sensors and actuators.

9.2.2 Variable-capacitance capacitor

The concept of the capacitor with a movable electrode is illustrated in Fig. 9.3. The two electrodes of area A are a nominal distance d apart. The movable electrode is designed to move small distances $\pm x$ from the nominal position d. This configuration produces a capacitor whose capacitance varies as a function of the displacement x, that is, $C = C(x)$.

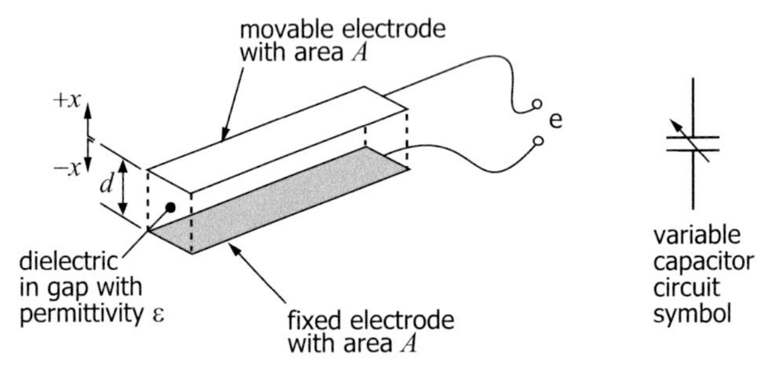

Fig. 9.3. A capacitor with a movable electrode has a variable capacitance $C(x)$.

The capacitance of this ideal capacitor is given by

$$C = \frac{\varepsilon A}{d + x}. \tag{9.5}$$

It follows that for positive x (gap increasing), capacitance decreases and for negative x (gap decreasing), capacitance increases.

Alternate configurations are possible. For example we can design the capacitor such that the motion of the movable plate is side to side, producing a variable capacitor in which the gap is fixed and capacitance varies with the changing overlap area, as illustrated in Fig. 9.4 (a). Or we can design the capacitor such that the electrodes are fixed and the dielectric material moves, as illustrated in Fig. 9.4 (b).

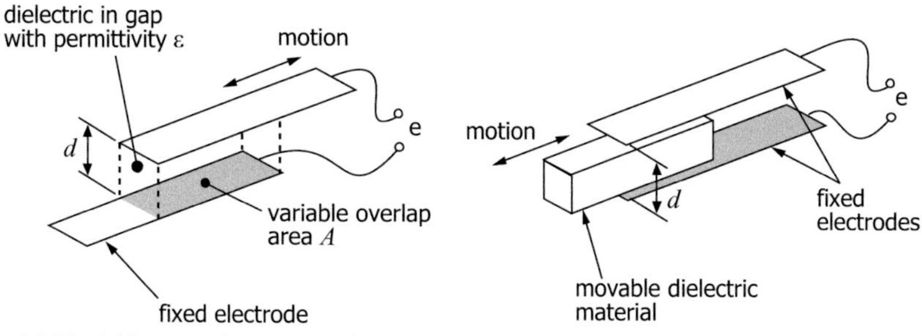

(a) Variable capacitance as a function of a changing area.

(b) Variable capacitance as a function of a changing permittivity.

Fig. 9.4. Alternate schemes for varying capacitance while still relating mechanical motion to energy stored

For each special configuration, we can write an expression for capacitance derived from Eq. (9.1) that expresses the variable capacitance in terms of the particular parameter (the gap, the area, or the permittivity) that varies, as we did in Eq. (9.5). However, since the majority of commercial MEMS capacitive transducers are designed with a variable gap, that's the system on which we'll focus our attention.

Turning our attention now to the variable-gap capacitor with constant d, ε and A, we can obtain an expression for the energy stored E by substituting Eq. (9.5) into Eq. (9.3) to obtain

$$E = \frac{q^2(d+x)}{2\varepsilon A}. \tag{9.6}$$

In the absence of other energy transfer mechanisms, conservation of energy requires that the power input to a system $\dot{W}$ equals the time rate of change of change of system energy. Taking the derivative of Eq. (9.6) with respect to time, and remembering that both the charge q and the displacement x are time-dependent, we obtain

$$\dot{W} = \frac{dE}{dt} = \frac{\partial E}{\partial x}\frac{dx}{dt} + \frac{\partial E}{\partial q}\frac{dq}{dt}. \tag{9.7}$$

Noting the derivative dx/dt is just the velocity v of the moving plate and the derivative dq/dt is just the current i through the capacitor, we can write

$$\dot{W} = \frac{dE}{dt} = \frac{\partial E}{\partial x}v + \frac{\partial E}{\partial q}i. \tag{9.8}$$

Because each of the terms on the right must be an expression of power, the first partial derivative represents a force (because force × velocity gives mechanical power) and the second partial derivative represents a voltage (because voltage × current gives electrical power). Thus the force of attraction between the two plates and the voltage drop across the capacitor are given by

$$\begin{aligned} \text{force}: f &= \frac{\partial E}{\partial x} = \frac{q^2}{2\varepsilon A} \\ \text{voltage}: e &= \frac{\partial E}{\partial q} = \frac{q(d+x)}{\varepsilon A} \end{aligned} . \tag{9.9}$$

Note that the voltage relationship is just a restatement of the basic constitutive behavior, given by Eq. (9.2), that $e = q/C$. The force relationship, however, leads to some new insights regarding transducer design. Substituting Ce for q in the force equation yields

$$f = \tfrac{1}{2}\varepsilon A \frac{e^2}{(d+x)^2} . \tag{9.10}$$

Note that the force scales with the inverse of gap distance. We therefore see the utility of creating a micro-sized device to make use of this force. Also, this model indicates that if we need a large force (for an actuator, for example), we can increase the input voltage e or the area A or the permittivity ε (by selecting a different dielectric material) or all of the above. We can also bring the plates closer together, making the sum $d+x$ smaller.

In practice, however, when the plates come too close together ($x \to -d$) or if the voltage is too large ($e \to \infty$), a large force ($F \to \infty$) slams the two plates together—a phenomenon known as **pull-in**. To avoid pull-in, the MEMS designer ensures that x is generally much smaller than d and that e is less than a limiting value called the *pull-in voltage*. We do not develop the pull-in analysis here, but details can be found in Senturia.

9.2.3 An overview of capacitive sensors and actuators

Regardless whether the variable capacitance is due to variable area, variable gap, or variable permittivity, once we have a movable component of the system, we can relate forces and motion (the mechanical domain) to electrical energy storage and voltage (the electrical domain) by means of the variable capacitance $C(x)$. Capacitive actuators are designed by considering the relationship between input voltages and output forces while capacitive sensors are designed by considering the relationship between input displacements to output voltages. This distinction is illustrated in Fig. 9.5. Sensors and actuators have in common the variable energy storage associated with the variable capacitance.

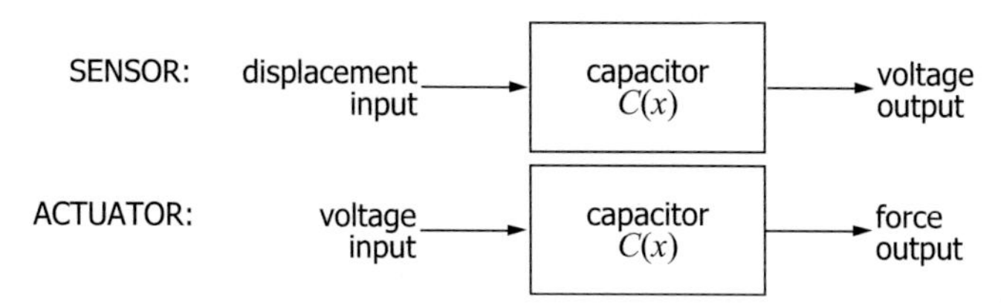

Fig. 9.5. Comparison capacitive transducers behaving as sensors and actuators

We can increase the sensitivity of a capacitive transducer by building electrodes with multiple "fingers" (or digits) and interleaving them, as illustrated in Fig. 9.6. Motion can be either parallel or transverse to the direction of the fingers.

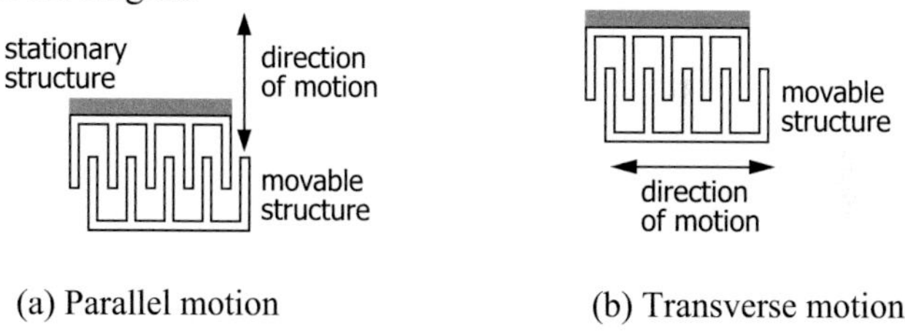

(a) Parallel motion (b) Transverse motion

Fig. 9.6. Interleaved capacitor "digits" to increase transducer sensitivity

Interleaving a number N digits multiplies the transducer sensitivity by N. Thus, in practical MEMS devices, the number of digits is made as large as practicable. Because the interleaved digits resemble a hair comb, the resulting structure is often called a **comb-drive**, illustrated in Fig. 9.7, which is Fig. 7.29 reproduced here for convenience. The capacitive energy storage occurs between the interleaved fingers of the comb.

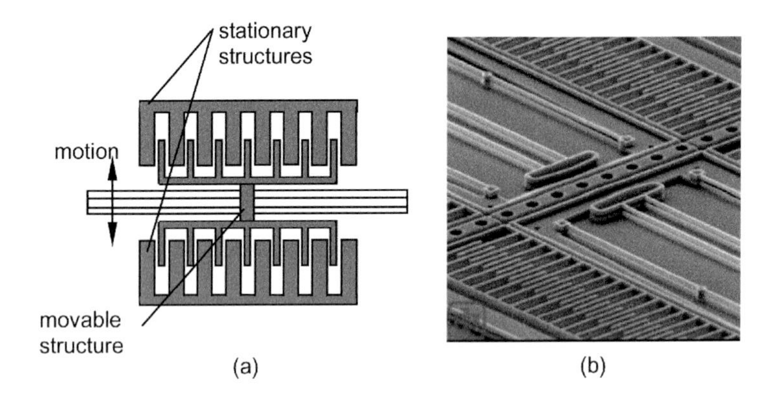

Fig. 9.7. Electrostatic comb-drives: (a) A schematic showing a typical layout. The center assembly is movable while the upper and lower structures are affixed to the substrate. (b) An SEM photo of an electrostatic comb-drive used to power a micro-motor. (Courtesy of Sandia National Laboratories, SUMMiT™ Technologies, www.mems.sandia.gov)

Sensors

The measurand for a capacitive sensor might be one of several mechanical quantities. In commercialized MEMS devices, acceleration and pressure are the most common. In each case the sensor is designed so that the measurand causes some portion of the capacitor assembly to move. The input motion alters the capacitance, producing a detectable change in output voltage. In the modeling sections to follow, we examine capacitive sensors in detail.

Actuators

An example of a conventional surface micromachined MEMS actuator is the linear comb-drive illustrated in Fig. 9.7. The center assembly is the moving part while the upper and lower structures (typically several hundreds of fingers each 10 microns long) are fixed to the substrate. Comb drives are used in optical switches and combined with other elements that can be used to produce rotary motion in micro-motors.

An example of an AC-excited comb drive is the MEMS resonator illustrated in Fig. 9.8. One of the most promising future markets for MEMS resonators is in the wireless communications market, where the performance of current quartz-based resonators is falling short of meeting industry needs. Though MEMS resonators show promise in this area, they have not yet reached volume-production phase due to unresolved technical issues such as poor temperature stability compared to quartz technology.

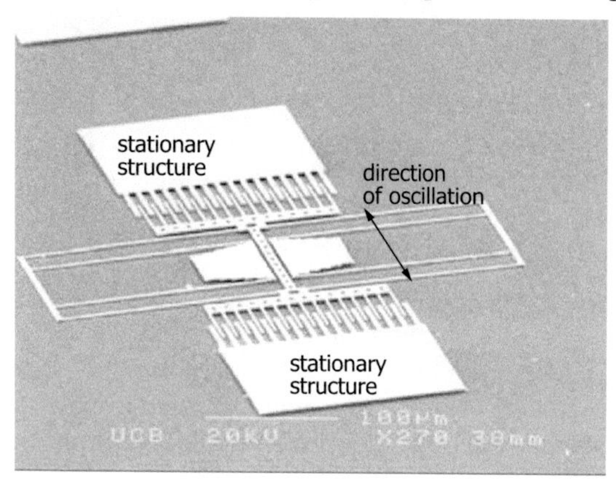

Fig. 9.8. Scanning electron microscope (SEM) image of an electrostatic resonator (From Gao et al. Courtesy Applied Physics Letters)

9.3 Modeling a capacitive sensor

9.3.1 Capacitive half-bridge

A commonly used approach in implementing a variable capacitance design for use as a sensor is to have an electrode that moves between two fixed electrodes, as shown in Fig. 9.9. Motion in one direction decreases the gap in that direction and simultaneously increases the gap in the opposite direction. With the three voltage connections shown, the assembly is called a **capacitive half-bridge**.

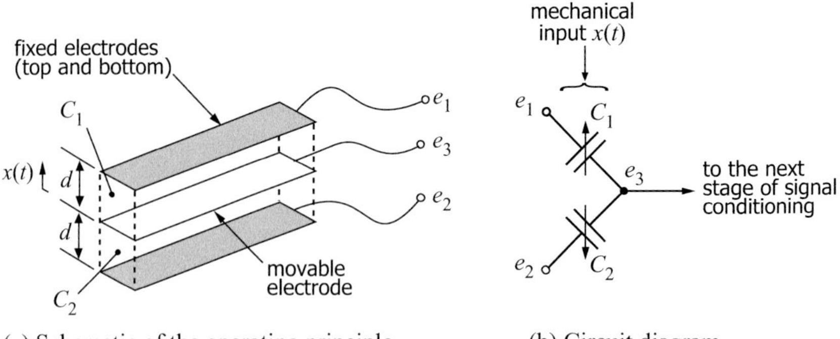

(a) Schematic of the operating principle (b) Circuit diagram

Fig. 9.9. Capacitive half-bridge

Voltages e_1 and e_2 are inputs; e_3 is the output to the next stage of signal conditioning. Both C_1 and C_2 are functions of the mechanical input $x(t)$. When the movable electrode is centered (its *nominal* position), the energy stored on both sides of the center electrode is identical—we'll label this nominal capacitance value C_0. When $x = 0$, both gaps have length d and $C_1 = C_2 = C_0$.

When the center electrode moves, suppose that x is small and positive (recall that $x << d$). With this motion, the gap for C_1 has length $d–x$ and its capacitance increases to $C_1 = C_0 + \Delta C_1$. Simultaneously, the gap for C_2 has length $d+x$ and its capacitance decreases to $C_2 = C_0 - \Delta C_2$. The magnitude of the two ΔC terms are approximately equal, as shown below.

$$\text{For } C_1: \Delta C_1 = C_1 - C_0, \qquad \text{For } C_2: \Delta C_2 = -C_2 + C_0. \tag{9.11}$$

Substituting for the capacitance terms,

$$\Delta C_1 = \frac{\varepsilon A}{d-x} - \frac{\varepsilon A}{d}, \qquad \Delta C_2 = -\frac{\varepsilon A}{d+x} + \frac{\varepsilon A}{d}. \tag{9.12}$$

Combining terms yields

$$\Delta C_1 = \frac{\varepsilon A x}{d^2 - xd}, \qquad \Delta C_2 = \frac{\varepsilon A x}{d^2 + xd}. \tag{9.13}$$

For $x << d$, the value of xd is negligible compared to d^2, and so the change in capacitance on either side of the half bridge for small x is given by

$$\Delta C(x) = \frac{\varepsilon A}{d^2} x. \tag{9.14}$$

To begin modeling the capacitive half-bridge (using nodal analysis), we assign currents i_1, i_2 and i_3 as shown in Fig. 9.10. We assume that a current i_3 is leaving the e_3 node to enter a latter stage of signal conditioning. This analysis will explain why capacitive half-bridges are commonly designed with e_1 and e_2 as periodic signals of equal magnitude and opposite sign.

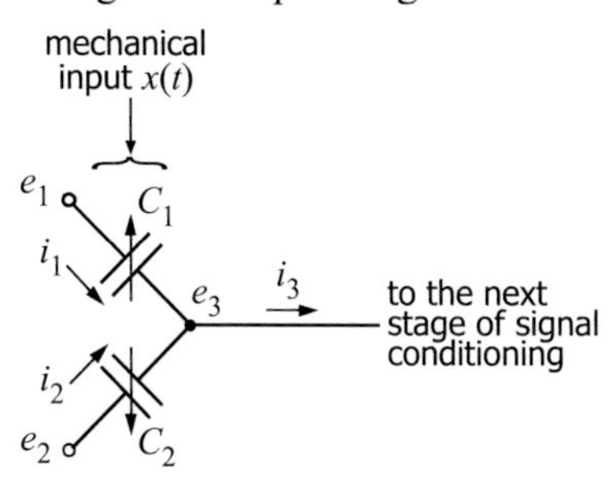

Fig. 9.10. Nodal analysis of a capacitive half-bridge

Kirchhoff's current law (KCL) written at the e_3 node is given by

$$i_3 = i_1 + i_2 . \tag{9.15}$$

Substituting dq/dt for current and then substituting for q in terms of capacitance and node voltages yields

$$i_3 = \frac{d}{dt}\left[C_1\left(e_1 - e_3\right)\right] + \frac{d}{dt}\left[C_2\left(e_2 - e_3\right)\right]. \tag{9.16}$$

Now we substitute for C_1 and C_2 from Eq. (9.11) and assume that both voltage sources are the same; that is, $e_1 = e_2 = e_S$. The result is

$$i_3 = \frac{d}{dt}\left[\left(C_0 + \Delta C\right)\left(e_S - e_3\right)\right] + \frac{d}{dt}\left[\left(C_0 - \Delta C\right)\left(e_S - e_3\right)\right]. \tag{9.17}$$

Combining terms yields

$$i_3 = \frac{d}{dt}\left[2C_0\left(e_S - e_3\right)\right]. \tag{9.18}$$

This results shows that by assuming the two input voltages are identical, the bridge output i_3 has no dependence on the variable capacitance ΔC and therefore no dependence on the position x of the movable plate—completely undermining our basic principle of operation! The solution is to make the two source voltages have identical magnitude but *opposite* polarity, that is, $e_1 = e_S$ and $e_2 = -e_S$.

With this new design, we can modify Eq. (9.17) to obtain

$$i_3 = \frac{d}{dt}[(C_0 + \Delta C)(e_S - e_3)] + \frac{d}{dt}[(C_0 - \Delta C)(-e_S - e_3)]. \tag{9.19}$$

Combining terms now yields

$$i_3 = 2\frac{d}{dt}(\Delta C e_S - C_0 e_3). \tag{9.20}$$

Taking the time derivative yields

$$i_3 = 2\left(e_S \frac{d}{dt}(\Delta C) + \Delta C \frac{d}{dt}(e_S) - C_0 \frac{d}{dt}(e_3) \right). \tag{9.21}$$

We obtain an expression for the derivative ΔC from Eq. (9.14). Substituting this expression we obtain

$$i_3 = 2\left(e_S \frac{\varepsilon A}{d^2}\dot{x} + x\frac{\varepsilon A}{d^2}\dot{e}_S - C_0 \dot{e}_3 \right), \tag{9.22}$$

where we replaced the time derivative with dot notation.

If the movable plate displacement x is nonzero but constant, then $\dot{x} = 0$ and the first term vanishes. If the voltage source is constant (DC), then $\dot{e}_S = 0$ and the second term vanishes. And the remaining third term does not depend on the variable differential capacitance ΔC. Thus, either the displacement x or the voltage source e_S has to continuously vary to obtain a useful bridge output. If we want our sensor to respond to a constant x input, it follows that we have to make the input voltage a varying (periodic) signal. Therefore, in many capacitive applications the two input voltages are sinusoids with equal magnitude but of opposite polarity (or, in other words, 180 degrees out of phase). In practice it is often easier to use a square wave instead of a sinusoid, but the rationale is the same. For analysis purposes as we proceed, let's assume a sinusoidal function such that

$$\begin{aligned} e_1 &= +e_S = +B\sin\omega t \\ e_2 &= -e_S = -B\sin\omega t \end{aligned}. \tag{9.23}$$

where ω is the frequency in rad/s of the AC signal. Recall that frequency f in Hz and radial frequency ω in rad/s are related by the expression $\omega = 2\pi f$.

One consequence of having a periodic input is that we have to be concerned with the frequency response of the system. We will have to select the frequency of the voltage inputs to the bridge, the range of frequencies, or *bandwidth*, over which the mechanical input will operate, and the frequency characteristics of the mechanical components of the sensor (its natural frequency ω_n and damping ratio ζ). These topics are addressed in subsequent sections.

We note that the coefficient "2" in Eq. (9.22) represents a doubling of the sensitivity of the sensor due to the half-bridge arrangement. Typical applications using a comb structure like that shown in Fig. 9.7 would therefore have a sensitivity of $2N$, where N is the number of half-bridges in the structure.

Now that we know something about the requirements for the input signals, we can continue building up the rest of the system.

9.3.2 Conditioning the signal from the half-bridge

The first step in conditioning the signal coming out of the capacitive half bridge is to amplify it so that the signal can be read "off-chip". A simple amplifier with feedback is shown in Fig. 9.11. We assume the op-amp is ideal and therefore behaves according to the two "rules" for op-amps with external feedback: 1) the inputs draw no current, and 2) the output attempts to do whatever is necessary to make the voltage difference between the inputs zero.

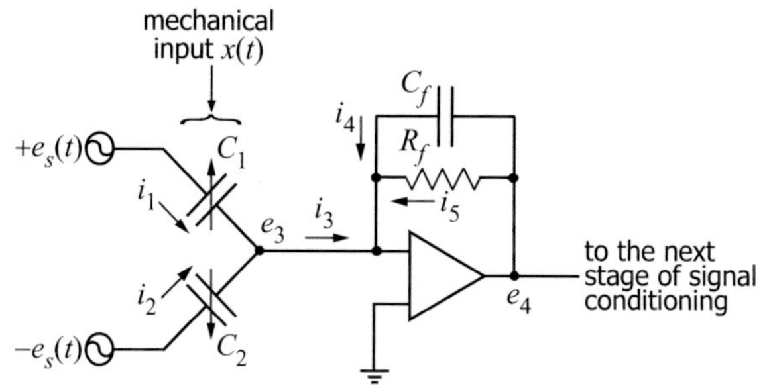

Fig. 9.11. Adding an amplifier

Using the first rule, we apply KCL at one input node to obtain

$$i_3 = i_4 + i_5. \tag{9.24}$$

Substituting for i_3 from Eq. (9.22) and using the usual constitutive laws for resistors and capacitors to substitute for i_4 and i_5 results in

$$2\left(e_S \frac{\varepsilon A}{d^2}\dot{x} + x\frac{\varepsilon A}{d^2}\dot{e}_S - C_0\dot{e}_3\right) = \frac{d}{dt}\left[C_f\left(e_4 - e_3\right)\right] + \frac{e_4 - e_3}{R_f}. \tag{9.25}$$

According to the second op-amp rule, having one input grounded (0 V) forces $e_3 \to 0$. (One of the purposes of the feedback capacitor C_f is to help enforce this zero condition on e_3.) Making this substitution and rearranging terms yields

$$2\frac{\varepsilon A}{d^2}\left(e_S\dot{x} + x\dot{e}_S\right) = C_f\dot{e}_4 + \frac{1}{R_f}e_4 . \tag{9.26}$$

This expression tells us that the output signal e_4 is a function of both the displacement of the center electrode x (which we want) as well as the velocity of the electrode $\dot{x}$ (which we do not want).

To make the velocity term negligible, we select a high frequency for the AC input signal e_S. To show the effect of the high frequency input, we examine the parenthetical term in Eq. (9.26). Substituting for e_s and its derivative using Eq. (9.23) yields

$$e_S\dot{x} + x\dot{e}_S = \dot{x}B\sin\omega t + x\omega B\cos\omega t . \tag{9.27}$$

We make a design decision here to select the frequency ω large enough so that $x\omega >> \dot{x}$. In this case, Eq. (9.26) simplifies to

$$2\frac{\varepsilon A}{d^2}x\dot{e}_S = C_f\dot{e}_4 + \frac{1}{R_f}e_4 . \tag{9.28}$$

We make a second design choice here by making the resistance R_f large enough such that the rightmost term has negligible amplitude compared to the other terms in the model to obtain

$$2\frac{\varepsilon A}{d^2}x\dot{e}_S = C_f\dot{e}_4 . \tag{9.29}$$

Recognizing that because the source e_s is a sinusoid at frequency ω, the output e_4 is a sinusoid at the same frequency, we can conclude that the amplitude ratio is given by

$$\left|\frac{e_4}{e_S}\right| = \frac{2\varepsilon A}{C_f d^2}x . \tag{9.30}$$

Thus by carefully selecting ω and R_f, we obtain a voltage signal from the amplifier $e_4(t)$ that has the same frequency as the input $e_S(t)$, is scaled by the factor $(2\varepsilon A)/(C_f d^2)$, and is multiplied by the magnitude of $x(t)$. This process is an example of *modulation*.

Modulation generally is the process of varying one waveform in relation to another waveform. The high-frequency waveform is called the *carrier*. The second waveform performs the *modulation*. The two most common types of modulation are frequency modulation (FM) and amplitude modulation (AM)—which are, of course, recognizable as the two types of modulation used in radio broadcasting.

In our capacitive sensor, we are using amplitude modulation. The low-frequency modulating signal $x(t)$ varies the amplitude of the high-frequency carrier $e_S(t)$. Since the displacement input $x(t)$ can itself be periodic, we have to determine its maximum frequency f_x and select our carrier frequency f_e so that $f_e >> f_x$.

We illustrate this modulation in Fig. 9.12. A high-frequency, constant amplitude carrier $e_S(t)$ is supplied to one input terminal of the capacitive half-bridge. An equal and opposite signal $-e_S(t)$ supplies the other terminal of the half-bridge. A mechanical measurand, for example acceleration or pressure, is applied to the sensor causing the central electrode comb to move relative to the fixed electrodes. This motion is $x(t)$. The small plots shown indicate the general shape of these signals. The output of the amplifier shows how the magnitude of the $x(t)$ has been imposed as an outer "envelope" on the carrier wave.

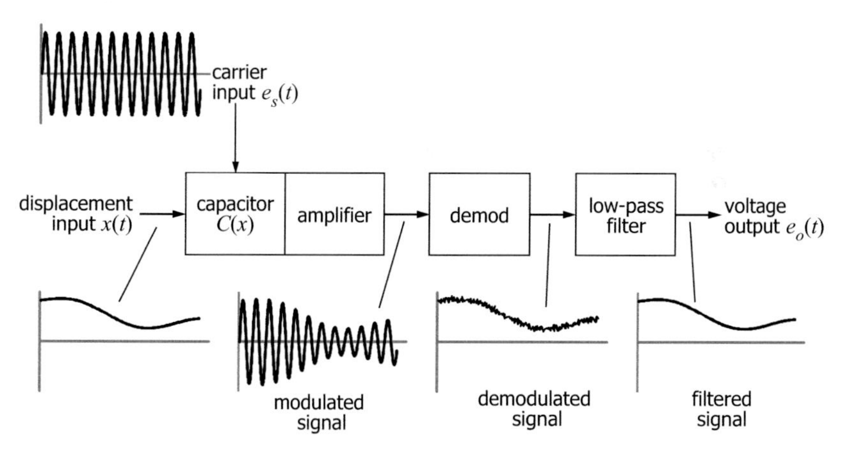

Fig. 9.12. Illustrating amplitude modulation, demodulation, and filtering

To extract the information we want from the shape of the outer envelop of the modulated signal, we subject the signal to *demodulation* and *filtering*. Both demodulation and filtering are purely signal conditioning components of the system, so we discuss their effects qualitatively rather than developing detailed mathematical models. The model depends on the type of circuit used and both demodulation and filtering have multiple "good" design alternatives.

Any of several demodulator circuit designs can be used, for example, a peak detector, a synchronous demodulator, or a track-and-hold circuit. All of these systems produce a demodulated signal, like that shown in Fig. 9.12, that includes a high-frequency "ripple"—a remnant of the high-frequency carrier wave.

To remove the ripple, the signal is passed through a low-pass filter. These circuits too take a wide variety of configurations, for example, a Butterworth filter, a Chebyshev filter, or a Bessel filter. For any low-pass filter, the designer selects a *cutoff frequency* f_c. Signal components with frequencies less than or equal to f_c "pass" through the filter with their amplitude unaffected (ideally). Signal components with frequencies greater than f_c are significantly attenuated (reduced) in amplitude. In our capacitive transducer, we select the cutoff frequency to be higher than the maximum displacement frequency f_x, but less than the carrier frequency f_e, that is, $f_x < f_c < f_e$.

The signal output from the filter is the desired output of the sensor: a voltage signal $e_o(t)$ that is directly proportional to the displacement input $x(t)$ that results from sensing a measurand. The interaction between a measurand (pressure or acceleration, for example) and the variable capacitance are modeled by considering the mechanical elements of sensor.

9.3.3 Mechanical subsystem

For illustrating and modeling the mechanical subsystem, we assume a comb structure like that shown in Fig. 9.13.

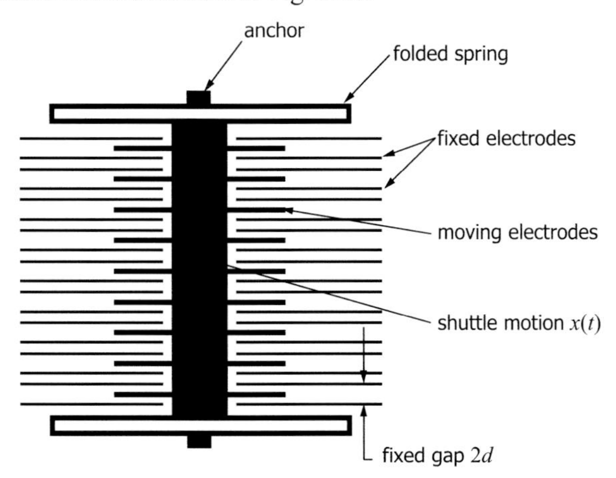

Fig. 9.13. Example of a mechanical subsystem in a capacitive sensor (Adapted from Senturia)

The "shuttle" is the movable portion of the assembly and includes the movable electrodes. Folded springs are anchored to the substrate at either end. In an accelerometer, the shuttle moves in response to the acceleration or deceleration of the system to which the sensor assembly is attached. In a pressure transducer, one of the anchors shown could be attached to a diaphragm subjected to pressure instead of being attached to the substrate. Under pressure, the diaphragm expands causing the shuttle to move. In either case, the measurand causes an input displacement $x(t)$ that, as we've seen, causes a change in capacitance $C(x)$.

The shuttle has total mass m. The springs have a net spring constant k designed to resist the force of attraction due to the separation of charge and to have sufficient force to restrict the motion such that $x \ll d$. Mechanical damping is also present due to the viscosity of the air between the electrodes. For illustration purposes, we'll use the simple velocity-dependent damping model $F = bv$, where b is a damping coefficient and v is the velocity of the shuttle. With these characteristics, we can model the entire mechanical subsystem as the simple "lumped-element" mass-spring-damper system shown in Fig. 9.14.

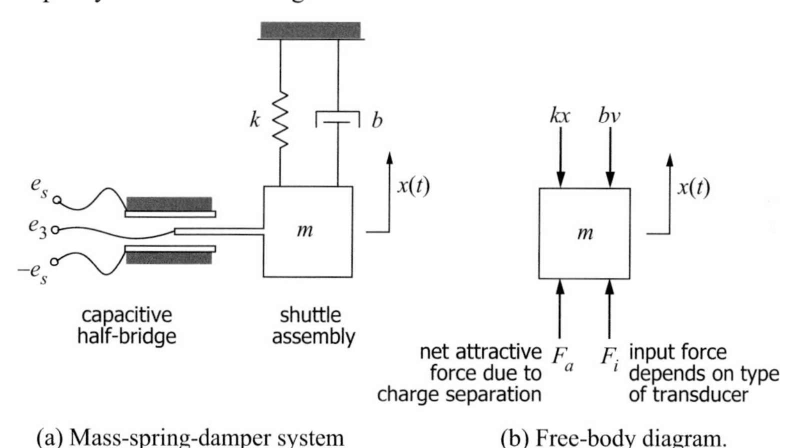

Fig. 9.14. Lumped-parameter model of the mechanical subsystem

The free-body diagram indicates the forces acting on the shuttle in motion. The spring force is modeled by kx and the damping force by bv. Force F_a represents the net force of attraction between electrodes due to the energy stored in all N capacitive half-bridges. Force F_i represents the net input force from the measurand (if appropriate). For example, a pressure transducer might include a diaphragm that presses with force F_i directly on the shuttle. An accelerometer, on the other hand, can be modeled without force F_i. The forces accelerating the object to which the accelerometer is attached result in a force from the MEMS substrate that acts directly on the fixed ends of the spring and damper. Thus the forces acting on the shuttle in an accelerometer are kx, bv, and F_a.

Newton's second law, applied to the free-body diagram, yields

$$F_a + F_i - kx - bv = m\ddot{x}\,. \tag{9.31}$$

where $\ddot{x}$ is acceleration. Rearranging terms in standard form yields the conventional second-order differential equation for a mass-spring-damper system given by

$$m\ddot{x} + b\dot{x} + kx = F_a + F_i\,. \tag{9.32}$$

Dividing by k yields

$$\frac{\ddot{x}}{k/m} + \frac{b}{k}\dot{x} + x = \frac{1}{k}\left(F_a + F_i\right), \tag{9.33}$$

which is equivalent to

$$\frac{\ddot{x}}{\omega_n^2} + \frac{2\zeta}{\omega_n}\dot{x} + x = \frac{1}{k}\left(F_a + F_i\right), \tag{9.34}$$

where

$\omega_n = \sqrt{k/m}$ is the natural frequency of the system in rad/s

$\zeta = b/\left(2\sqrt{km}\right)$ is the damping ratio of the system (unitless)

The natural frequency of the sensor is a critical design value, because if any input to the system has a frequency in the neighborhood of the natural frequency of the lightly-damped mechanical system, it will cause resonance (as discussed in Chapter 6), exciting the shuttle to move with a much greater amplitude than the system can tolerate. Thus we design the stiffness k and the mass m such that the natural frequency is much greater than the maximum possible frequency of the measurand f_x. That is, if f_n is the natural mechanical frequency in Hz, we design the system so that $f_x << f_n$.

It turns out that the attractive force due to the separation of charge can be neglected in our design as long as we select a high enough frequency for the electrical input. This is a good thing, because otherwise the shuttle would be acted on by these forces turning our sensor into an actuator and thereby disrupting out ability to sense the input measurand! The following analysis shows why this is case.

Recall from Eq. (9.8) that the force of attraction is given by the partial derivative of the stored energy E with respect to displacement x. For the case of the half-bridge capacitor, the stored energy is given by

$$E = \frac{q_1^2}{2C_1} + \frac{q_2^2}{2C_2}. \tag{9.35}$$

Substituting for C_1 and C_2 and collecting terms yields

$$E = \frac{d^2}{2\varepsilon A}\left(\frac{q_1^2}{d+x} + \frac{q_2^2}{d-x}\right). \tag{9.36}$$

The total force of attraction acting on the movable electrode is given by

$$F_a = \frac{\partial E}{\partial x} = \frac{d^2}{2\varepsilon A}\frac{\partial}{\partial x}\left(\frac{q_1^2}{d+x} + \frac{q_2^2}{d-x}\right). \tag{9.37}$$

Taking the partial derivative yields terms in the denominators involving d^2, xd, and x^2. Recalling once again that $x << d$, the terms xd and x^2 are negligible compare to d^2. This yields

$$F_a = \frac{1}{2\varepsilon A}\left(q_2^2 - q_1^2\right). \tag{9.38}$$

Substituting $q_1 = C_1 e_s$ and $q_2 = -C_2 e_s$ yields

$$F_a = \frac{1}{2\varepsilon A}\left(C_2^2 - C_1^2\right)e_s^2. \tag{9.39}$$

Substituting for e_s yields our final model for the force

$$F_a = \frac{1}{2\varepsilon A}\left(C_2^2 - C_1^2\right)\left(B\sin\omega t\right)^2. \tag{9.40}$$

The critical feature of this model is that the force F_a is a sinusoid oscillating at the same frequency ω as the electrical source voltage. We select this frequency to be much higher than the natural frequency of the mechanical subsystem. The frequency response characteristics of the mechanical, second-order system are such that signals with frequencies much higher than the natural frequency are highly attenuated (or reduced). Thus, the attractive force has a negligible affect on the motion of the shuttle mass. We conclude, therefore, that $f_x << f_n << f_e$.

9.4 Device case study: Capacitive accelerometer

To examine some of the practical aspects of the analysis presented in the previous section, we'll use the ADXL50 accelerometer manufactured by Analog Devices. We first introduced this sensor in our discussion of block diagrams in Chapter 6. The figure from that chapter is reproduced below.

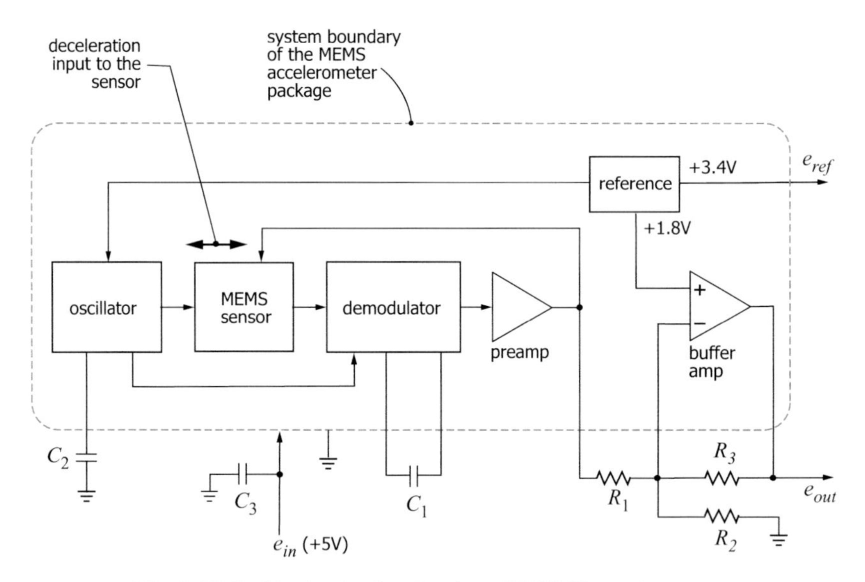

Fig. 9.15. Inside the Analog Devices ADXL50 accelerometer

When the reader first glimpsed this block diagram in Chapter 6, the purpose of the various blocks was probably not obvious. Some of the mystery should now be resolved:

- The block labeled "MEMS sensor" is a capacitive half-bridge structure.
- The oscillator generates periodic voltage inputs to the bridge, square waves at 1 GHz.
- The oscillator signal is also used for synchronous demodulation.
- The buffer amp and associated resistors are used to set the output scale and zero-offset level, thereby determining the accelerometer sensitivity. Sensitivity is factory-calibrated to 19 mV/g (one "g" is 9.81 m/s^2 or 32.2 ft/s^2).
- Capacitor C_1 on the demodulator sets the measurement bandwidth.

- A low-pass filter is not included in the package. A customer provides their own filter suited to the needs of the application.

The package is fabricated on a single monolithic integrated circuit, depicted below in (a). The IC is installed in a package, with the cover removed, as shown in (b).

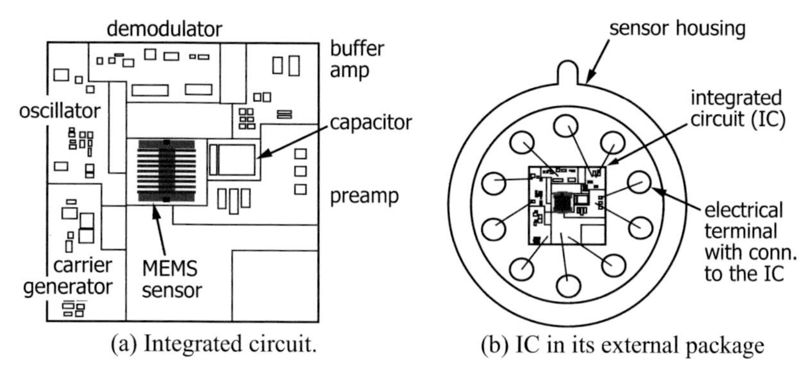

(a) Integrated circuit. (b) IC in its external package

Fig. 9.16. The integrated chip for the ADXL50 accelerometer

The full-scale measurement range is 650 g at frequencies from DC to 10 kHz. Summarizing the frequency characteristics of this sensor:

- The maximum frequency of the measurand is 10 kHz.
- The natural frequency of the mechanical subsystem is 24 kHz.
- The frequency of the carrier signal is 1 GHz.

Thus the system satisfies the requirements we noted earlier, namely that that $f_x << f_n << f_e$.

References and suggested reading

Beeby S, Ensell G, Kraft M, White N (2004) MEMS Mechanical Sensors, Artech House, Inc., Norwood, MA

Busch-Vishniac I (1998) Electromechanical Sensors and Actuators, Springer, New York,

Gao D, Howe RT, Maboudian R (2003) High-selectivity Etching of Polycrystalline 3C-SiC Films using HBr-based Transformer Coupled Plasma. Applied Physics Letters, 82:1742-1744

Madou MJ (2002) Fundamentals of Microfabrication 2nd edn., CRC Press, New York

Maluf N, Williams K (2004) An Introduction to Microelectromechanical Systems Engineering, 2nd edn., Artech House, Inc., Norwood, MA

"Microsystems Science, Technology and Components Dimensions." *Sandi National Laboratories.* 2005. http://www.mems.sandia.gov/
Senturia S (2001) Microsystem Design, Kluwer Academic Publishers, Boston

Questions and problems

9.1 What is the basic principle behind capacitive actuation? capacitive sensing?
9.2 What is pull-in? How can it be avoided?
9.3 Why are electrostatic comb structures typically designed with large number of fingers?
9.4 What are some typical mechanical inputs to a capacitive sensor? Are there other inputs to a typical capacitive sensor? What are they? What does the electrical output look like?
9.5 What is a capacitive half-bridge? What are its various inputs and outputs?
9.6 Why are the source voltages to a typical capacitive sensor of opposite polarity
9.7 What types of signal conditioning are performed on the output of capacitive sensors?
9.8 What is modulation? What is its purpose in capacitive sensing?
9.9 Under what conditions can the attractive force due to the separation of charge can be neglected in a capacitive sensor?
9.10 Write an expression for the variable capacitance of a capacitor, similar to Eq. (9.5), but in which the area A is the varying quantity instead of the gap. See Fig. 9.4(a). Express this capacitance as a function of the distance x the plate moves side to side, assuming a rectangular plate with width w.
9.11 Based on the previous problem, write an expression for the energy stored and develop an expression for the force of attraction as a function of the side-to-side displacement x that changes the overlap area A. What conclusions can you draw regarding what aspects of the design affect the magnitude of the force?
9.12 Using Eq. (9.27) we concluded that $x\omega >> \dot{x}$, where ω is the frequency of the carrier signal, x is the position of the movable electrode, and $\dot{x}$ is its velocity. Using the following parameters, determine if the Analog Devices ADXL50 satisfies this requirement. The maximum value of displacement is 4 μm, the sensor bandwidth is 10 kHz, and the carrier frequency is 1 GHz.

9.13 If the mass of the Analog Devices ADXL50 shuttle is 162×10^{-12} kg, and the system's natural frequency is 24 kHz (per the manufacturer's spec sheet), determine the stiffness of the "spring" structure that connects the proof mass to the substrate.

Chapter 10. Piezoelectric transducers

10.1 Introduction

In Chapter 5 we investigated the properties of thin films deposited onto substrates. In so doing, we found that differences between the thermal expansion coefficients of the thin film and the substrate could lead to a stress in the thin film, which in turn could bend the substrate. Now imagine that we can actively control the level of stress in the thin film. This would allow us to bend the substrate at will. This is essentially what happens in many piezoelectric actuators, the big difference being that the controlled stress is not thermal mismatch stress, but rather a thin film stress resulting from the application of an electric field. This actuation principle is used in applications such as inkjet print heads (Fig. 10.1), micropumps, microvalves and hard disk drive controls to name a few.

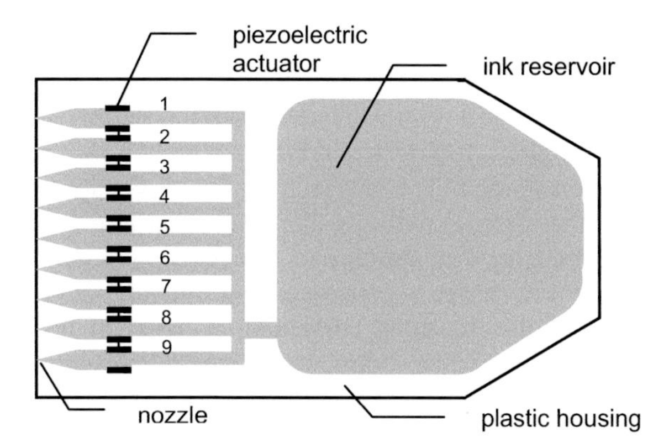

Fig. 10.1. An inkjet print head is a common MEMS device that makes use of piezoelectric actuation. The actuators squeeze the nozzles, causing them to eject ink.

The reverse process is also possible; that is, a piezoelectric material being deformed results in an electric field within the material. Hence, piezo-

T.M. Adams, R.A. Layton, *Introductory MEMS: Fabrication and Applications*,
DOI 10.1007/978-0-387-09511-0_10,

electric materials are used for both actuation and sensing. Acceleration and pressure are the most commonly sensed quantities using piezoelectric materials.

In Chapter 7 we investigated piezoelectric actuation and sensing in a mainly qualitative manor. In this chapter we will explore the nature of piezoelectric materials and the piezoelectric effect in more detail, as well as develop the necessary equations to model such materials. We will also delve a bit deeper into solid mechanics so that we can relate deflection and electric field to each other in a MEMS transducer. Ultimately our goal will be to predict the deflection of a piezoelectric MEMS actuator under the action of an applied voltage. Another way to look at it is that we will develop a physical model that represents the inside of the block in the block diagram shown in Fig. 10.2.

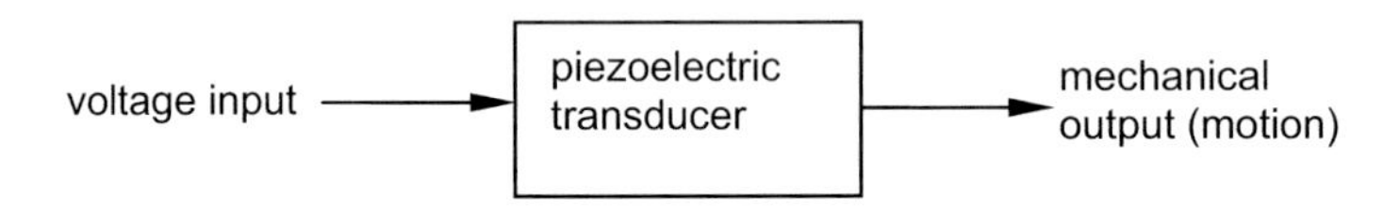

Fig. 10.2. System inputs and outputs for a piezoelectric transducer behaving as an actuator

10.2 Modeling piezoelectric materials

As we have seen, **piezoelectric materials** exhibit a non-zero change in electric field when deformed. Conversely, when a voltage is applied to such a material, the material deforms. For a material to exhibit the piezoelectric effect it must have an appropriate molecular structure as far as the distribution of electric charge is concerned. Specifically, the material cannot be centrosymmetric, meaning containing a center of inversion. Molecular structures with complete symmetry in regard to electric charge may deform easily enough, but the deformation will not be accompanied by any change in its internal electric field. Hence, typical piezoelectric materials tend to be crystals formed of ionic bonds with moderately to highly complicated geometries. These include naturally occurring crystals such as quartz, and man-made materials such as lead zirconium titanate (PZT). Certain polymers also exhibit piezoelectricity.

As an example consider the two materials shown in Fig. 10.3. The figure shows one view of the quartz unit cell as well as a sodium chloride crystal. When the differing charges of the ions in the quartz of Fig. 10.3 (a) are rearranged due to an applied stress as shown, more positive charge (Si^{4+} ions) is concentrated at the bottom whereas more negative charge (O^{2-} ions) is concentrated at the top. This produces a non-zero electric field. For the ions of a sodium chloride unit cell, however, the applied stress essentially brings electrically neutral sheets of atoms closer together, and therefore it is not accompanied any change in electric field. We see then that quartz is a piezoelectric material whereas NaCl is not.

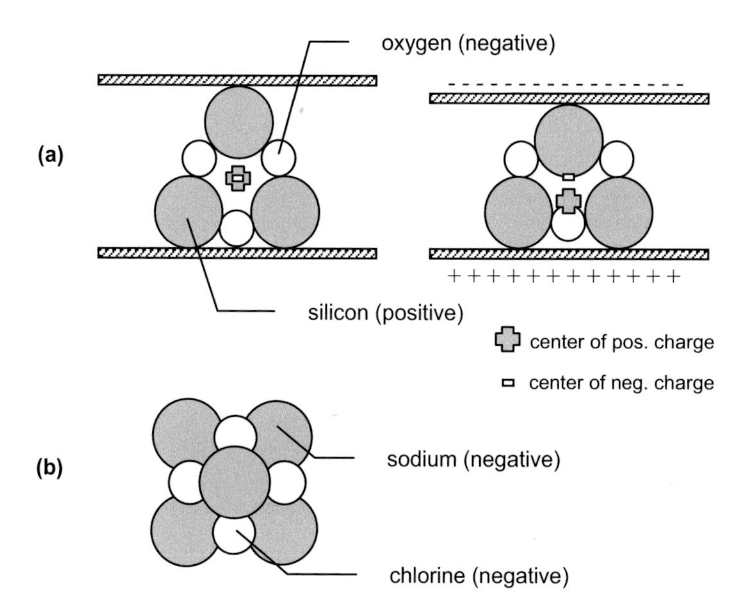

Fig. 10.3 (a) Quartz's crystalline structure is such that it exhibits piezoelectric properties. (b) The sodium chloride crystalline structure does not exhibit the piezoelectric effect.

In Fig. 10.3 (a) the resulting electric field in the quartz is shown in the y-direction, the same direction as the deformation. However, the quartz unit cell is such that electric fields can form in directions other than the deformation; i.e. in the x-direction and/or the z-direction (in and out of the page). As piezoelectric materials cannot be centrosymmetric, this is the case more often than not. In short, piezoelectric materials exhibit high degrees of anisotropy, and piezoelectricity is necessarily a three dimensional phenomenon. What's more, piezoelectricity also couples the two energy domains of electricity and solid mechanics. Modeling piezoelectric materials therefore requires a little more mathematical machinery than other actuation types.

A large number of formulations relating piezoelectric parameters exist, depending on the quantities of interest and the phenomena present. One of the most accessible formulations involves the **piezoelectric charge coefficients** labeled d_{ij}, first introduced in Chapter 7. Both subscripts on d_{ij} refer to the x, y, and z directions by taking on the values of 1, 2 and 3, respectively. The i subscript refers to axis on which the electric phenomenon is happening, whereas the j subscript is the direction where the mechanical quantity is happening. For example, d_{31} can be used to relate electric field in the z direction to stress in the x direction. Typical units for d_{if} are C/N, or equivalently, m/V. Which dimensions/units one uses depend on the desired formulation.

Depending on the piezoelectric material, many of the components of d_{ij} are related to each other, and others may drop out completely. The only non-zero components of d_{ij} for zinc oxide (ZnO), for example, are d_{31}, d_{32}, and d_{33}. Furthermore, $d_{32} = d_{31}$. Many of the piezoelectric materials that can be deposited as thin films, ZnO and lead zirconate titanate (PZT) for example, exhibit this same behavior. Others such as quartz behave much differently.

Figure 10.4 shows a piezoelectric material sandwiched between two electrodes. We will assume that the material behaves like ZnO in terms of its piezoelectric charge coefficients, with the z direction oriented perpendicular to the plan of the material. For this set up, an applied voltage in the z direction, e_z, creates and electric field equal to e_z/t in that direction. This in turn results in strains in the x, y, and z directions of, respectively,

$$\frac{\Delta L}{L} = d_{31} \frac{e_z}{t}, \tag{10.1}$$

$$\frac{\Delta W}{W} = d_{31} \frac{e_z}{t}, \tag{10.2}$$

and

$$\frac{\Delta t}{t} = d_{33} \frac{e_z}{t} . \qquad (10.3)^1$$

From knowledge of the strains the deformations are easily calculated.

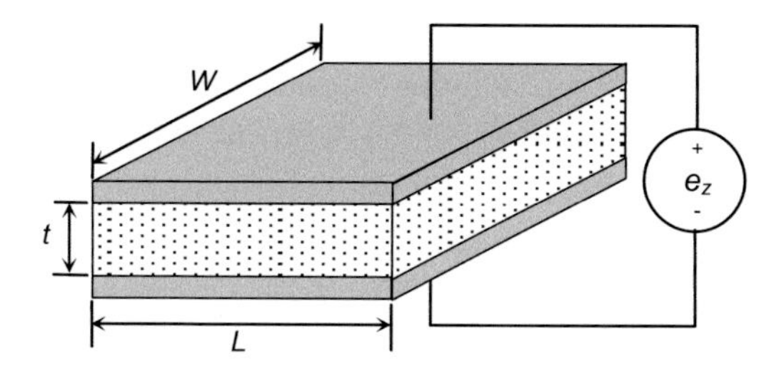

Fig. 10.4. Piezoelectric material between two electrodes

Though one may readily understand the mechanism by which piezoelectric actuation is accomplished from Fig. 10.4, actual MEMS actuators usually don't look like this. This is because the amount of deformation obtainable from an arrangement like that in Fig. 10.4 is much too small to be useful in most applications. A more common arrangement has a piezoelectric material cleverly arranged such that it induces stress in other components, which in turn deform by much greater amounts. One such arrangement is a thin film of piezoelectric material atop a thin membrane, as shown in Fig. 10.5. A voltage applied to the thin film causes a stress in the film, which then bends the membrane to which it is attached.

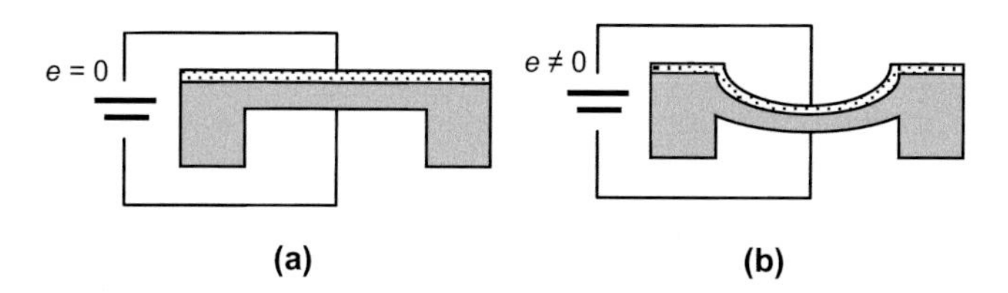

Fig. 10.5 A simple piezoelectric actuator design. (a) A thin film of piezoelectric material on top of a membrane before a voltage is applied. (b) The thin film experiences stress due to an applied electric field, causing the membrane to bow.

[1] The thickness t appearing in the denominator in both the left hand side and the right hand side of Eq. (10.3) may seem extraneous. However, it is included to remind you that the actual quantities being related by d_{ij} are strain and electric field.

The formulation of piezoelectricity given in Eqs. (10.1) to (10.3) is one special case of the many possible relationships between electric field and the various components of the stress or strain tensors first seen in Chapter 5. In order to develop some of the more general cases, let us start by considering stress and strain by themselves.

In Chapter 5 we saw that stress and strain in a non-isotropic material were related via a tensor containing information about how the moduli of elasticity varied in different directions. For a general material this relation is given by

$$\sigma_i = C_{ij}\varepsilon_j \tag{5.24}$$

where the indices i and j vary from 1 to 6 and the Einstein summation convention is observed. (Remember that indices 4-6 for σ and ε refer to shear quantities.) The inverse of this relation is given by

$$\varepsilon_i = S_{ij}\sigma_j \tag{10.4}$$

where S_{ij} are the **compliance coefficients**, the components of the inverse matrix of **stiffness coefficients** C_{ij}. Compliance coefficients have the dimensions of length squared per force, a typical unit being Pa^{-1}.

If only mechanical stress and strain are present, then the two preceding equations suffice to model a non-isotropic material. When piezoelectricity is thrown into the mix, these equations are modified to include the additional effect of the applied or the resulting electric field. For the strain formulation of Eq. (10.4) we have

$$\varepsilon_i = S^E{}_{ij}\sigma_j + d_{ji}E_j \tag{10.5}$$

where E_j are the components of the electric field vector $\boldsymbol{E}$, and d_{ji} are the piezoelectric constants introduced previously. As E_j is a vector, we need only sum over 1-3 for the second term. (Correspondingly, d_{ji} is at most a represented by a 3×6 matrix.) The superscript on S_{ij} in Eq. (10.5) signifies that the compliance coefficients are measured at constant electric field.

An alternate formulation comes from modifying Eq. (5.24) directly with stress as the dependent variable rather than strain:

$$\sigma_i = C^E{}_{ij}\varepsilon_j - e_{ji}E_j \tag{10.6}$$

where e_{ji} are different set of piezoelectric constants with dimensions of strain per applied electric field, a typical unit being C/m^2. Again, as E_j is a vector, we need only sum over 1-3 for the second term, and the $C^E{}_{ij}$ components are measured at constant electric field.

Equations (10.5) and (10.6) are two of the most common formulations of the general piezoelectric effect. At least eight formulations are possible depending on the choice of dependent and independent variables, with the additional corresponding piezoelectric constants defined accordingly.

10.3 Mechanical modeling of beams and plates

10.3.1 Distributed parameter modeling

In many of the preceding chapters we saw how one might model a MEMS device essentially as a "black box", or a number of black boxes. These black boxes have distinct inputs and outputs, but their contents are unknown and/or unimportant for the analysis at hand. For example, an electrostatic actuator a may treated as an electrical switch that is either in the on or off position (output) for a given voltage (input). For that matter, so can the piezoelectric actuator shown in Fig. 10.5. Despite what we may or may not know about how the switches themselves work, we are ultimately interested in the input/output relation.

This type of model is often called a **lumped element model**. One of things we give up upon adopting a lumped element approach is knowledge of the *spatial variations* of quantities within the "lumps". For example, if we would like to know how much a membrane has deflected *at a given location*, then such a model does not suffice. Rather, we must apply the necessary physical concepts with at least one continuously varying spatial coordinate as a variable. Only in this way can distributions of stress, displacement, and/or other parameters be predicted. Designers often refer to this idea as **distributed parameter modeling** in order to emphasize the goal of acquiring spatial information.

Despite the two approaches looking much different from each other, the results of distributed parameter modeling are often used in lumped element modeling. For example, one possible lumped element representation of the membrane of Fig. 10.6 is a spring, with the spring deflection being analogous to the center deflection of the membrane. For the spring and membrane deflection to be equal for the same applied force, however, the spring constant must have the correct value. Distributed parameter modeling helps us determine this value.[2]

In this section we will explore the idea of distributed parameter modeling specifically applied to bending in beams and plates. Though we develop these ideas here specifically to be used with piezoelectric devices, the ubiquity of beam and plate structures in MEMS devices allows the application of these ideas in numerous other situations as well.

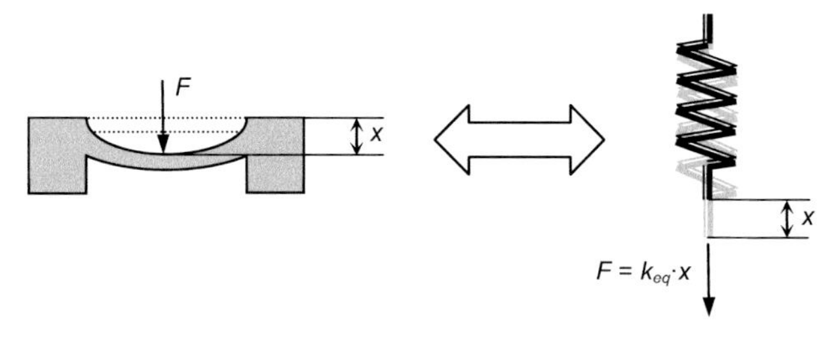

Fig. 10.6. The concept of an equivalent spring constant is one example of how distributed parameter modeling can be linked to lumped element modeling.

10.3.2 Statics

Before looking at beams and plates directly, let us first review the principles of **statics**, or the field of mechanics in which nothing moves.[3] As such the working equations for statics are quite simple. They are given by reduced forms of the conservation of linear and angular momentum:

$$\sum \vec{F} = \vec{0} \tag{10.7}$$

$$\sum \vec{M}_p = \vec{0} \tag{10.8}$$

where F is force and M_p is the moment or torque taken about some point p. Eqs. (10.7) and (10.8) apply to any static system, whether it be a particle, a rigid body, an elastic body, or a material point within a continuous body. The point p about which the moment in Eq. (10.8) is taken is completely

[2] We actually used this very idea of an equivalent spring constant when modeling a capacitive accelerometer back in Chapter 9. Specifically, we used one value of spring constant to model a number of spings working together.

[3] Naturally the structures at which we are looking move. However, the structures move so little and/or over time frames that are so small that their accelerations can be ignored. Strictly speaking, statics involves things that don't accelerate.

arbitrary. We can therefore pick any point p that is convenient for us when summing moments.

Loads and reactions:

When a mechanical component is subjected to an externally applied force, the force is called the **load**. Loads can be applied at specifics points, or distributed over a length or area. Such loads are referred to as **point loads** and **distributed** loads, respectively. (Fig. 10.7.) The applied pressure in a piezoresistive pressure sensor and the thin film stress in piezoelectric actuator both form distributed loads.

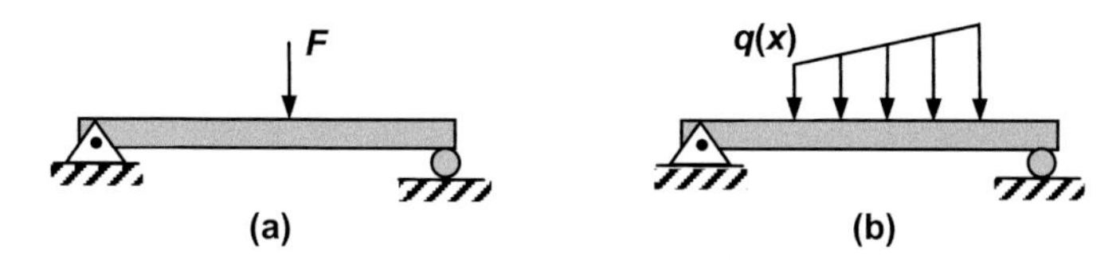

Fig. 10.7: Types of traverse loads: (a) Point load. (b) Distributed load.

The structures we will analyze can be supported in a number of different ways. When we isolate a structure for static analysis, we "remove" the support and replace it with the forces and/or moments it supplies to the structure. Such forces/moments are called **reactions**. When trying to figure out whether a reaction consists of forces, moments, or both, it is useful to think about the way in which the support *restrains the motion* of the structure. For example, if a support keeps a beam from moving up and down, then a reaction force develops in the vertical direction. If a support keeps a beam from rotating about an axis, then a moment reaction develops about that axis. Figure 10.8 shows some of the most common supports and their corresponding reactions.

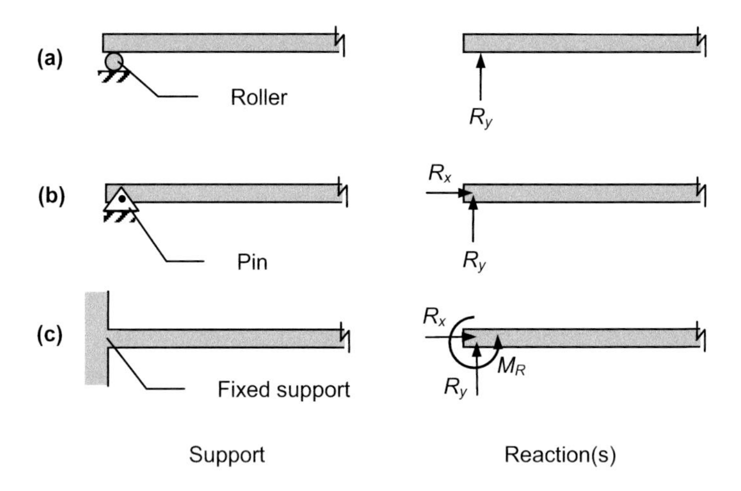

Fig. 10.8: Common support types and their reactions

Of particular interest in the preceding figure is the fixed support. For such a support, the support and the mechanical component itself are formed from one continuous piece of material. If we were to make a "cut" anywhere along the structure, these same types of reactions would therefore appear. These reactions at the various points along the structure are appropriately deemed *internal reactions*. Finding the spatial variation of internal reactions forms the first step in distributed parameter modeling for mechanical components. Example 10.1 illustrates this procedure.

Example 10.1

Internal reactions of a MEMS cantilever

A point load acts at the end of a MEMS cantilever.

(a) Find the reactions at the support, and
(b) the internal reactions as a function of x.

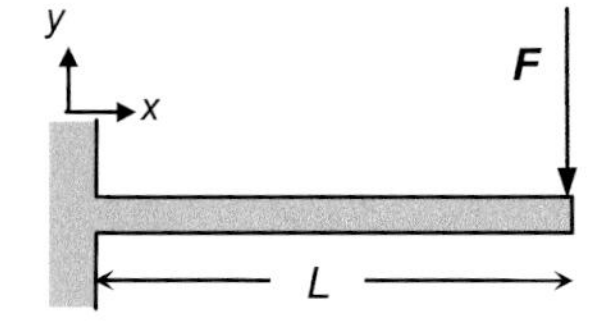

(a) The first step is to isolate the beam from the wall for analysis. Removing the fixed support at the wall, we have a system that looks like this:

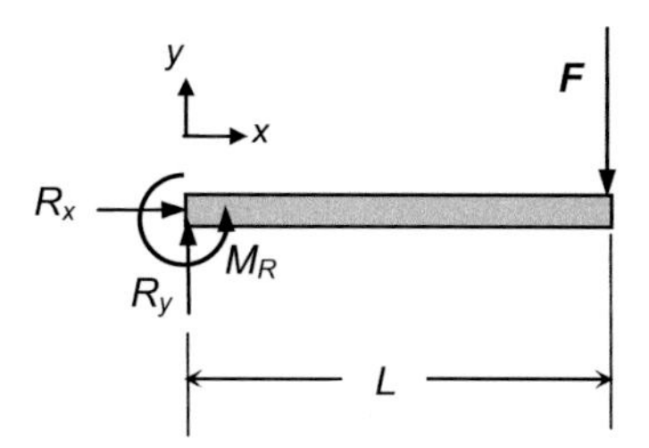

Equilibrium requires that the sum of the x components of force be zero. Hence,

$$\sum F_x = 0$$

$$R_x = 0$$

The y-direction forces must also sum to zero:

$$\sum F_y = 0$$

$$R_y - F = 0$$

$$R_y = F$$

The moments taken about any point must be zero as well. For convenience, let's sum the moments about the left end of the beam. About this point, only the force F creates a torque. We will call counterclockwise moments positive.

$$\sum M_{x=0} = 0$$

$$M_R - LF = 0$$

$$M_R = LF$$

(b) To find the internal reactions, we "cut" the beam at an arbitrary location x from the left edge. The removed portion of the beam to the right supplies the left hand portion with the same kind of reactions as a fixed support. We label the internal reactions as a normal force P (normal to the cut), a shear force V (tangent to the cut) and an internal bending moment, M_x.

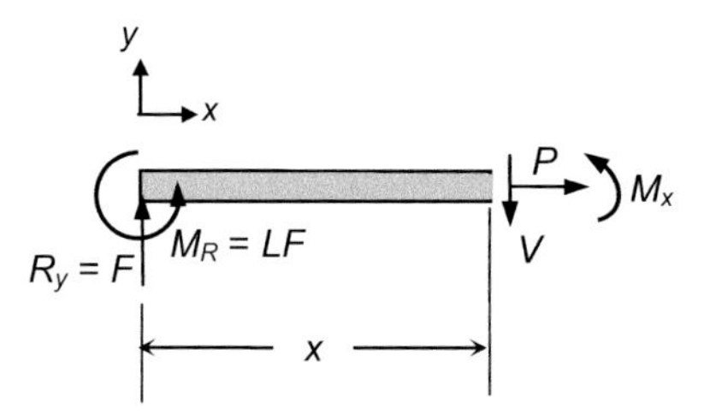

Applying equilibrium in the x-direction,

$$\sum F_x = 0$$

$$P = 0$$

For the y-direction,

$$\sum F_y = 0$$

$$F - V = 0$$

$$V = F$$

Once again summing the moments about the left end of the beam with counterclockwise being positive,

$$\sum M_{x=0} = 0$$

$$LF + M_x - xV = 0$$

$$M_x = (x - L)F$$

For this example we see that the application of a point load at the end of a cantilever beam results in a constant internal shear force V equal to F at any beam location x. The internal bending moment, however, varies as a function of x. The largest magnitude of this moment is $-LF$ at x=0, and the smallest is zero occurring at x=L. These variations are shown below.

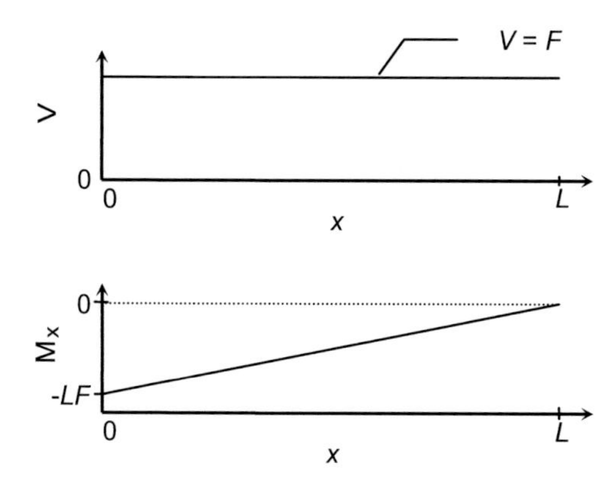

An equally valid analysis can be performed by analyzing the portion of the beam to the right of the cut, as shown in the figure below. From Newton's third law, the internal reactions are drawn in the opposite directions.

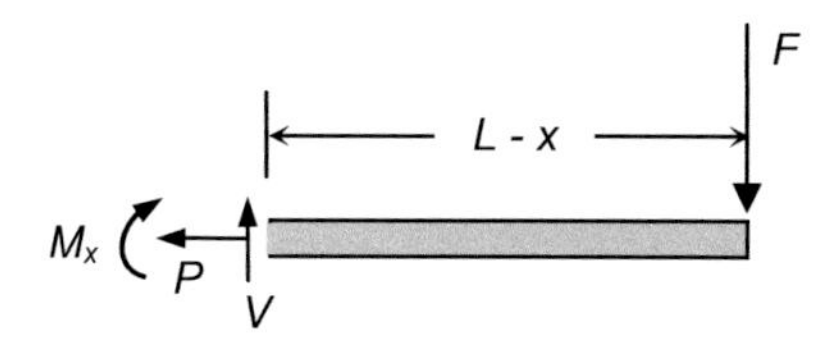

It is left as an exercise to show that the magnitudes of the internal reactions remain the same when analyzing the above system. ◄

In Example 10.1 we get our first taste of distributed parameter modeling in finding the spatial variations if internal reactions. Example 10.1 also shows us the need to come up with a sign convention for these internal reactions. The convention used in this chapter is given in Fig. 10.9, in which an arbitrary section of beam is shown with cuts being made on both the left and right hand faces. Positive values of normal reaction forces point away from the section, which is consistent with sign convention given in Chapter 5 for normal stress. Positive values of shear force tend to rotate the beam segment clockwise, and positive values of bending moment tend to bend the ends of the segment upwards.

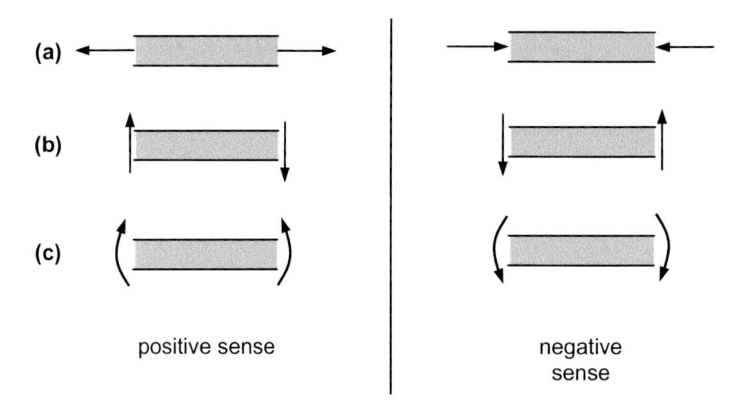

Fig. 10.9: Sign convention for (a) normal reaction force, (b) internal shear and (c) internal bending moment

If we have a traversely loaded beam with only a continuously varying load $q(x)$ (load per unit length), the load, internal shear force and internal bending moment are related to each other via the following equations:

$$\frac{dV}{dx} = -q \tag{10.9}$$

$$\frac{dM}{dx} = V \tag{10.10}$$

The derivation of these relations is the subject of Problem 10.9.

10.3.3 Bending in beams

As we have just seen, loads externally applied to beams cause internal moments, and we know from experience that moments result in bending. Think of bending a meter stick, for example. In order to predict beam deflection we must therefore first relate internal moments to bending.

Consider a short section of beam subject to a negatively directed bending moment, M_O, as shown in Fig. 10.10 (a). This applied moment will result in the beam bending downward as shown in Fig. 10.10 (b).

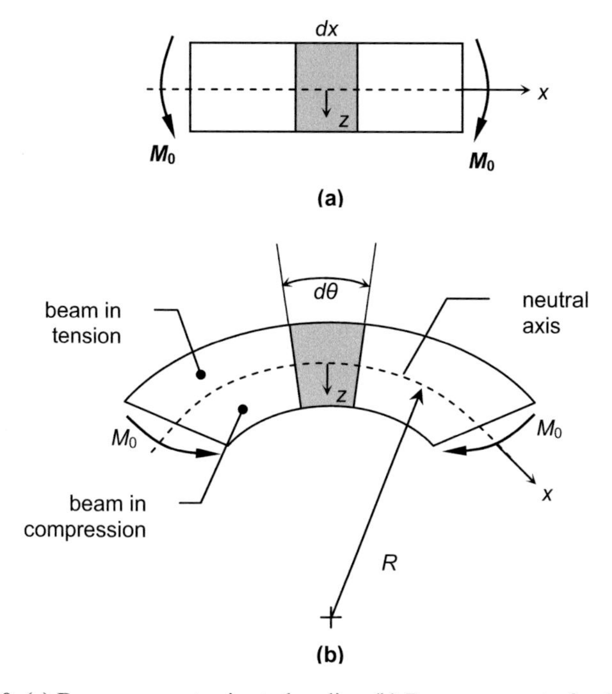

Fig. 10.10. (a) Beam segment prior to bending (b) Beam segment after bending

First we focus on a differential width of beam, dx. Prior to bending, the entire thickness of beam defined by dx has a length of, well, dx. After bending, however, we see that top portion of the segment has *lengthened* in the x-direction and the bottom portion has *shortened.* This also means that the top portion of the beam is in tension and the bottom part is in compression. There is also a special location somewhere in the middle where no deformation occurs called the **neutral axis**. Hence we see that the length of the segment after deformation, dx^*, is a function of the traverse location within the beam. The neutral axis serves as a convenient origin for the traverse coordinate, z, which we'll assign to be positive in the downward direction.

From the geometry of Fig. 10.10 the length of the beam element after bending, dx^*, is related to the local radius of curvature, R, the traverse coordinate, z, and the bending angle $d\theta$, by

$$dx^* = (R - z)d\theta\,. \tag{10.11}$$

Equation (10.11) gives the correct deformation for both positive and negative values of z. At the neutral axis where z=0, the length dx^* is the same as *dx*, the length before bending. Eq. (10.11) therefore gives us a way to calculate the *x*-direction strain:

$$\varepsilon_x = \frac{dx^* - dx}{dx} = \frac{(R-z)d\theta - Rd\theta}{Rd\theta} \tag{10.12}$$
$$= -\frac{z}{R}$$

Assuming that the stress/strain relation at any location z is uniaxial, the normal stress in the x-direction is, then[4]

$$\sigma_x = E\varepsilon_x = -\frac{zE}{R}. \tag{10.13}$$

The next step is to relate the normal stress to the local internal bending moment. To do this, consider a beam segment sliced at some arbitrary x location as shown in Fig. 10.11.

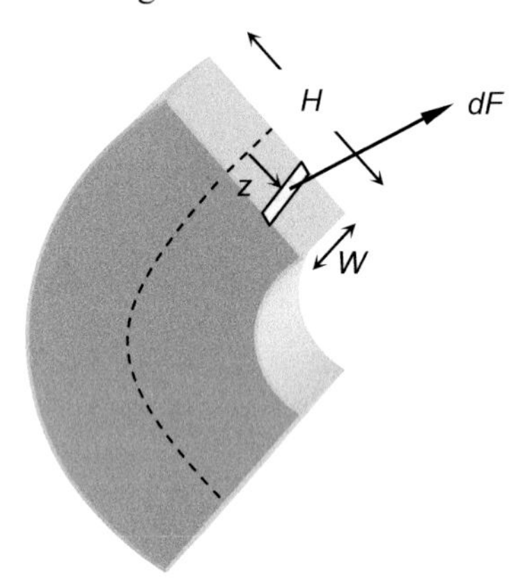

Fig. 10.11. Slice of a bent beam segment. The force *dF* is due to normal stress.

The infinitesimal force *dF* in the figure is due to the normal stress acting on the area defined by *dz* and the width of the beam, *W*:

$$dF = \sigma_x W dz = -\frac{zE}{R} W dz. \tag{10.14}$$

[4] This assumption amounts to modeling the beam as a number of infinitesimally thin sheets of material stacked on top of each other. In so doing we have ignored Poisson contraction in the z direction.

The corresponding moment dM about the neutral axis is

$$dM = zdF = -\frac{z^2 WE}{R} dz\ . \tag{10.15}$$

The total internal bending moment is the integral of this term evaluated across the entire cross section.

$$M = \int_{-H/2}^{H/2} -\frac{z^2 WE}{R} dz = -\frac{WH^3}{12}\frac{E}{R} \tag{10.16}$$

You might recognize the term $WH^3/12$ as the second moment of area, I, of a rectangle about its centroid. Should we have had a differently shaped beam cross section, we would have obtained a similar result with I for the new cross section shape instead. Generalizing we have

$$M = -\frac{IE}{R}\ . \tag{10.17}$$

Now that we have a relationship between the local radius of curvature and internal bending moment, all that remains is to relate the radius of curvature to beam deflection. This is readily accomplished by considering the geometry given in Fig. 10.12.

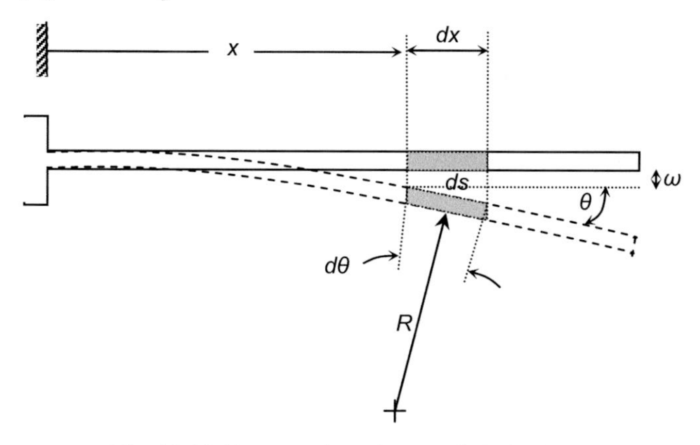

Fig. 10.12. Beam undergoing small deflection

If we assume small deflections in the beam, this is a relatively easy task. When the beam in Fig. 10.12 deflects under the action of an applied load (not shown),

$$ds = Rd\theta\ . \tag{10.18}$$

For small deflections $dx \approx ds$, hence $dx \approx Rd\theta$ and

$$\frac{d\theta}{dx} \approx \frac{1}{R}. \tag{10.19}$$

For small deflections we can also write $\tan(\theta) \approx \theta$. In terms of deflection this is

$$\frac{d\omega}{dx} \approx \theta. \tag{10.20}$$

Substituting into of Eq. (10.19),

$$\frac{d^2\omega}{dx^2} = \frac{1}{R}. \tag{10.21}$$

Eliminating R from Eqs. (10.17) and (10.21),

$$\frac{d^2\omega}{dx^2} = -\frac{M}{EI}. \tag{10.22}$$

Eq. (10.22) is called the **beam deflection equation**. It is highly useful in that it allows us to relate the internal reactions in a loaded beam, something we can get from simple statics, to the local deflection in the beam. To solve the beam deflection equation, the following steps are generally followed:

1. Solve for the support reactions first for a given load first *as if there were no deflection.*
2. Find the internal reactions (V and/or M) as a function of x, again *as if there were no deflection.*
3. Use the internal reactions found in 2. along with the beam deflection equation to find the deflection as a function of x.

Example 10.2 illustrates this method.

Example 10.2

Deflection in a MEMS cantilever

A point load acts at the end of a cantilever, causing it to deflect.

(a) Find the deflection ω as a function of x.

(b) Find the equivalent spring constant k_{eq} for the beam based on the tip deflection (ω at $x = L$) and the point load force, F.

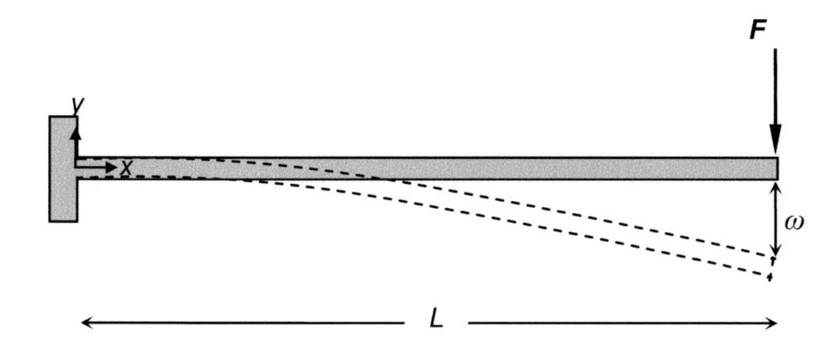

(a) Let's use our previously outlined method of attack. First we solve for the support reactions as if there were no deflection.

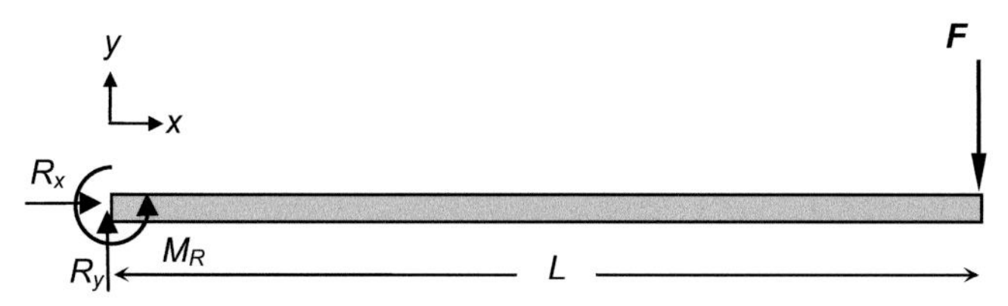

The situation is identical to Example 10.1, hence we can use the results from that problem. For the reactions at the support those are

$$R_x = 0,$$

$$R_y = F$$

and

$$M_R = -LF.$$

For the internal reactions,

$$P = 0,$$

$$V = F$$

and

$$M_x = (x - L)F\ .$$

Next we make use the beam deflection equation using the internal bending moment for which we just solved:

$$\frac{d^2\omega}{dx^2} = -\frac{M}{EI} = -\frac{(x - L)F}{EI}\ .$$

Integrating once with respect to x yields

$$\frac{d\omega}{dx} = \frac{FL}{EI}x - \frac{F}{2EI}x^2 + C_1,$$

where C_1 is a constant of integration. A second integration gives ω as

$$\omega - \frac{FL}{2EI}x^2 - \frac{F}{6EI}x^3 + C_1x + C_2,$$

where C_2 is a second constant of integration.

The constants of integration in this case are found by applying boundary conditions at $x = 0$. The deflection at $x = 0$ must be zero, so that C_2 is zero. The fixed support requires that the slope of the deflection, $d\omega/dx$, is also zero at $x = 0$. Thus, $C_1 = 0$ and the deflection as a function of x is given by

$$\omega = \frac{FL}{2EI}x^2 - \frac{F}{6EI}x^3$$

(b) The equivalent spring constant k_{eq} based on the tip deflection is defined by

$$F = k_{eq}\omega_{tip},$$

where F is a point load applied to the tip and ω_{tip} is ω at $x = L$. From part (a) the tip deflection is found to be

$$\omega_{tip} = \frac{FL}{2EI}L^2 - \frac{F}{6EI}L^3 = \frac{FL^3}{3EI}$$

Substituting and solving for k_{eq},

$$k_{eq} = \frac{F}{\omega_{tip}} = \frac{3EI}{L^3}. \blacktriangleleft$$

10.3.4 Bending in plates

We next extend the previously derived equations to plate bending. Only the main results will be given, without rigorous derivations.

As an example consider a plate of thickness H shown in Fig. 10.13. A distributed load such as an applied pressure has caused the plate to deflect out of the x-y plane.

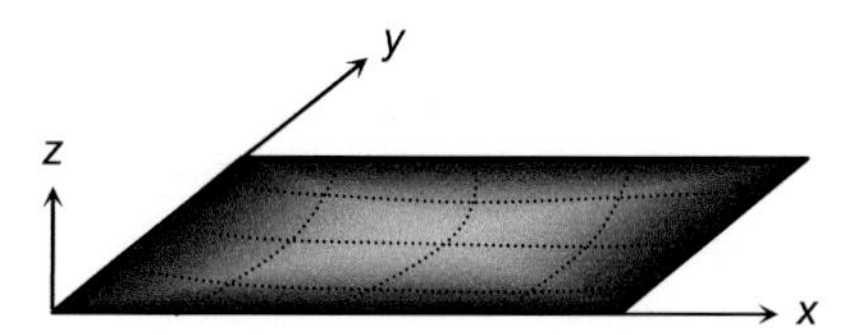

Fig. 10.13. Deflection in a flat plate

For plate bending there are two radii of curvature, one associated with the x-direction and one with the y-direction. Each of these is related to the local deflection in an analogous way as for bending in beams.

$$\frac{1}{R_x} = \frac{\partial^2 \omega}{\partial x^2} \tag{10.23}$$

$$\frac{1}{R_y} = \frac{\partial^2 \omega}{\partial y^2} \tag{10.24}$$

Likewise the x and y-direction strains are related to their respective radius of curvature and the coordinate z, though z is now the distance measured from the neutral *surface* rather than the neutral axis:

$$\varepsilon_x = -\frac{z}{R_x} \tag{10.25}$$

$$\varepsilon_y = -\frac{z}{R_y} \tag{10.26}$$

The stress/strain relations for the x and y-directions, respectively, are given by

$$\sigma_x = -\frac{E}{1-\nu^2} z \left(\frac{1}{R_x} + \frac{\nu}{R_y} \right) \tag{10.27}$$

and

$$\sigma_y = -\frac{E}{1-\nu^2} z \left(\frac{1}{R_y} + \frac{\nu}{R_x} \right). \tag{10.28}$$

In the preceding equations you should recognize that a stress and a strain are related to each other via a modulus of elasticity. Here that modulus is given by $E/(1-v^2)$, and is called the *plate modulus*. It looks much like the biaxial modulus, but should not be confused with it.[5]

Finally, the analogue of the beam deflection equation is given by

$$D\left(\frac{\partial^4 \omega}{\partial x^4}+\frac{\partial^4 \omega}{\partial x^2 \partial y^2}+\frac{\partial^4 \omega}{\partial y^4}\right)=P(x,y) \tag{10.29}$$

where $P(x,y)$ is applied the load per unit area and D is a parameter called the flexural rigidity, given by

$$D=\frac{1}{12}\left(\frac{EH}{1-v^2}\right). \tag{10.30}$$

Solving Eq. (10.29) tends to be quite a bit more involved than solving the beam deflection equation. It sometimes can be accomplished by assuming a polynomial form for the solution or by using eigenvalue expansions. For complicated geometries, however, the likelihood of generating a closed form solution is greatly diminished. Furthermore, it should be noted that none of the equations developed in this section are valid when the deflections are too large for small angle approximations to be used. In these later cases, modeling the mechanical behavior is often accomplished using energy methods or through sophisticated numerical means, such as finite element techniques.

10.4 Case study: Cantilever piezoelectric actuator

One of the uses of the equations developed in the previous section is to calculate the stress that results from loading a MEMS device. If we first assume we have an initially stress-free device, the applied load is used to calculate deflection as a function of position, which in turn is used to find the local radius or radii of curvature. Next, the curvature(s) is used to find the strain and then the stress as a function of the traverse coordinate z. (This is the subject of Problem 10.12.)

---

[5] The plate modulus can be derived from Eqs. (5.33) and (5.34) by assuming that plate motion is sufficiently restrained so that stress applied in one direction causes a compensatory stress in the perpendicular direction resulting from zero *additional* strain in the second direction.

In the case of a piezoelectric actuator, the above process can be used somewhat in reverse. That is, the applied electric field causes a known stress within the actuator. The known stress is then related to bending moment, which is then used to find the local deflection. This is the process we will employ in modeling the actuator in shown Fig. 10.14.

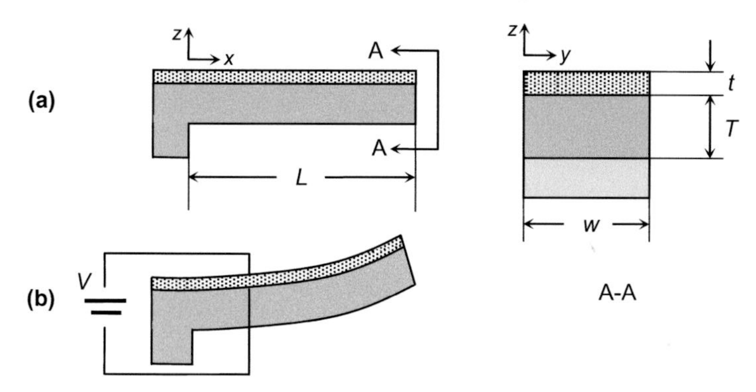

Fig. 10.14. The modeled piezoelectric actuator: (a) A thin film of piezoelectric material is deposited onto a cantilever beam. (b) Stress in the thin film induced by an applied electric field causes the beam to bend.

For the simple actuator shown in Fig. 10.14, a thin film of a piezoelectric material of thickness t is deposited onto a cantilever beam of greater thickness T. When a voltage V is applied in the traverse direction across the piezoelectric material, it results in an electric field in that direction.[6] We will assume that the piezoelectric material is of the class that behaves as does zinc oxide in terms of its piezoelectric constants, and furthermore that the z-direction corresponds to the traverse direction. Hence, the applied field would like to cause deformations in both the z-direction (the 33 mode) and in the plane itself (the 31 mode). The thicker diaphragm, however, forces the thin film to experience no in-plane deformation. An in-plane stress therefore develops within the thin film to compensate. This stress bends the cantilever, which serves as the actuation event. In the case shown in Fig. 10.14, the in-plane stress is tensile, resulting in the beam bending up.

To quantify the deformation, we would first like to know the stress resulting from the applied field. We therefore use the stress formulation of the piezoelectric effect represented by the equations in Eq. (10.6). Assum-

[6] We have switched notation for voltage from e to V in this analysis in order to avoid confusion between voltage and the piezoelectric constants given by e_{ij}.

ing that the film is stress-free prior to the applied electric field, the strains are all zero. Since the thicker diaphragm keeps the in-plane strains at zero even after the field is applied, the first two equations represented by Eq. (10.6) reduce to

$$\sigma_1 = C_{31}\varepsilon_3 - e_{31}E_3 \tag{10.31}$$

and

$$\sigma_2 = C_{32}\varepsilon_3 - e_{32}E_3 . \tag{10.32}$$

The film is not restrained in the z-direction, keeping the stress at zero. Hence the third equation of Eq. (10.6) becomes

$$\sigma_3 = 0 = C_{33}\varepsilon_3 - e_{33}E_3 . \tag{10.33}$$

Solving Eq. (10.33) for strain and making use of the fact that the electric field is given by $E_3 = V/t$,

$$\varepsilon_3 = \frac{e_{33}}{C_{33}}\frac{V}{t} . \tag{10.34}$$

Symmetry arguments for all materials require that $C_{31} = C_{13}$ and $C_{32} = C_{23}$. Additionally, for piezoelectric materials behaving like ZnO, $C_{23} = C_{13}$. Favoring the notation x, y and z over the indices 1-3 and substituting Eq. (10.34) into Eqs. (10.31) and (10.33) yields

$$\sigma_x = \sigma_y = \left(C_{13} - \frac{e_{33}}{C_{33}} \right) \frac{V}{t} . \tag{10.35}$$

The in-plane stresses given by Eq. (10.35) are those which cause internal bending moments, and thus actuator deflection. Since we have assumed that the thin film is much smaller than the beam thickness, the internal bending moment within the beam caused by this stress can be approximated as

$$M = \frac{T}{2}\sigma_y tw . \tag{10.36}$$

Substituting this moment into the beam deflection equation, Eq. (10.22),

$$\frac{d^2\omega}{dx^2} = -\frac{T\sigma_y tw}{2EI}$$

$$= -\left(C_{13} - \frac{e_{33}}{C_{33}} \right) \frac{TwV}{2EI} . \tag{10.37}$$

Equation (10.37) can be solved by direct integration in a similar fashion as in Example 10.2. Since the modeled actuator is a cantilever, we will also assume the same boundary conditions as in Example 10.2, namely that the deflection and its first derivative are zero at $x = 0$. The result is

$$\omega(x) = -\left(C_{13} - \frac{e_{33}}{C_{33}} \right) \frac{TwV}{4EI} x^2 . \tag{10.38}$$

Finally, the beam cross section is rectangular so that the second moment of area is given by $I = wT^3/12$. Equation (10.38) becomes

$$\omega(x) = -\left(C_{13} - \frac{e_{33}}{C_{33}} \right) \frac{3V}{ET^2} x^2 . \tag{10.39}$$

Tip deflection is found by simply substituting $x = L$:

$$\omega_{tip} = -\left(C_{13} - \frac{e_{33}}{C_{33}} \right) \frac{3V}{ET^2} L^2 . \tag{10.40}$$

Equation (10.40) gives an approximation of the tip deflection of a cantilever-type piezoelectric actuator as a function of geometry and applied voltage. Many restrictions apply, of course, including the use of a certain class of piezoelectric material, that the piezoelectric material is much thinner than the beam material, and that the deflections are small. In the event that the deflections are large, some of the geometric approximations would need to be relaxed, resulting in a more complicated derivation. And if the deflections become too large, linear elastic theory may no longer provide adequate constitutive relations. Advanced energy methods would most likely need to be employed to obtain analytical solutions, and/or numerical simulations.

Another restriction that applies to the models in this chapter is that the structures are initially stress-free. This restriction is the probably one of the most suspect. As we saw in the fabrication portion of the text, residual stresses are a common byproduct of the microfabrication process. To compensate for this, the analyses can be modified to include initial stress within the components, resulting in more complicated derivations still. The values for the residual stresses for use in the analysis may be measured after the device has been fabricated, or in some cases predicted based on the details of the particular fabrication method employed. More common, however, is to attempt to reduce the stress during fabrication itself. And so, as is quite common in the world of MEMS, the design and analysis of a MEMS device cannot be divorced from the details of how it is made.

References and suggested reading

Craig R (1999) Mechanics of Materials 2nd edn. John Wiley and Sons, Canada

Deckert M (2007) Design of a MEMS Power Generating Device Using a Shape Memory Alloy and Piezoelectric Thin Film. Master's thesis, Rose-Hulman Institute of Technology

Fung YC (1965) Foundations of Solid Mechanics. Prentice-Hall, Englewood Cliffs, NJ

Gere JM, Timoshenko SP (1997) Mechanics of Material, 4th edn. PWS Publishing, Boston

Gopal P, Spaldin N., (2005) Polarization, piezoelectric constants and elastic constants of ZnO, MgO and CdO. arXiv:cond-mat/0507217v2

Krushenisky C (1993) Inkjet Printers: No Longer the Third Wheel in the Printer Market. Hardware 4:2

Malvern LE (1969) Introduction to the Mechanics of a Continuous Medium. Prentice-Hall, Englewood Cliffs, NJ

Madou MJ (2002) Fundamentals of Microfabrication, The Art and Science of Miniaturization., 2nd Ed., CRC Press, New York

Maluf M, Williams K (2004) An Introduction to Microelectromechanical Systems Engineering, 2nd edn. Artech House, Norwood, MA

Senturia S (2001) Microsystem Design, Kluwer Academic Publishers, Boston

Questions and problems

10.1 The two types of the piezoelectric constants given in this chapter relate electric field to strain (d_{ij}) and electric field to stress (e_{ij}). Find one other type of piezoelectric constant and give the physical quantities that it relates to each other. In what modeling situation might this formulation be useful?

10.2 Explain the piezoelectric effect and its origins.

10.3 Would you expect single crystal silicon to exhibit piezoelectricity? Why or why not?

10.4 Explain the difference between lumped element modeling and distributed parameter modeling. When do you think each approach might be preferred over the other?

10.5 Describe the general operating principle of actuators that employ thin films of piezoelectric material. Why are these arrangements preferred over actuators whose moving parts are made entirely of piezoelectric materials?

10.6 What are reactions? What are internal reactions?

10.7 Where would you expect the internal normal stress in a bent beam to be the larger at the top or at the neutral axis? Why?

10.8 Using the system consisting of the portion of the cantilever to the right of the cut in Example 10.1, show that the magnitudes of the internal reactions match those found in the example.

10.9 Show that for a *continuously* distributed traverse load $q(x)$ the following relations apply:

$$\frac{dV}{dx} = -q$$

$$\frac{dM}{dx} = V$$

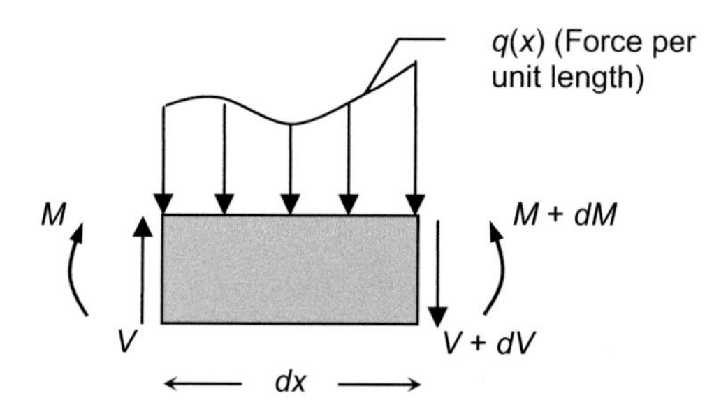

(Hint: For an infinitesimally small segment, an arbitrary distributed loading function $q(x)$ can be considered constant across dx.)

10.10 Find the equivalent spring constant for the cantilever-type actuator modeled in Example 10.2 in terms of its dimensions and elastic material properties.

10.11 Use Eqs. (10.21), (10.22) and (10.36) to derive the formula for calculating stress using the Disk Method, given by Eq. (5.39). (Hints: Eliminate $d^2\omega/dx^2$ from the equations, and remember that the Disk Method assumes *biaxial* stress/strain, not uniaxial.)

10.12 A silicon cantilever of length 500 μm, width 50 μm and thickness 2 μm is subjected to a uniform distributed traverse load $q(x)$ = constant = q_0. Assume E_{Si} = 160 GPa.

a. Find an expression for the tip deflection.

b. Find an expression for the stress experienced by the top and bottom surfaces of the beam.

c. The yield stress for silicon can be taken as 7 GPa. (See section 5.2.3.) What load q_0 in N/m can the cantilever experience before this stress is realized somewhere in the beam?

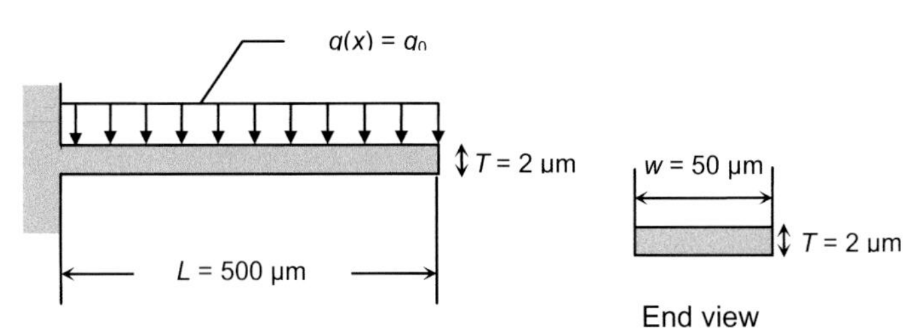

10.13 A piezoelectric actuator made of polysilicon has the geometry shown in Fig. 10.14 with a thin film of aluminum nitride (AlN), which is a piezoelectric material in the same class as ZnO. Use the following parameters:

- $L = 300$ μm
- $w = 50$ μm
- $T = 8$ μm
- $t = 0.3$ μm
- Si properties: $E_{si} = 160$ Gpa
- AlN properties: $C_{13} = 100$ GPa, $C_{13} = 390$ GPa $e_{33} = -0.58$ C/m^2, $e_{33} = 1.55$ C/m^2

Determine:

a. The applied voltage required to produce a tip delfection of 2 μm

b. The sensitivity of the actuator

Chapter 11. Thermal transducers

11.1 Introduction

There are a number of MEMS devices that can be classified as being thermal transducers due to the involved energy going through the thermal mode at some point. In particular, thermal actuators have mechanical components that move when heated, and the temperature of thermal sensors changes in response to stimuli from their environment. Often the physical quantity of interest in a thermal sensor is temperature itself, though several other physical quantities can be measured indirectly via such sensors, particularly velocity and other flow quantities. Qualitative descriptions of several of thermal actuators and sensors were given in Chapter 7.

In the present chapter we will develop three different models for a hot-arm actuator, a simple block diagram for which is given in Fig. 11.1. As in the previous three chapters, much of this chapter is devoted to developing the necessary machinery to physically understand what is inside of the block. The models are presented in order of increasing complexity, each one building on the previous. Energy in the thermal mode, however, behaves much differently than other forms of energy, especially in the way that it is transported. As such, a brief introduction to the field of thermal energy transport, or **heat transfer**, is presented first.

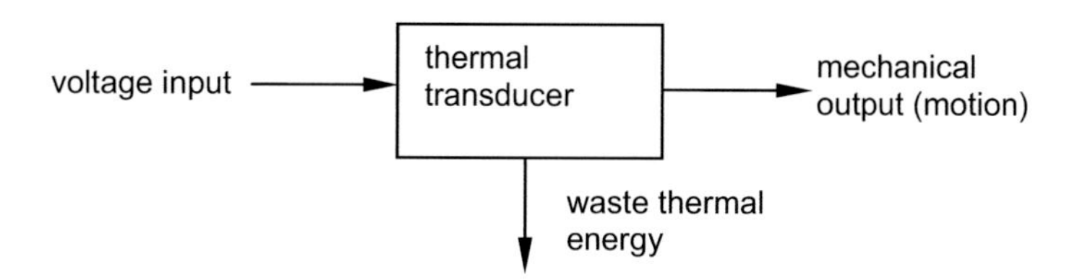

Fig. 11.1. System inputs and outputs for a thermal transducer behaving as an actuator

T.M. Adams, R.A. Layton, *Introductory MEMS: Fabrication and Applications*,
DOI 10.1007/978-0-387-09511-0_11, © Springer Science+Business Media, LLC 2010

11.2 Basic heat transfer

Energy can be transported via several mechanisms including mechanical work, the movement of electric charge, the movement of mass, and many other forms. When energy is transported specifically due to spatially occurring temperature differences, it is called heat transfer. As a consequence of the Second Law of Thermodynamics, thermal energy always travels from a region of high temperature to a region of low temperature.[1] The converse is impossible; i.e., thermal energy can not spontaneously go from regions of low temperature to regions of high temperature. Thus, many MEMS thermal actuators have no sensor counterpart, and *vice versa*.

There are three ways in which thermal energy can be transported known as the three modes of heat transfer. These include **conduction**, **convection** and **radiation**. To get an idea of how the three modes of heat transfer work, let us imagine that thermal energy at a high temperature is represented by the water in a large pond. We are interested in filling a bathtub some distance away with this thermal energy. Naturally the water in the bathtub must be at a lower temperature than the pond.

One way of accomplishing this is with a bucket brigade, each person handing a bucket of thermal energy to the next person in the line, while otherwise remaining still. (Fig. 11.2.) This is what happens in **conduction**. The people in the brigade represent molecules of the medium through which the thermal energy travels. Other than the exchange of energy between the molecules, the molecules remain stationary. That is to say, there is no *bulk motion* of the medium. As such, conduction is the dominant form of heat transfer within solids. Conduction can also occur in stationary liquids and gases.

Fig. 11.2 A conduction analogy

[1] The Second Law is also responsible for the waste thermal energy in Fig. 11.1 being a non-zero quantity. Since this output is not the transducer output of interest, it is often not calculated *per se*. Minimizing its value, however, does lead to increased device performance.

When conduction is accompanied by bulk motion of the medium, we have **convection**. To continue with our analogy, in convection buckets of thermal energy are first conducted to members of the bucket brigade, but now the members of the brigade are free to run with this energy en route to the bathtub. (Fig. 11.3.) This extra form of energy transfer, sometimes called *advection*, accounts for the larger rates of heat transfer in convection over conduction. Convection is the dominant form of heat transfer in flowing liquids and gases, and thus is very important in microfluidic devices.

Fig. 11.3 A convection analogy

Thermal radiation travels in the form of electromagnetic waves. In specific, thermal radiation is that part of the electromagnetic spectrum emitted by a body due to its temperature. Thermal radiation includes portions of the ultraviolet, all of the visible range and much of the infrared regions of the spectrum. As it is an electromagnetic wave, no medium is required for thermal radiation to occur. As our last analogy, thermal radiation corresponds to spraying thermal energy directly from the pond into the bathtub. (Fig. 11.4.)

Fig. 11.4. A radiation analogy

In many cases one mode of heat transfer dominates over the other two, and hence only that mode need be considered. In other cases two or all three modes must be considered simultaneously. A notable example is that in which convection within a gas and radiation exchange between surfaces bounding that gas are both on the same order of magnitude. The small length scales in MEMS applications, however, tend to lead to a dominance of one mode over others. For example, at large scales temperature gradients within air lead to spatial differences in density and give rise to buoyancy forces. These buoyant forces in turn lead to air motion causing a form of convection called *natural convection.* The small air gap sizes encountered in many MEMS devices, however, may suppress the buoyant forces in the air. As a result, the air remains stagnant and the mode of heat transfer within it becomes conduction instead.

We next turn our attention to the simple forms of the working equations for the three modes. We begin with conduction, which is probably the most important heat transfer mechanism in MEMS applications.

11.2.1 Conduction

In Fig. 11.5 a slab of material with thickness d has two surfaces of surface area A maintained at constant temperatures T_1 and T_2. With $T_1 > T_2$, the *rate* of thermal energy transport through the slab is given by

$$\dot{Q} = -\kappa A \frac{T_2 - T_1}{d} \tag{11.1}$$

Where $\dot{Q}$ is the rate of conduction heat transfer (energy per unit time) and κ is the **thermal conductivity** of the slab. The negative sign indicates that the energy flows in the direction of decreasing temperature. The thermal conductivity is a material property that measures the ease with which thermal energy is conducted through a material. For metals this number is rather high, with κ_{copper}, for example, being on the order of 400 W/m·°C. Insulating materials such as Styrofoam have low thermal conductivities on the order of tens of mW/m·°C. For silicon $\kappa_{Silicon} \approx 148$ W/m·K.

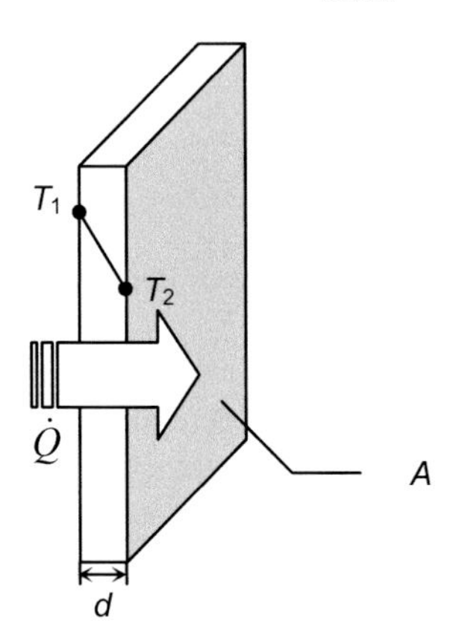

Fig. 11.5. One dimensional conduction through a slab

Equation (11.1) can be arranged into a more intuitive form:

$$\dot{Q} = \frac{T_1 - T_2}{\frac{d}{\kappa A}} = \frac{T_1 - T_2}{R_{th}} \tag{11.2}$$

Equation (11.2) is a thermal analogue to Ohm's law for a simple resistor. Heat transfer rate is the analogue of current, and its driving potential is temperature, the counterpart of voltage. Rather than electrical resistance, we have **thermal resistance** instead, represented by the group $d/\kappa A$. The typical unit of thermal resistance is °C/W, also called a *thermal ohm* and represented by the symbol ॐ.

In addition to being the analogue of electrical resistance, thermal resistance has the exact same form as electrical resistance. In thermal resistance, electrical conductivity has simply been replaced by thermal conductivity. As such, the same trends apply to thermal resistance as to electrical resistance. For example, thin bodies (small *d*) offer less resistance to heat transfer than do thick. Large cross sectional areas result in smaller resistances than small areas, as do large conductivities. (Fig 11.6.) These geometric trends are exploited explicitly in hot-arm actuators.

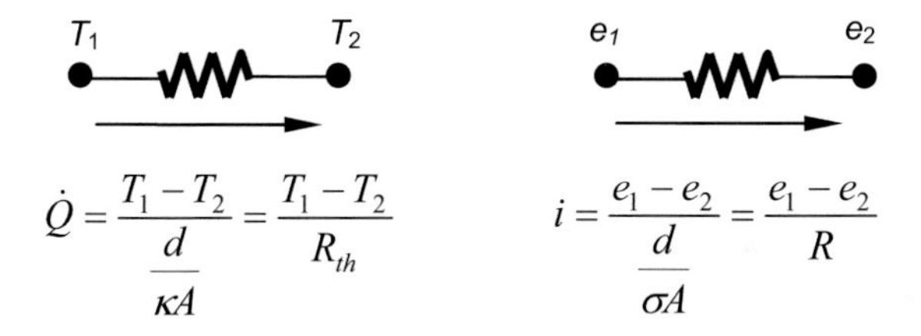

Fig 11.6 Thermal resistance analogy

Equation (11.2) holds for one dimensional conduction for which no internal sources of energy are converted to thermal energy; for example, via chemical reactions or electrical power dissipation. A more general equation for conduction applicable for these cases is given by letting the dimension *d* shrink to zero:

$$\dot{Q} = -\kappa A \frac{dT}{dx} \tag{11.3}$$

In cases where we have conduction in multiple directions, we let the area shrink to zero as well, and the derivative of Eq. (11.3) becomes the temperature gradient:

$$\vec{q} = -\kappa\vec{\nabla}T\,, \tag{11.4}$$

where $\vec{q}$ is the heat flux vector (rate of heat transfer per unit area).

11.2.2 Convection

Convection is the mode of heat transfer encountered in flowing fluids. There are two major types of convection. In the first type we force the fluid to move by some external means such as a fan or a pump. This is known as *forced convection*. In the second type buoyancy forces move the fluid due to temperature-induced density gradients, giving the appearance that the fluid moves by itself. This is known as *natural* or *free convection*, and is the dominant form of heat transfer in otherwise stagnant gases. (You have heard that "Hot air rises," yes?)

In quantifying the heat transfer rate in convection we refer to the temperature of a solid surface past which the fluid is flowing (or *in* which for pipe flow), and the temperature of the fluid far from the surface. For a surface of area A maintained at temperature T_s the rate of heat transfer due to convection is given by

$$\dot{Q} = hA(T_s - T_\infty) \tag{11.5}$$

where $\dot{Q}$ is the rate of convective heat transfer from the surface to the fluid, h is the **convective heat transfer coefficient** and T_∞ is the temperature of the fluid far from the surface. (Fig. 11.7.) Equation (11.5) is often referred to as Newton's law of cooling.

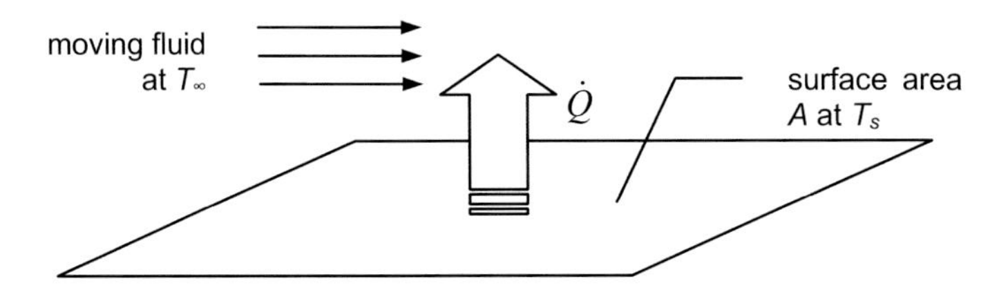

Fig. 11.7. Convection from a solid surface to a moving fluid

Unlike the thermal conductivity κ, the convective heat transfer coefficient is not a material property, but rather a function of the fluid, fluid velocity, surface-fluid geometry and many other things. Determining h is a rich science in itself and beyond the scope of our discussion here. For now, representative values of h are given in Table 11.1.

Table 11.1. Representative values of convective heat transfer coefficients

Convection	Fluid	h (W/m^2 °C = 1/(ॐ˙ m^2)
Natural	Gas	2-20
Natural	Liquid	10-1000
Forced	Gas	20-250
Forced	Liquid	100-20,000

The convection relation can be recast into a resistance form similar to Eq. (11.2):

$$\dot{Q} = \frac{T_s - T_\infty}{R_{th}} = \frac{T_s - T_\infty}{\frac{1}{hA}} \tag{11.6}$$

And so we see that the thermal resistance due to convection is given by $1/hA$.

11.2.3 Radiation

All bodies emit electromagnetic radiation. That part of the radiation that is due to a body's temperature is termed thermal radiation. The amount of thermal radiation emitted by real surfaces is compared to an idealized, perfect emitter of thermal radiation called a *blackbody*. Surfaces approaching blackbody behavior need not appear black to the eye, as thermal radiation includes portions of the ultraviolet and infrared regions of the electromagnetic spectrum.

Figure 11.8 shows a surface of area A maintained at *absolute temperature* T_s. If the surface were black (that is, a blackbody) the rate of heat transfer would be given by[2]

$$\dot{Q} = \sigma A T_s^4 \tag{11.7}$$

where σ is the Stephan-Boltzmann constant with a value of 5.67×10^{-8} W/m^2·K^4 in the SI system. For a real surface the rate of heat transfer per unit area is less, and is given by

[2] The derivation of the rate of heat transfer per unit area for a blackbody (often referred to as *blackbody emissive power*) can be achieved by an appeal to statistical mechanics or quantum mechanics. The derivation is lengthy and tedious, and beyond the scope of our interest here.

$$\dot{Q} = \varepsilon \sigma A T_s^4, \tag{11.8}$$

where ε is the **emissivity** of the surface, a dimensionless quantity ranging from zero to one.[3] An emissivity of one corresponds to a blackbody. In general the emissivity of a surface is a function of wavelength, direction, temperature and other surface material properties. In practice, however, an overall average value of ε for all directions, wavelengths and temperatures is sought. Table 11.2 gives some representative values of ε for various surfaces.

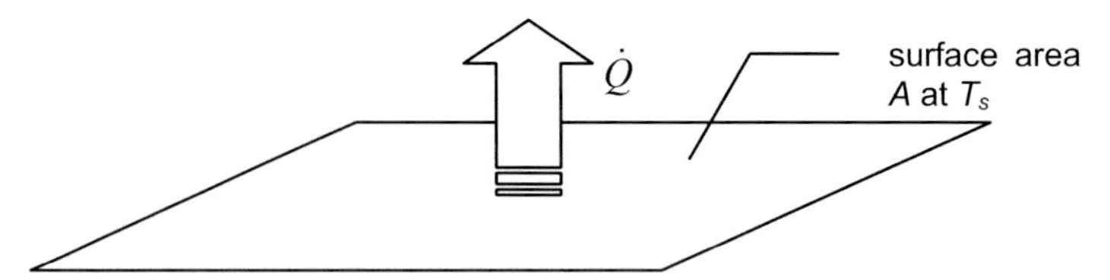

Fig. 11.8. Radiation from a solid surface

Table 11.2. Representative values of emissivities for several surfaces at 300 K

Convection	Emissivity, ε
Aluminum, polished	0.04
Black paint	0.98
Skin	0.90
Si	0.67

Equations (11.7) and (11.8) give the amount of heat transfer per unit time emitted from a perfect emitter of thermal radiation and a real surface respectively. In heat transfer in general, however, we are usually more concerned with the amount of heat transfer between things at different temperatures. In radiation this amounts to the net exchange between two surfaces at different temperatures. The shape of the surfaces and their relative orientation therefore introduce a strong dependence on geometry. However, a simple result for a very common configuration that can be used in many situations is given by

[3] The multidisciplinary aspect of MEMS once again poses potential confusion in regards to notation, as ε also refers to strain in solid mechanics, and σ to normal stress. One should be able to know which is which from context, however.

$$\dot{Q} = \varepsilon\sigma A\left(T_s^{\ 4} - T_{surr}^{\ \ \ 4}\right). \tag{11.9}$$

In Eq. (11.9) T_{surr} is the temperature of a *surface* that completely surrounds another small surface maintained at T_s. A common example where Eq. (11.9) would apply is in the net radiation exchange between an object in a room and the surface composed of the walls, ceiling and floor.

Example 11.1

Time to warm up a micro-olive

A frozen olive initially at a temperature of T_0 is dropped into a martini at a temperature T_∞. The martini is then stirred with a flamingo swizzle stick. We are interested in how the olive temperature changes with time, most notably how long it takes to warm up to T_∞.

Taking the convective heat transfer coefficient between the olive and the martini to be h_{conv} = 100 W/(m^2·°C) and the thermo-physical properties of an olive to be the same as water, find

a. the time to warm a spherical olive with a 2.0-cm diameter to within 2% of the final temperature, and
b. the time to warm a spherical olive with a 10-μm diameter to within 2% of the final temperature.

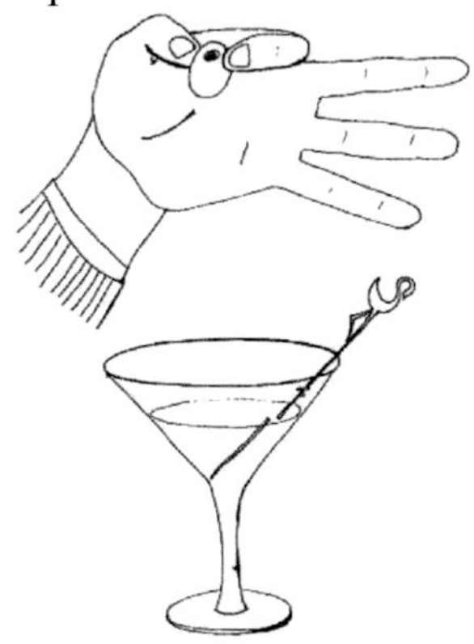

Solution

The *conservation of energy* principle will help us derive an equation describing the temperature of the olive as a function of time. In words, this equation is

[Rate of energy transport into the olive] = [Rate of energy increase of olive]

The mode of heat transfer to the olive is convection. Hence the left hand side of this equation is given by

$$\dot{Q}_{in} = hA(T_\infty - T_{olive}),$$

where A is the surface area of the olive and we have assumed that the temperature of the olive is uniform throughout, allowing us to use $T_s = T_{olive}$. (Note that since this is transport *into* the olive we use T_∞-T_{olive} rather than T_{olive}-T_∞.)

The rate of energy increase of the olive is given by the mass multiplied by the specific heat and the rate of temperature increase:

$$\frac{dE_{olive}}{dt} = m_{olive} c_p \frac{dT_{olive}}{dt}.$$

Substituting these expressions into conservation of energy and recognizing that mass is the density multiplied by the volume we have

$$h_{conv} A(T_\infty - T_{olive}) = \rho V c_p \frac{dT_{olive}}{dt},$$

where ρ and V are the olive density and volume, respectively. Rearranging,

$$\frac{dT_{olive}}{dt} + \frac{h_{conv} A}{\rho V c_p} T_{olive} = \frac{h_{conv} A}{\rho V c_p} T_\infty$$

We recognize this as the standard form of a first order linear differential equation for T_{olive} in time, the solution of which is given by

$$\frac{T_{olive} - T_\infty}{T_0 - T_\infty} = e^{-\frac{t}{\tau}}.$$

where τ is the time constant, here given by

$$\tau = \frac{\rho V c_p}{h_{conv} A}.$$

The exponential form of the solution suggests that olive never reaches T_∞ but only asymptotically approaches it. This, of course, is a limitation of our simple model.

In such first order systems, 98% of the final steady state response has been achieved by four time constants. Here, the olive temperature has increased by 98% of the initial temperature difference. For the 2.0-cm olive, the time constant is

$$\tau = \frac{(850\frac{\text{kg}}{\text{m}^3})(\frac{4}{3}\pi \cdot (0.01\,\text{m})^2)(1780\frac{\text{J}}{\text{kg}\cdot\text{K}})}{(100\frac{\text{W}}{\text{m}^2\cdot\text{K}})(4\pi \cdot (0.01)^2)} = 50.4\,\text{s},$$

where we have used the properties of water for the olive. And so within 4×50.4 = 202 s (about 3 ½ minutes) the 2.0-cm olive has warmed to within 2% of its final steady state temperature.

Recalculating for the micro-olive, however, gives a much different result:

$$\tau = \frac{(850\frac{\text{kg}}{\text{m}^3})(\frac{4}{3}\pi \cdot (5\times10^{-6}\,\text{m})^2)(1780\frac{\text{J}}{\text{kg}\cdot\text{K}})}{(100\frac{\text{W}}{\text{m}^2\cdot\text{K}})(4\pi \cdot (5\times10^{-6})^2)} = 0.025\,\text{s},$$

so that within only 4×0.025 s = 0.10 s the olive has warmed to the same temperature! ◄

Example 11.1 shows a favorable scaling associated with micro-scale heat transfer; that is, much faster fast response times than at the macro-scale. In the example, however, we assumed that the entire olive was characterized by a single temperature and ignored any potential temperature gradients within. Was this a valid assumption?

To answer this question we calculate a dimensionless quantity called the **Biot number**, which represents the relative resistance to internal conduction within a body to the external resistance to convection.

$$Bi = \frac{\text{Resistance to internal conduction}}{\text{Resistance to external convection}} \tag{11.10}$$

Using the simple expressions for conduction and convection resistance given by Eqs. (11.2) and (11.6),

$$Bi = \frac{L_{char}/(\kappa A)}{1/(h_{conv}A)} = \frac{h_{conv}L_{char}}{\kappa}, \tag{11.11}$$

where L_{char} is some characteristic dimension of the body in question, usually taken to be the volume divided by the surface area. When $Bi < 0.1$, there is little resistance to internal conduction so that temperature gradients do not form, allowing us to treat the body in question as a "lump" all at one temperature. It is left as an exercise to show that the Biot numbers for the two olives in Example 11.1 are 0.95 and 0.00048 for the 2-cm olive and the 10-μm olive, respectively. This validates our model in the case of the micro-olive, but casts doubt on the case of the larger olive model.

From Eq. (11.11) we see that Bi scales with L. This leads to another major favorable scaling of heat transfer, namely that temperature gradients don't occur as readily in the microworld. This in turn leads to less thermally-induced stress, the ability to heat MEMS components to high temperatures without breaking and the ability of MEMS components to endure high rates of thermal cycling with high reliability.

11.3 Case study: Hot-arm actuator

We saw in Chapter 5 that thermal effects can lead to stress in MEMS elements, causing them to bend. Then in Chapter 7 we explored ways of exploiting this phenomenon when we a looked at the operating principles of a number of thermal actuators. In the current section we will model one of those actuators, a hot-arm actuator, in various levels of detail.

We first review the operating principle of a hot arm actuator. In such an actuator, the differences in geometry of its different components cause them to expand at different rates when heated. The resulting stress causes actuator deflection.

Figure 11.9 gives a schematic of a typical hot arm actuator. Voltage is applied across the two contact pads at the left of the figure, both of which are anchored to the substrate. Current flows through the long thin part at the top of the figure, the *hot arm*, and then the thicker part at the bottom, the *cold arm*. The larger electrical resistance of the hot arm causes it to dissipate more thermal energy than the cold arm, and therefore to experience a larger increase in temperature. This larger temperature coaxes the hot arm to expand more than the cold arm. However, its motion is restricted by its being attached to the cold arm, inducing a mechanical stress that causes the whole assembly to bend.

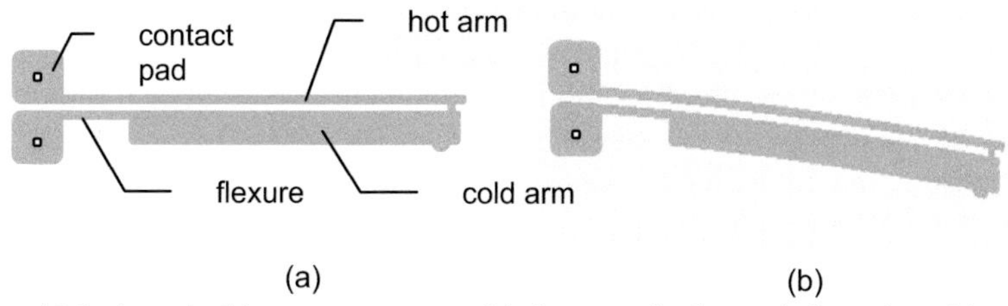

Fig. 11.9. A typical hot arm actuator: (a) Actuator in the undeflected position; (b) Actuator after a voltage is applied across the contact pads.

The flexure is the thin piece that connects the cold arm to the contact pad. Like the hot arm, the thin flexure also heats up, but it is usually too short for the temperature to increase significantly over the cold arm. Rather, the flexure ideally operates as a low-stiffness extension of the otherwise stiff cold arm, allowing the entire actuator to bend more than it would without it. If it is too short, however, the effective stiffness of the actuator will remain high. On the other hand, making it too long results in too much heating, thereby reducing the effectiveness of the hot arm. The flexure's ideal dimensions for optimum performance are therefore important questions in the design of a hot arm actuator. What's more, the length of the flexure tends to affect the maximum force and maximum deflection in a nonlinear fashion.

Many things make modeling a hot-arm actuator perhaps the most challenging modeling task we've undertaken so far. First of all, three energy domains are present - electrical, thermal and mechanical. Secondly, the geometry is somewhat complicated, with two differently sized arms of different shapes affixed to each other. And after they bend, that geometry changes. Last but not least, physical parameters are not uniform over space, necessitating a distributed parameter model for the most accuracy.

In the sections that follow we introduce three models for the actuator, each one increasing in complexity over the previous. We start with a lumped approach, move on to a more detailed distributed parameter model, and finish with a full-blown computer model using the commercial software package COMSOL.

11.3.1 Lumped element model

For this model we will assume that only two temperatures characterize the entire device, one for the hot arm and one for the cold arm. Furthermore, we will neglect the contributions of the portions of the hot and cold arms connecting them to the contact pads and the structure that connects the two arms to each other. This will allow us to assume that electric current and thermal energy flow through arms of equal length that vary only in their cross sectional areas, greatly simplifying the geometry, and also removing many of the previously discussed complications of the flexure. Lastly we will assume that all material properties are constant. With this last assumption, the hot and cold arms each have one electrical resistance. Figure 11.10 gives a schematic of the simplified actuator.

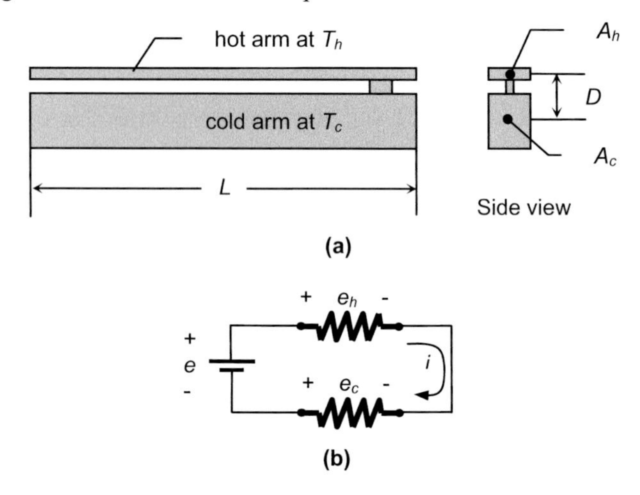

Fig. 11.10. (a) Geometry of the simplified actuator; (b) Electrical circuit with hot and cold arms as resistors

For the simplified actuator, the hot arm and the cold arm have electrical resistances of, respectively,

$$R_h = \rho \frac{L}{A_h} \tag{11.12}$$

and

$$R_c = \rho \frac{L}{A_c}, \tag{11.13}$$

where ρ is the electrical resistivity of the actuator material. The two resistors are connected in series, so that the electrical current flowing through each is the same. Hence Ohm's law requires

$$\frac{e}{R_h + R_c} = \frac{e_h}{R_h} = \frac{e_c}{R_c}, \tag{11.14}$$

where e is the total voltage applied to the actuator and e_h and e_c are the voltage drops across the hot and cold arms, respectively. Solving for e_h and substituting Eqs.(11.12) and (11.13),

$$e_h = \frac{R_h}{R_h + R_c} e = \frac{\rho(L/A_h)}{\rho(L/A_h) + \rho(L/A_c)} e \tag{11.15}$$

$$= \frac{A_c}{A_h + A_c} e .$$

As we have already seen, time scales in microscale heat transfer situations are very small. As such, the transient response of the actuator occurs very quickly and a steady state is reached almost immediately after the voltage is applied. At steady state conservation of energy requires that all of the electrical energy dissipated in the hot arm be removed via convection from the hot arm surface area:

$$\dot{W}_{in,elec} = \dot{Q}_{conv,out} . \tag{11.16}$$

Recognizing that dissipated power is e_h^2/R_h and substituting the appropriate relation for convective heat transfer

$$\frac{e_h^2}{R_h} = h_{conv} A_{h,surf} (T_h - T_{amb}). \tag{11.17}$$

where $A_{h,surf}$ is the surface area of the hot arm.[4] Making use of Eqs. (11.12) and (11.15) and substituting the appropriate expression for $A_{h,surf}$,

$$\frac{A_h}{\rho L}\left(\frac{A_c}{A_h + A_c}\right)^2 e^2 = h_{conv} \mathscr{P}_h L (T_h - T_{amb}). \tag{11.18}$$

where $\mathscr{P}_h$ is the perimeter of the hot arm cross section. Solving for T_h gives

[4] The appropriate area for convective heat transfer is not the cross sectional area A_h, but rather the area over which the solid surface makes contact with the surrounding air, $\mathscr{P}_h L$. One of the most common mistakes students new to heat transfer make, as well as one of the most fatal, is not to visualize the area over which heat transfer is taking place, thereby confusing expressions for it.

$$T_h = \frac{A_h}{h_{conv}\mathcal{P}_h \rho L^2}\left(\frac{A_c}{A_h + A_c}\right)^2 e^2 + T_{amb}. \tag{11.19}$$

The analysis is identical for the cold arm. Hence the expression for the cold arm temperature is given by

$$T_c = \frac{A_c}{h_{conv}\mathcal{P}_c \rho L^2}\left(\frac{A_h}{A_h + A_c}\right)^2 e^2 + T_{amb}. \tag{11.20}$$

Equations (11.19) and (11.20) give first order approximations for the temperatures of the hot and cold arms, respectively. Now that we have these, we can move on to a solid mechanics model.

For a first attempt at a solid mechanics model we will assume that the actuator behaves in a matter quite similar to the piezoelectric actuator modeled in Chapter 11. The thicker cold arm restrains the expansion of the thinner cold arm, inducing a stress in the hot arm. That stress in turn creates a bending moment in the cold arm, causing it to deflect. (Fig. 11.11.)

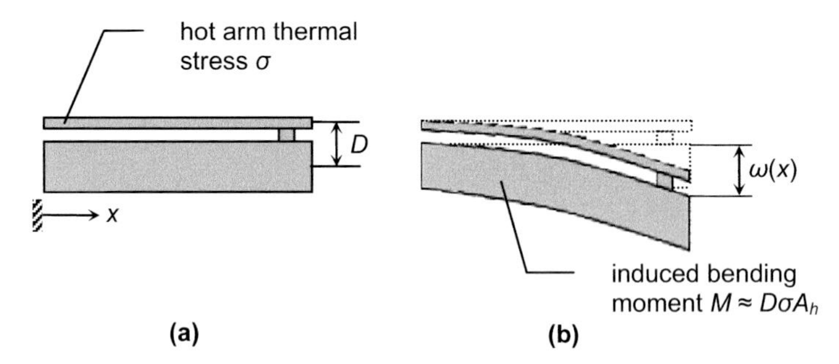

Fig. 11.11. A thermal stress of σ in the hot arm (a) creates a bending moment in the cold arm (b) of approximately $M \approx D\sigma A_h$, causing deflection.

For a hot arm initially at ambient temperature T_{amb}, heating to the temperature T_h should induce a thermal strain of

$$\varepsilon_h = \alpha_{T,h}(T_h - T_{amb}). \tag{11.21}$$

The cold arm experiences a thermal strain of

$$\varepsilon_c = \alpha_{T,c}(T_c - T_{amb}). \tag{11.22}$$

Since the cold arm is thicker than the hot arm to which it is attached, as a first order estimate we can assume that the actual strain in both arms is that of the cold arm. Assuming uniaxial stress/strain, this causes a stress in the hot arm equal to Young Modulus multiplied by the difference in thermal strains of the two arms:

$$\sigma = E(\varepsilon_c - \varepsilon_h) = E\left[\alpha_T(T_c - T_{amb}) - \alpha_T(T_h - T_{amb})\right] \tag{11.23}$$
$$= E\alpha_T(T_c - T_h).$$

The induced bending moment in the cold arm is now approximated as

$$M = D\sigma A_h, \tag{11.24}$$

D being the distance from the neutral axis of the cold arm to the centerline of the hot arm.

The beam deflection equation, Eq. (10.22), is now used with the moment found in Eq. (11.24):

$$\frac{d^2\omega}{dx^2} = -\frac{M}{EI} = -\frac{D\sigma A_h}{EI}. \tag{11.25}$$

For boundary conditions appropriate for a cantilever, namely that deflection and its first derivative are both zero at $x = 0$, integrating Eq. (11.25) twice gives

$$\omega(x) = -\frac{D\sigma A_h x^2}{2EI}. \tag{11.26}$$

Finally, substituting expressions for σ, T_h and T_c, we arrive at

$$\omega(x) = \frac{D\alpha_T A_h e^2 x^2}{2h_{conv}\rho L^2 I}\left[\frac{A_h}{P_h}\left(\frac{A_c}{A_h + A_c}\right)^2 - \frac{A_c}{P_c}\left(\frac{A_h}{A_h + A_c}\right)^2\right]. \tag{11.27}$$

Equation (11.27) is a first order approximation for the deflection of the actuator only. Many of the assumptions that went into its derivation are suspect, namely that the actuator has only two temperatures, that the cold arm is thick enough compared to the hot arm to validate the bending moment as calculated by Eq. (11.24), etc. And so a valid question is, “Of what use is Eq. (11.27)?”

The answer is plenty. The model with its assumptions gives a closed form analytic solution, from which it is easy to discern trends, even if the numbers that come out of it do not agree particularly well with experiment. For example, Eq. (11.27) shows us that larger deflections are effected by larger voltages and coefficients of thermal expansion. Both of these trends are to be expected. However, Eq. (11.27) also suggests that larger electrical resistivities produce smaller deflections, and that the deflection does not depend on Young's Modulus, both of which are trends that are not obvious. The model also begins to give us some insight on how the relative cross sectional areas of the two arms along with the geometry in general affects the deflection.

Furthermore, when more detailed models are attempted, how do we know whether their results are reasonable? The "ballpark" estimates given by simpler models give us a way of gauging this. What's more, it has been the authors' experience that such compact models often do a better job of predicting actual numbers than one would initially think. Besides, one of the sources of uncertainty in Eq. (11.27) is what value to use for h_{conv}, which is a problem that does not go away even when using the most detailed and sophisticated numerical computer models.

11.3.2 Distributed parameter model

For a model with a bit more detail, we relax the assumption that the actuator is characterized by only two temperatures. One way to do this is to go to three temperatures, one for the entire length of the hot arm from its contact pad to the actuator tip, one for the thick section of the cold arm, and then one more for the section of the cold arm connecting its thicker portion to its contact pad. The analysis for this three temperature model would proceed in the same manner as for the two temperature model, hopefully providing more accurate temperatures.

In reality, we know that the actuator is characterized by neither two nor three temperatures, but rather by a continuously varying spatial temperature distribution. A higher order model would therefore be a distributed parameter model in which we solve for the temperature dependence on location, and then use the results to find strain and the consequential deflection.

For this model we will set up one equation to be used for solving for the distribution of temperature along the hot arm and another for the cold arm. We will assume that the flow of thermal energy is one dimensional along the direction of the electric current flow. As such, it is convenient to use one coordinate for position along the hot arm, x_h, and another for the cold arm, x_c. (Fig. 11.12.)

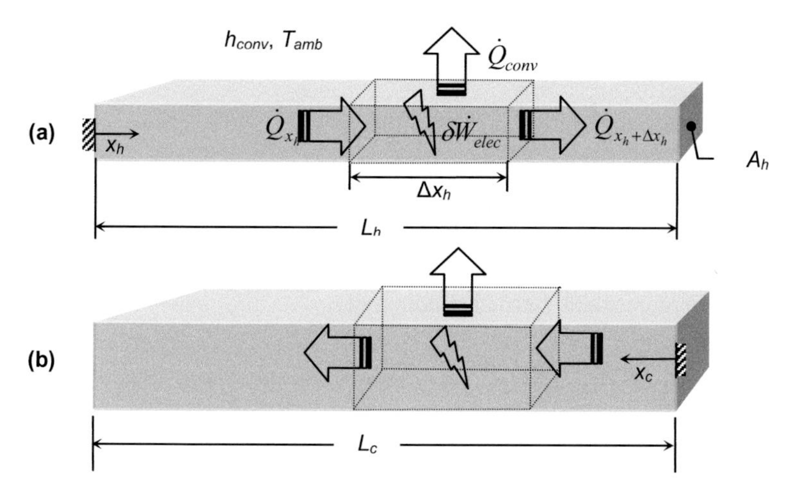

Fig. 11.12. Systems used to derive temperature distributions for the (a) hot arm and (b) the cold arm.

Figure 11.12 also shows a small element of length Δx_h within the hot arm. Thermal energy is conducted to and from the element on the x_h face and the $x_h+\Delta x_h$ face, respectively, and convected to the surrounding air from the surface area defined by $\mathscr{P}_h\Delta x$. Assuming constant electrical resistivity, the electrical energy being dissipated by the element is the fraction $\Delta x/L$ of the energy dissipated by the entire hot arm. With these assumptions, conservation of energy for the element is given by

$$\dot{Q}_{x_h} - \dot{Q}_{x_h+\Delta x_h} - \dot{Q}_{conv} = \delta\dot{W}_{elec} \tag{11.28}$$

$$\dot{Q}_{x_h} - \dot{Q}_{x_h+\Delta x_h} - h_{conv}\mathscr{P}_h\Delta x_h\left(T(x_h) - T_{amb}\right) = \frac{\Delta x_h}{L_h}\frac{e_h^{\;2}}{R_h}. \tag{11.29}$$

Dividing by Δx,

$$\frac{\dot{Q}_{x_h} - \dot{Q}_{x_h+\Delta x_h}}{\Delta x_h} - h_{conv}\mathscr{P}_h\left(T(x_h) - T_{amb}\right) = \frac{1}{L_h}\frac{e_h^{\;2}}{R_h}. \tag{11.30}$$

In the limit as $\Delta x \rightarrow 0$,

$$-\frac{d\dot{Q}_{x_h}}{dx_h} - h_{conv}\mathscr{P}_h\left(T(x_h) - T_{amb}\right) = \frac{1}{L_h}\frac{e_h^{\;2}}{R_h}. \tag{11.31}$$

Using Eq. (11.3) for the conduction term and derived expressions for e_h and R_h from the previous model

$$-\frac{d}{dx_h}\left(-\kappa\frac{dT}{dx_h}\right)-h_{conv}\mathcal{P}_h\left(T(x_h)-T_{amb}\right)^2 \tag{11.32}$$

$$=\frac{1}{L_h}\frac{A_h}{\rho L_h}\left(\frac{A_c}{A_h+A_c}\right)^2 e\,.$$

Finally, for a constant thermal conductivity,

$$\kappa\frac{d^2T}{dx_h^{\ 2}}-h_{conv}\mathcal{P}_h\left(T(x_h)-T_{amb}\right)=\frac{A_h}{\rho L_h^{\ 2}}\left(\frac{A_c}{A_h+A_c}\right)^2 e^2\,. \tag{11.33}$$

The solution of Eq. (11.33) is contingent upon the appropriate boundary conditions. Since the equation is second order, we require two boundary conditions involving T. One commonly used boundary condition comes from assuming that the contact pad end of the hot arm remains at ambient temperature, since the surface area of the pad is large enough to dissipate energy effectively to the surrounding air. This amounts to

$$T(x_h=0)=T_{amb}. \tag{11.34}$$

The second boundary condition is a bit trickier. At the other end of the hot arm, the temperature must be continuous. That is, the temperature of the hot arm must be the same as the temperature of the cold arm at that point. Also, the conduction out of the hot arm must all go into the cold arm. In equation form,

$$T(x_h=L_h)=T(x_c=0) \tag{11.35}$$

and

$$-\kappa A_h\frac{dT(x_h=L_h)}{dx_h}=-\left(-\kappa A_c\frac{dT(x_c=0)}{dx_c}\right). \tag{11.36}$$

This gives us one too many boundary conditions, however. We must choose one or the other. But once we do, we still don't know the value of temperature at $x_c = 0$, or its first derivative either.

The remedy to this comes from setting up the differential equation for the cold arm and applying boundary conditions at its two ends. The analysis is similar to that for the hot arm, resulting in a differential equation for conservation of energy given by

$$\kappa \frac{d^2T}{dx_c^{\ 2}} - h_{conv}\mathcal{P}_c\left(T(x_c) - T_{amb}\right) = \frac{1}{L}\frac{A_c}{\rho L_c}\left(\frac{A_h}{A_h + A_c}\right)^2 e^2 . \tag{11.37}$$

We choose either Eq. (11.35) or Eq. (11.36) as the boundary condition for $x_c = 0$, whichever one is left over from the hot arm. The remaining boundary condition is most likely ambient temperature at the contact pad:

$$T(x{=}L_c) = T_{amb}. \tag{11.38}$$

Together Eqs. (11.33)-(11.38) form a complete set of equations and unknowns. Equations (11.33) and (11.37) are second order, non-homogenous, linear differential equations, for which there are many analytical solution methods available. Numerical schemes can also be employed.

However Eqs. (11.33)-(11.38) are solved, they give us an estimate for the temperature distribution only. Ultimately we are interested in the deflection of the actuator. The next step is to use the knowledge gained form the temperature distributions to find the thermal strain. If the entire hot arm were at one temperature, we could use Eq. (11.21) to calculate the strain as it heats up from T_{amb}:

$$\varepsilon_h = \alpha_T (T_h - T_{amb}) . \tag{11.21}$$

However, there is not just one T_h for the entire hot arm, but rather a continuously varying temperature. We must therefore use an integral form of Eq. (11.21):

$$\varepsilon_h = \int_{x_h=0}^{L_h} \alpha_T \frac{d(T - T_{amb})}{dx_h} dx_h = \int_{x_h=0}^{L_h} \alpha_T \frac{dT}{dx_h} dx_h \,,. \tag{11.39}$$

where dT/dx_h is known from the thermal analysis. The analogous expression for the cold arm is

$$\varepsilon_h = \int_{x_c=0}^{L_c} \alpha_T \frac{dT}{dx_c} dx_c . \tag{11.40}$$

Once these strains are calculated, the strain in the hot arm can be approximated in a similar fashion as in our first model,

$$\sigma = E(\varepsilon_c - \varepsilon_h), \tag{11.41}$$

which in turn can be used to estimate the deflection using Eq. (11.26):

$$\omega(x_h) = -\frac{D\sigma A_h x_h^{\ 2}}{2EI}. \tag{11.42}$$

As you no doubt have become aware, the currently discussed model is complicated, and it is no trivial matter to solve the equations involved. What's more, even though we now have a distributed parameter model for temperature, we have assumed that the electrical resistivity is constant throughout. In reality, the resistivity is a function of temperature, and for the large temperature changes typically encountered in hot arm actuators, this assumption usually needs to be revised. As we will see, adding this improvement to the model will greatly increase the complexity of the resulting equations. Thus, we will only outline the model here.

Variable resistivity can be accounted for by using an equation of the form

$$\rho(T) = \rho_0 + a(T - T_0), \tag{11.43}$$

where a is a constant and ρ_0 is the resistivity at some reference temperature T_0. Simple substitution of Eq. (11.43) into Eqs. (11.33) and (11.37) is not possible, however, since the varying resistivity no longer results in a linear drop in voltage across the arms. Rather, differential forms of the electrical energy dissipation terms (e/R^2 terms) need to be used. For the hot arm, this becomes

$$\dot{Q}_{x_h} - \dot{Q}_{x_h+\Delta x_h} - h_{conv}\mathcal{P}_h \Delta x_h \left(T(x_h) - T_{amb}\right) \tag{11.44}$$
$$= \frac{\left(e_x - (e_{x+\Delta x})\right)^2}{\frac{\rho(T)\Delta x_h}{A_h}} = -\frac{A_h \Delta e^2}{\rho(T)\Delta x_h}.$$

Dividing by Δx_h and taking the limit as $\Delta x_h \rightarrow 0$

$$-\frac{d\dot{Q}_{x_h}}{dx_h} - h_{conv}\mathcal{P}_h\left(T(x_h) - T_{amb}\right) = -\frac{A_h}{\rho(T)}\left(\frac{de}{dx_h}\right)^2. \tag{11.45}$$

Substituting Eq. (11.3) for $\dot{Q}_{x_h}$ and Eq. (11.43) for ρ,

$$\kappa\frac{d^2T}{dx_h^{\ 2}} - h_{conv}\mathcal{P}_h\left(T(x_h) - T_{amb}\right) = -\frac{A_h}{\rho_0 + a(T - T_0)}\left(\frac{de}{dx_h}\right)^2. \tag{11.46}$$

Again, the analysis is the same for the cold arm:

$$\kappa \frac{d^2T}{dx_c^{\ 2}} - h_{conv}\mathcal{P}_c\left(T(x_c) - T_{amb}\right) = -\frac{A_c}{\rho_0 + a(T - T_0)}\left(\frac{de}{dx_c}\right)^2. \qquad (11.47)$$

From the two last equations we see that solution procedure has become significantly more difficult than in the constant property case. Specifically, the values of the terms on the right hand side cannot be known without an additional appeal to the electrical equations. That is, the thermal and the electrical domains are coupled. What's more, we will see that the electrical equations are not so simple either.

To complete the equation set we need Ohm's Law in differential form. For a differential voltage drop somewhere within the hot arm, the equivalent of $e = iR$ is

$$de = i\frac{\rho(T)dx_h}{A_h}, \qquad (11.48)$$

Solving for the current i and using Eq. (11.43) for ρ,

$$i = \frac{A_h}{\rho_0 + a(T - T_0)}\frac{de}{dx_h}. \qquad (11.49)$$

Equivalently for the cold arm,

$$i = \frac{A_c}{\rho_0 + a(T - T_0)}\frac{de}{dx_c}. \qquad (11.50)$$

Since the model is for steady state and the arms are in series, the value of the current through each as given by Eqs. (11.49) and (11.50) must be the same constant.

Finally, we look at the total voltage applied across the actuator. This is no longer a simple sum of two discrete voltage drops across the two arms, but an integral:

$$e = \int_{x_h=0}^{L_h}\frac{de}{dx_h}dx_h + \int_{x_c=0}^{L_c}\frac{de}{dx_c}dx_c. \qquad (11.51)$$

Equations (11.46)-(11.47) and (11.49)-(11.51), together with the previously given boundary conditions for temperature, form a complete equations set that can be solved for the distributions of temperature and voltage. Due to the nonlinearity of the equations and their high degree of coupling, however, this is an onerous task indeed. We would almost certainly need to resort to computational schemes for the solution. And imagine the added complexity of also accounting for a variable thermal conductivity!

Even after the ardor of solving for the temperature distribution, we still need to add the model for calculating the actuator deflection. Given the effort required to solve for the temperature distribution, it seems counterproductive also to continue with the simplified deflection model employed so far. That is to say, if we must exert large amounts of effort to solve for a better estimate of temperature, we might as well exert equal amounts of effort into improving the solid mechanics model as well. This is the type of model we will explore in the next section.

11.3.3 FEA model

Finite element analysis (FEA) computer codes are powerful tools used to simulate physical systems of high complexity. In such codes, the differential forms of the conservation laws along with the appropriate constitutive relations are solved over a region of space defined by the user. In relation to our current example, an FEA code can be used to solve the differential form of conservation of energy (the conservation law) assuming Fourier's law of conduction (the constitutive relation) in order to find the temperature distribution within the heat actuator. Quite complex geometries are often possible, including three dimensional ones, which is one reason why such software packages have become so popular.

The basic idea of FEA (also called FEM for *finite element methods*) is to approximate the solutions to partial differential equations by solving for numerical values of physical parameters at discrete points in space. To do this, the partial differential equations are reduced to either algebraic equations or ordinary differential equations. The network connecting all the discrete points used in an analysis is called a **mesh**.

Figure 11.13 shows a screenshot of a hot-arm actuator drawn in COMSOL Multiphysics®, a commercially available FEA package with MEMS modules capable of modeling physical systems and processes coupled across several energy domains. The dimensions of the actuator are given in the figure caption. The thickness throughout is 2 μm.

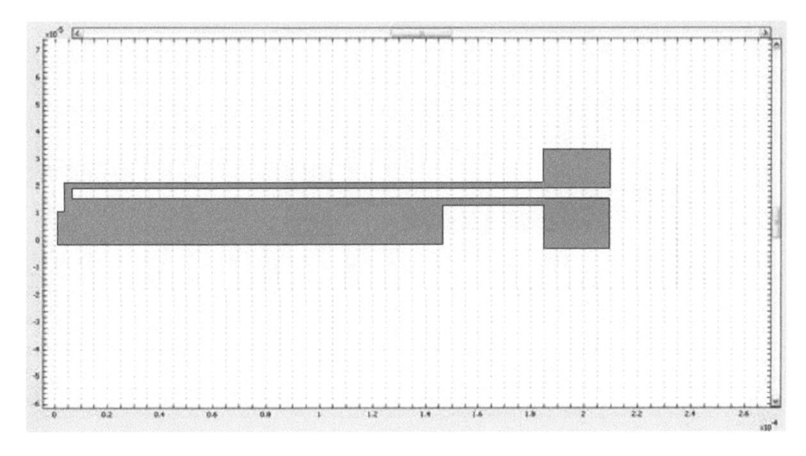

Fig. 11.13. Hot-arm actuator drawn in COMSOL Multiphysics®. The lengths of the hot arm, cold arm and flexure are 185 μm, 145 μm and 40 μm, respectively. The hot arm and flexure are both 2 μm wide, whereas the cold arm is 16 μm wide. The spacing between the hot and cold arms is 2μm. All components are 2 μm thick. (COMSOL and COMSOL Multiphysics are registered trademarks of COMSOL AB)

Figures 11.14 and 11.5 give the spatial variation of electric potential and temperature as calculated by COMSOL Multiphysics®, respectively. The actuator material is assumed to be copper with a variable electrical resistivity. The assumed boundary conditions for heat transfer are constant temperatures of 300 K at the contact pads and convection heat transfer to air at 293 K, with a heat transfer coefficient of 240 $W/m^2 \cdot °C$ for all other surfaces. There are 2148 elements in the mesh. The applied voltage is 1 V.

Note from Fig. 11.14 that roughly half the total voltage drop has occurred midway through the hot arm, a trend which is to be expected based on the design of the actuator. Correspondingly in Fig. 11.15 we see that the largest temperature of 482 K is realized about midway down the hot arm. The thin flexure also experiences a large temperature of approximately 460 K. The majority of the cold arm, however, is within 40 degrees of the contact pad temperature of 300 K.

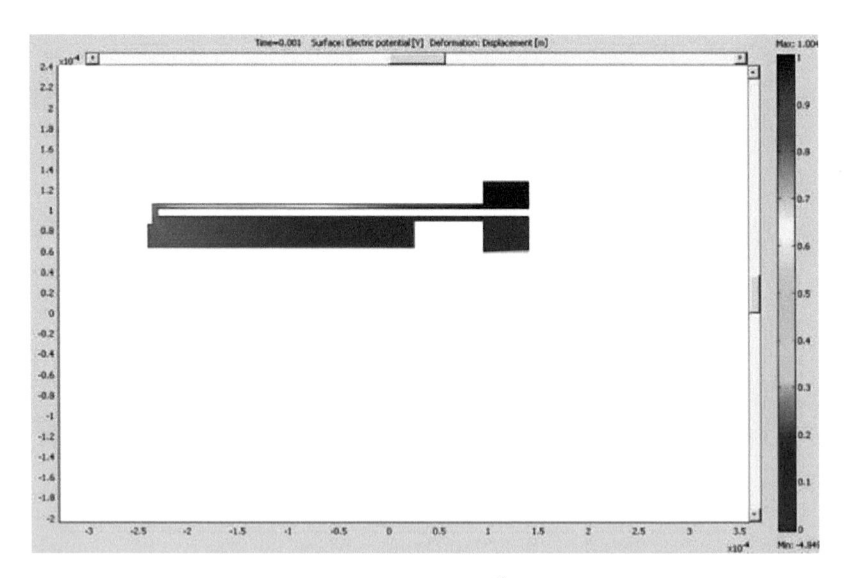

Fig. 11.14. Output from COMSOL Multiphysics® spatial variation of electric potential. (COMSOL and COMSOL Multiphysics are registered trademarks of COMSOL AB)

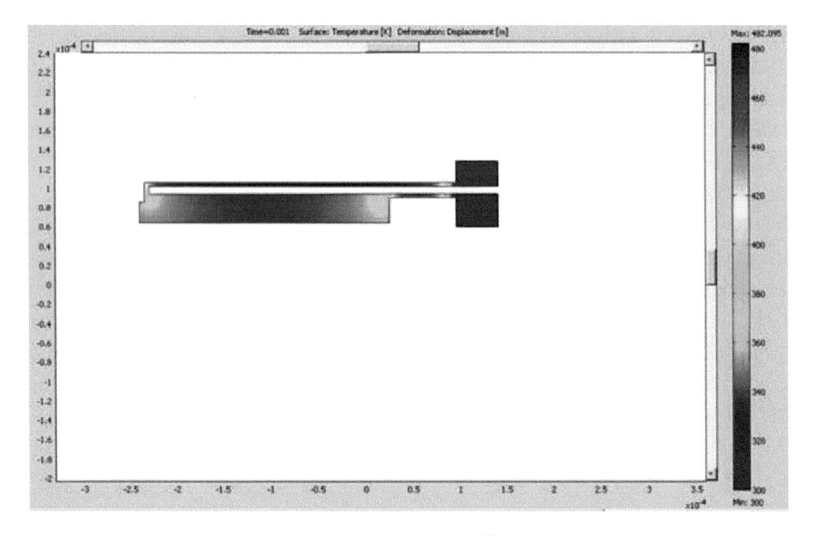

Fig. 11.15. Output from COMSOL Multiphysics® giving spatial variation of temperature. (COMSOL and COMSOL Multiphysics are registered trademarks of COMSOL AB)

Part of the power of a software package such as COMSOL Multiphysics® in modeling MEMS devices comes in its ability to solve equations in different energy domains simultaneously. In sections 11.3.1 and 11.3.2, in contrast, we assumed throughout that the thermal domain and the mechanical domain were decoupled. That is, once the temperature distribution was known, we used it to find thermal strain. We then used elastic theory to relate the strain to stress, and finally to the resulting deflection. In reality, however, the geometry of the actuator changes as it deforms, which affects the thermal distribution, which in turn affects the strain, which affects the geometry, and so on.

FEA software can take this type of coupling into account with relative ease. Indeed, in the previous two figures the equations governing the distribution of electric potential and temperature are coupled via the temperature-dependent electrical resistivity. The results of an analysis *additionally* coupling the mechanical energy domains are given in Fig. 11.16, in which the actuator deflection is found. The assumed boundary conditions for the solid mechanics equations are zero deflection and zero slope where the hot arm and the flexure meet the contact pads. The boundary conditions for electric potential and temperature are the same as in Figs. 11.14 and 11.15.

The scale on the left hand side of Fig. 11.16 has units of 10^{-4} m, showing a total tip deflection of about 22 μm. Notice that coupling the mechanical and thermal energy domains has resulted in a reduction of the maximum temperature from 482 K in Fig. 11.15 to 475 K here.

In this section we have seen the power of doing detailed distributed parameter modeling in several different energy domains at once. Though COMSOL Multiphysics® and similar software packages are impressive in their breadth and number of features, their use in no way guarantees accurate results. Indeed, the problem with computer codes is that they do exactly what you tell them. Enter a poor choice for a convective heat transfer coefficient as a boundary condition, or ignore the effects of thermal radiation when they may be appreciable, and the output from FEA codes may look pretty, but have very little meaning. It is therefore usually desirable to do modeling at several different levels as outlined in this chapter, so that the important trends and variables are discerned well before investing the time and effort into large scale numerical models.

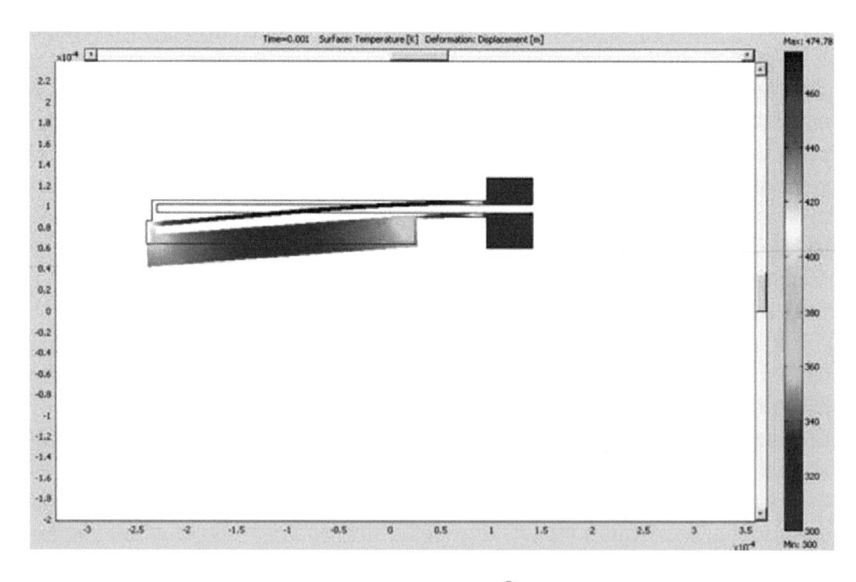

Fig. 11.16. Output from COMSOL Multiphysics® giving deflection of the hot-arm actuator. . (COMSOL and COMSOL Multiphysics are registered trademarks of COMSOL AB)

Essay: Effect of Scale on Thermal Properties

Michael S. Moorhead
Department of Mechanical Engineering, Rose-Hulman Institute of Technology

In the study of heat transfer in MEMS devices, we are often interested in conduction, as it is the mechanism of transfer of thermal energy within a solid. This mode of heat transfer is summarized by Fourier's Law, which states that the rate of heat transfer is proportional to the temperature gradient in the material (Eqs 11.3 and 11.4). The constant of proportionality in this relationship is known as the thermal conductivity, κ. Materials that transport heat very effectively are good thermal conductors and have a high value of κ. Gold (Au), for instance, has a thermal conductivity of κ = 317 W/m·K. Materials which do not transport heat very effectively, on the other hand, are good thermal insulators and have a low value of κ. Silicon dioxide (SiO_2) has a thermal conductivity of only κ = 1.38 W/m·K.

Just how is heat transported in solids? There are actually two answers to that question. The first is via electrons. Some electrons in a material are not bound to a particular atom. These are known as free electrons. As they move about in a solid, they carry *both* electrical charge *and* thermal energy with them. It is therefore no coincidence that many of the best electrical conductors are also excellent thermal conductors (e.g. gold). The second answer is *phonons*. Phonons are vibrations in the crystal lattice that makes up a solid. The best analogy is to think of the atoms in Figures 2.2 and 2.3 as billiard balls that are connected by Slinkies®. If you were to perturb one of the atoms, you would send a wave of energy flowing through the lattice. These vibrations have the ability to transport thermal energy. The portion of thermal conductivity due to phonons helps us understand why certain materials (e.g. diamond) are very good electrical insulators, while also being very good thermal conductors (κ = 2300 W/m·K). From this point on, we will refer to both electrons and phonons simply as energy carriers.

What does this have to do with size? Energy carriers are characterized by the mean free path, λ. The mean free path is the average distance that an energy carrier will travel before encountering something else (i.e. another energy carrier or a boundary). In bulk materials, the mean free path may be on the order of $\lambda \sim$ 10-100 nm. In such a circumstance, it is a good bet that an energy carrier will collide with another energy carrier before it encounters anything else. This is one reason why thermal conductivity in bulk materials is an *intensive property*. That is, it doesn't matter how big the material is. One cubic centimeter of gold has the same thermal conductivity as cubic meter of gold. In the realm of micro and nano-fabrication, however, it is not uncommon to have length scales that are smaller than the mean free path of the energy carriers. For example, Intel's current 65 nm chip architecture uses a SiO_2 film which is only 1.2 nm thick. In such a case, it is very likely that an energy carrier will encounter a boundary before it is able to transfer its energy to another energy carrier. This has the effect of reducing the thermal conductivity of the material. In some cases, this decrease has been very dramatic.

Measuring the thermal conductivity at these small scales is no easy task. Traditional methods of determining thermal conductivity involve measuring temperature gradients through a material with a known heat transfer rate. Unfortunately, there are no nano-thermometers that we can use to make such measurements. Researchers have therefore resorted to laser-based techniques and sophisticated computer models to help determine thermal conductivity in thin films and multi-layered materials. For the foreseeable future, it won't be as simple as looking it up in the back of the book.

(Note: Given room temperature bulk thermal conductivities obtained from Incropera et al, Fundamentals of Heat and Mass Transfer, 6th Ed.)

References and suggested reading

Comtois JH, Bright VM (1997) Applications for surface-micromachined polysilicon thermal actuators and arrays. Sensors and Actuators, 58.1:19-25

Incropera FP, DeWitt DP (2007) Fundamentals of Heat and Mass Transfer, 6th edn. Wiley, Hoboken, NJ

Lee BJ, Zhang, ZM, Early EA, DeWitt DP, Tsai BK (2005) Modeling radiative properties of silicon with coatings and comparison with reflectance measurements. J. Thermophysics and Heat Transfer, 19.1:558-65

Madou MJ (2002) Fundamentals of Microfabrication, The Art and Science of Miniaturization, 2nd edn. CRC Press, New York

Mankame ND, Ananthasuresh GK (2001) Comprehensive thermal modeling an characterization of an electro-thermal-compliant microactuator. J. Micromech. Microeng. 11:1-11

Read BC, Bright VM, Comtois JH (1995) Mechanical and optical characterization of thermal microactuators fabricated in a CMOS process. In: Proc. SPIE, vol. 2642 pp. 22-32

Yan D, Khajepour A, Mansour R (2003) Modeling of two-hot-arm horizontal thermal actuator. J. Micromech. Microeng. 13:312–322

Questions and problems

11.1 What are the three modes of heat transfer?

11.2 List two ways radiation is different from the other modes of heat transfer.

11.3 In general, which mode of heat transfer is more efficient, convection or conduction? Why?

11.4 Microwave ovens emit radiation. Is it *thermal* radiation? Why or why not?

11.5 Explain the difference between natural and forced convection.

11.6 List two advantages of small scale heat transfer.

11.7 In the thermal resistance model, what is the analog of voltage? current? resistance?

11.8 What is the interpretation of the Biot number? Do you expect small or large values of Biot number for most MEMS devices?

11.9 Explain the operating principle behind a hot-arm actuator. Specifically address how the thermal stresses come about, their sign (tension and/or compression) and how this leads to motion.

11.10 When is a lumped element method valid from a heat transfer perspective?

11.11 Give two advantages of closed form, analytic models over numerical models. Give two advantages of numerical models over closed form, analytic models.

11.12 Give one way you might account for non-constant properties within a hot-arm actuator while still using the lumped element model in 11.3.1.

11.13 Find a micro-device from the literature that incorporates the thermal energy mode and describe it in qualitative terms. The device should make use of thermal energy in some fashion, even if it is neither an actuator nor a temperature sensor. Be sure to cite your source(s).

11.14 A 500-μm thick silicon wafer 75 mm in diameter (a 3-inch wafer) rests on top of a hot plate in the MEMS lab. The temperature of the hot plate is measure to be 500°C. The top surface of the wafer is cooled in ambient air for which the convective heat transfer coefficient is h = 10 W/(m·°C) and T_{air} = 20°C. Assuming the system has reached steady state,

a. derive an expression for the thermal energy conducted through the wafer in terms of the temperature of the top wafer surface, T_2,

b. derive an expression for thermal energy convected away from the top surface of the wafer in terms of the temperature of the top wafer surface, T_2, and

c. use your answers to a. and b. to calculate the temperature of the top wafer surface, T_2. Is your answer surprising?

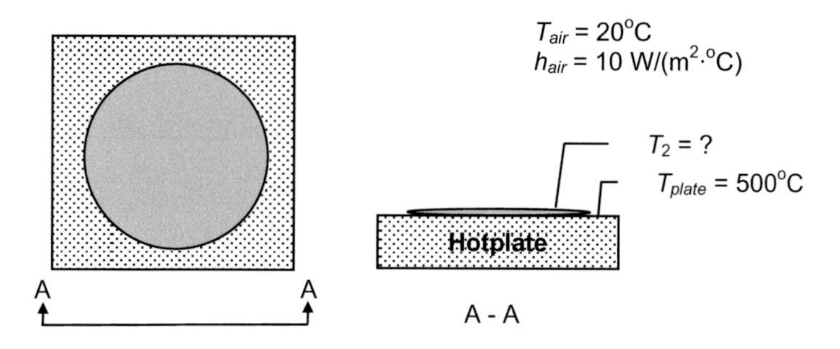

11.15 Based on the model in section 11.3.1, do you expect the sensitivity to be linear? Why or why not?

11.16 Calculate the sensitivity of the hot-arm actuator modeled in section 11.3.3.

11.17 Consider the hot-arm actuator shown in the figure. If the convective heat transfer coefficient can be taken to be $h_{conv} = 1.5$ W/m^2-°C,

a. calculate Biot numbers for the hot arm and the cold arm if the actuator material is copper.

b. Repeat a. for a polyimide actuator.

c. Based on your results from a. and b., how do you feel about using the lumped element model given in section 11.3.1 for the two materials, respectively? Why? What if the convective heat transfer coefficient were $h_{conv} = 150$ W/m^2·°C?

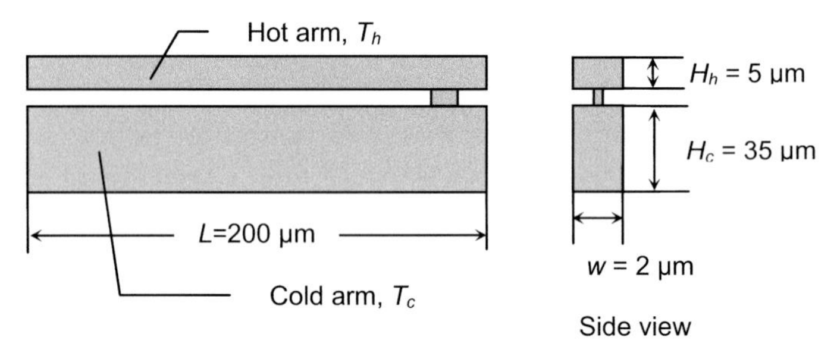

11.18 For the actuator geometry given in Problem 11.17 approximate the temperatures of the hot arm and the cold arm and the resulting tip deflection using the model of section 11.3.1. Assume the material is aluminum and use the following parameters:

- $e = 0.2$ V
- $\rho = 5.961\times10^{-8}$ Ω·m
- surrounding air is at $T_\infty = 25$°C with a convective heat transfer coefficient of $h_{conv} = 12{,}800$ W/m^2·°C

11.19 Reconsider Problem 11.18. To account for the temperature dependence of resistivity, let $\rho_c = 2.873\times10^{-8}$ Ω·m and $\rho_h = 9.049\times10^{-8}$ Ω·m for the cold arm and hot arm, respectively. Other parameters are the same.

a. find the electrical resistance (in Ω) of the hot arm and the cold arm.

b. Find the electrical power dissipated by the hot arm and the cold arm their respective temperatures. (Hint: Equations (11.19) and (11.20) are *not* valid here, but Eqs. (11.16) and (11.17) are.)

c. Find the tip deflection. (Hint: Equation (11.27) is *not* valid here, but Eqs. (11.23) and (11.26) are.)

11.20 One of the most common ways to increase heat transfer from a surface is to make the surface area bigger. This can be accomplished by adding *fins* which are usually nothing more than long, skinny extensions of the surface itself. Thermal energy is conducted to the base of the fin, a little of that energy is convected away from the fin surface, the remaining energy is conducted further down the fin, and so on.

a. The maximum thermal energy that can be dissipated by a fin would occur if the fin had an infinite thermal conductivity. Write an expression for that maximum heat transfer rate. (Hint: What is the temperature of the fin surface area?)

b. In reality, the temperature of a fin gets incrementally smaller as you traverse the fin, and so the fin dissipates less thermal energy than the maximum value. Derive a differential equation similar to Eq. (11.33) that could be solved to find the temperature $T(x)$ for the cylindrical fin in the figure.

c. The ratio of the actual heat transfer rate dissipated by a fin divided by the maximum (found in part a.) is defined as the fin efficiency:

$$\eta = \frac{\dot{Q}}{\dot{Q}_{\max}}.$$

For a cylindrical fin like the one shown here,

$$\eta = \frac{\tanh(aL)}{aL}, \text{ where } a = \sqrt{\frac{4h}{\kappa_{fin} D}}.$$

Calculate the fin efficiency for a polysilicon fin with L = 100mm, D = 5 mm and h_{conv} = 10 W/m^2·°C, then repeat for with L = 100μm, D = 5 μm and h_{conv} = 10 W/m^2·°C. Based on your answers to b., do you think anyone is interested in microfins?

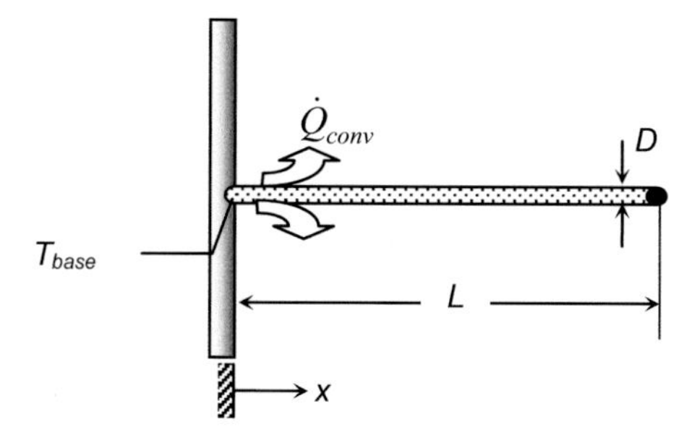

11.21 Consider a thin cylindrical silicon rod (κ=148 W/m·°C) maintained at a constant, uniform temperature of T_{rod} = 100°C. The rod is surrounded by flowing air for which h_{conv}=15 W/m^{2}·°C and T_{∞}=20°C.

a. Find the rate of heat transfer from the rod to the air in W.

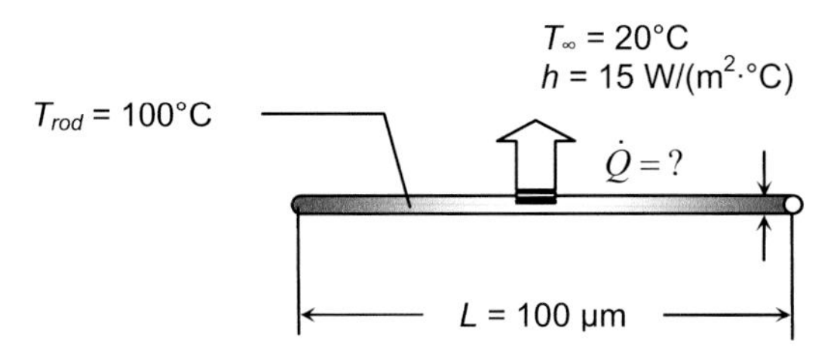

b. Consider a similar silicon rod that is surrounded by an insulating material while its left and right ends are maintained at constant temperatures T_1 and T_2. If T_1=100°C what would T_2 need to be in order to achieve the same heat transfer rate as in part a?

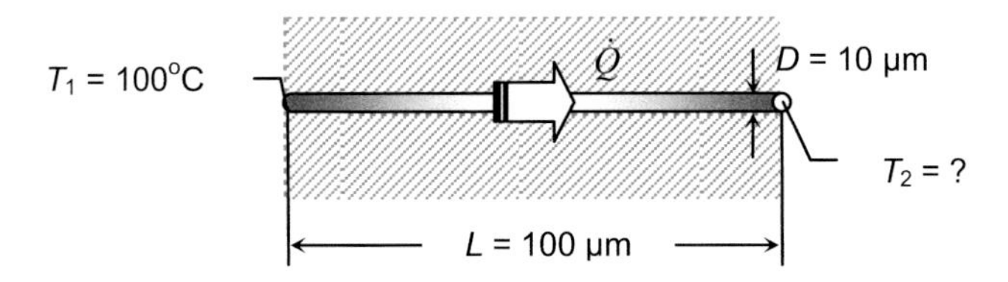

11.22 Consider the hot arm of a hot-arm actuator that not only convects to the surrounding air at a temperature of T_{amb}, but also exchanges thermal radiation with a large surface that encloses the actuator, such as the device packaging surface. If the actuator material has a constant surface emissivity of ε and the device packaging is at a constant temperature of T_{surr}, derive the equivalent of Eq. (11.33), the solution of which could be used to find the temperature distribution of the hot arm.

Chapter 12. Introduction to microfluidics

12.1 Introduction

Utilization of MEMS devices that employ fluids offers many advantages over macroscale devices, including enhanced portability and performance, as well as allowing for increased integration and automation. The array of microscale devices that incorporate fluids is large, and includes inkjet printer heads, devices used for DNA amplification, a variety of sensors, and a number of emerging technologies such as the use of microneedles for drug delivery. The use of fluids at the microscale is also becoming increasingly important in thermal management, particularly through the use of microfluidic devices to cool electronic components. Power generation via microscale turbomachinery and even microrocket engines are currently under development as well. Figure 12.1 gives yet another application, the use of controlled fluid flow to grow crystals.

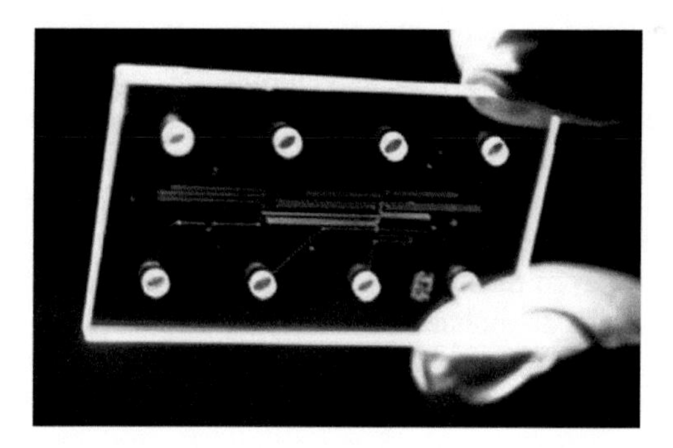

Fig. 12.1. Designed at NASA's Marshall Space Flight Center, this microfluidics chip was developed to grow biological crystals aboard the International Space Station. The lines on the chip are the microchannels that connect the various ports, which appear as holes in the figure. Microvalves control the various processes. (Courtesy of NASA)

T.M. Adams, R.A. Layton, *Introductory MEMS: Fabrication and Applications*,
DOI 10.1007/978-0-387-09511-0_12,

Currently, the most commonly encountered microfluidic devices are probably those used in chemistry and the life sciences for doing separations and analysis. These devices can take a small fluid sample, move it through a microchannel, and then separate out the important components. These components are consequently detected and quantified. Researchers have been working to incorporate detection apparati on the same device as the microfluidics components in developing what has become know as a "lab-on-a-chip". Such devices incorporate the features of a traditional lab in a device that can fit in the palm of your hand, allowing the lab to go to the sample rather than the other way around. In addition to this portability, a lab-on-a-chip has other advantages, including smaller required sample sizes, disposability in the case of hazardous samples, and greatly decreased analysis times.

In this chapter we will we will develop the analytical tools necessary to describe two of the most important aspects of microfluidics as used in chemical and biological detection: **electro-osmotic flow** and **electrophoretic separation**. In electro-osmotic flow an applied electric field is used to make a fluid move, and in electrophoretic separation, an electric field is responsible for separating the constituents of a fluid sample. Obviously the field of microfluidics encompasses much more than this. It is hoped, however, that by thoroughly examining a couple of the most important aspects of current microfluidic devices, the reader will develop a solid foundation with which s/he can begin the study of this intriguing field.

Fluids behave much differently than solids, and the same physical concepts used in previous chapters to model MEMS devices will therefore look different when applied to fluidic systems. We therefore devote a significant portion of the chapter to developing the necessary analysis tools needed to understand fluid mechanics. This makes the current chapter a little more technical and more equation-based than others. What's more, electro-osmotic flow and electrophoretic separation both require the use of an electric field, thereby coupling the mechanical and electrical energy domains. And so, even those readers already familiar with fluid mechanics may find themselves in uncharted territory.

Nonetheless, it is hoped that if nothing else this last chapter truly conveys how multidisciplinary MEMS is, and how the traditional lines delineating our so-called fields are in fact artificial boundaries that must be redrawn time and again in order to broaden the frontiers of technology.

12.2 Basics of fluid mechanics

A fluid is defined as any substance that continuously deforms under the action of a **shear stress**. A shear stress in turn is a force per unit area acting *tangent* to a surface. If you place your hand on the water surface of an aquarium and then move it tangent to the surface, you are applying a shear stress to the water. The water near your hand will move, but the water far from your hand remains stationary. Thus, the water is deforming and it will continue to do so as long as you apply the shear stress. It is a fluid. If you tried the same thing with the surface of a table, however, you would get a much different result. Either your hand would move along a stationary table, or, if you pushed really hard, the table would move as an entire unit. We see from this thought experiment that both gases and liquids can be fluids, but solids cannot.

We will model our fluids as **continuums**, which are substances for which it makes sense to talk about properties having a value *at a point*. Put another way, if we keep chopping our continuum in half and then in half again forever (like taking a limit is calculus) there will be some material left at the end. Though not stated explicitly, we made this same assumption when looking at solid mechanics back in Chapter 5.

In reality we know that fluids are not continuums, but made of millions of molecules of finite size. If we keep chopping a fluid in half, eventually we get to the point where we notice that there is a molecule of fluid here and a molecule there, but nothing in between them. Figure 12.2 illustrates this point.

Figure 12.2 shows the density of air as a function of volume at standard atmospheric conditions. When the volume gets too small, the density varies wildly. Thus, in order to get away with the continuum assumption, then, we have to stay at length scales sufficiently above this point, which is beat estimated using the **mean free path** of the molecules.[1]

[1] In some microfluidic devices with extremely small dimensions, especially those utilizing gases, the continuum model may not apply. If we stick with liquids in most common microfluidic devices, however, we are OK.

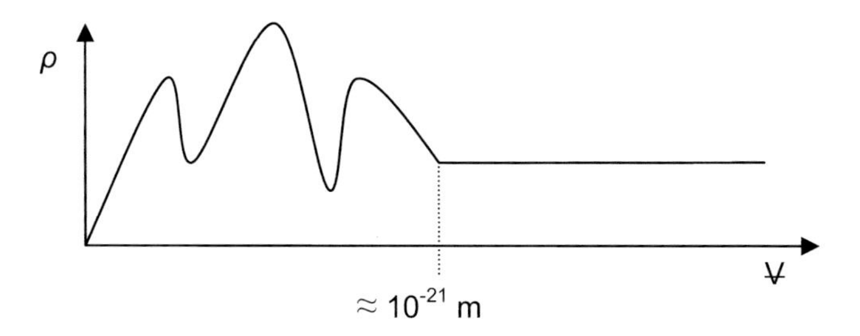

Fig. 12.2. Breakdown of the continuum assumption for air

12.2.1 Viscosity and flow regimes

Imagine a fluid flowing in a circular pipe in which we inject a stream of dye as in Fig. 12.3. For sufficiently small fluid velocities, the dye remains relatively smooth and uniform in its path, as in Fig. 12.3 (a). If the velocity is increase somewhat, the dye steam will start to wiggle around a bit, as in Fig. 12.3 (b). When the velocity really gets going, the dye follows a seemingly random path as in Fig. 12.3 (c).

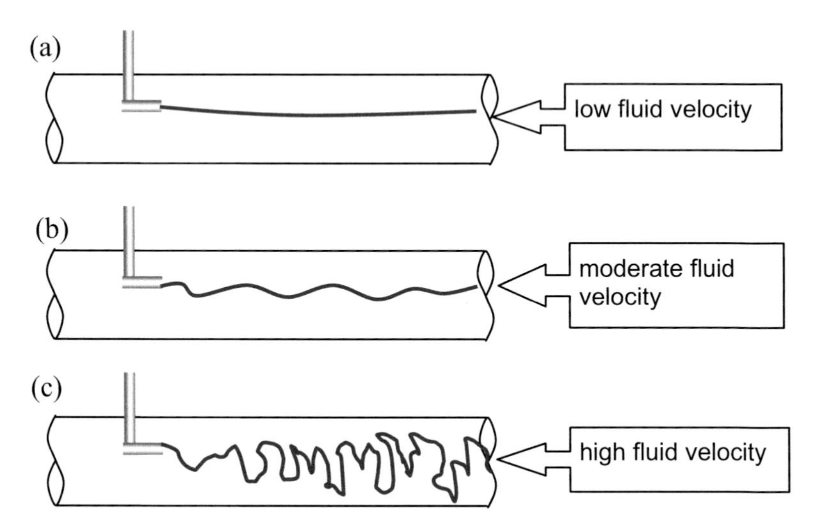

Fig. 12.3. Flow regimes in pipe flow

If we repeat this experiment using a smaller channel, we will find that the velocities where the various dye patterns are observed are larger. If we repeat the experiment yet again using thicker (more viscous) fluids, we find the required velocities are again larger.

Osborne Reynolds actually did this experiment many years ago and found that he could predict when occurrence of these various patterns based on a dimensionless number which now bears his name. The **Reynolds number** for a circular channel is given by

$$Re = \frac{\rho v D}{\eta} \tag{12.1}$$

where v is the average fluid velocity, D is the channel diameter and η is the fluid **viscosity**. For circular channels with Reynolds numbers less than about 2,300, patterns as in Fig. 12.4 (a) are observed. This is referred to as **laminar flow**. From Reynolds numbers around 2300 to 4000, the flow looks like that in Fig. 12.4 (b), and is called **transition flow**. For sufficiently high Reynolds numbers, we observe **turbulent flow** as illustrated in Fig. 12.4 (c).

When channel cross sections are not round, D is usually replaced by a quantity called the **hydraulic diameter**. The hydraulic diameter is given by

$$D_h = \frac{4A}{\mathscr{P}_w}, \tag{12.2}$$

where A is the cross sectional area and $\mathscr{P}_w$ is the wetted perimeter. As many microfabrication processes result in channel geometries other than circular, Eq. (12.2) is used frequently.

The Reynolds number is sometimes interpreted as the ratio of inertia forces to viscous forces. Thus, viscosity dominates in laminar flow, and inertia in turbulent flow. In the macroworld, most practical flows of interest are turbulent. Due to the small dimensions encountered in microfluidics, however, most flows tend to be laminar. It is therefore necessary for us to learn a bit more about viscosity.

Qualitatively, viscosity is a measure of the "stickiness" or the "thickness" of a fluid. Honey, for example, is more viscid than water. Quantitatively, viscosity relates the shear stress within a fluid to its rate of deformation, or *strain rate*. Thus, viscosity plays a similar role in fluid mechanics as the various moduli of elasticity do in solid mechanics with strain rate replacing strain. We can't relate stress to strain in a fluid, of course, since fluids are substances that continuously deform as long as shear stress is present. In any case, kinematics can be used to show that a fluid's deformation rate is related to its velocity gradient. And so practically speaking, *viscosity relates shear stress to velocity gradient.*

Consider the simple case of two infinite plates bounding a viscous fluid as in Fig. 12.4. The bottom plate is stationary whereas the top plate moves to the right with a constant velocity of v_0.

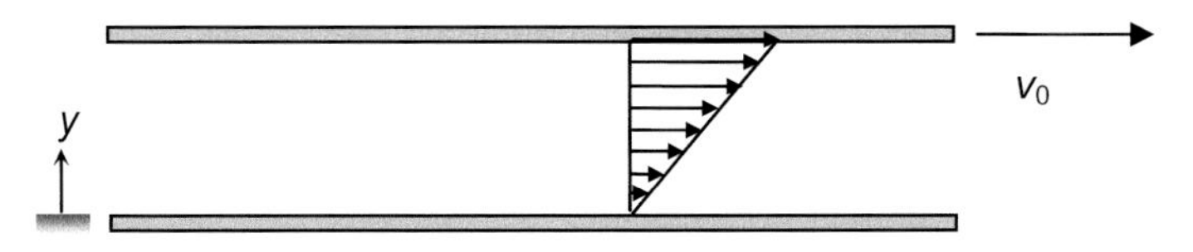

Fig. 12.4. Viscous flow between two plates

At the two solid surfaces, there is no *relative* velocity between the fluid and surfaces themselves. This is known as the **no slip boundary condition** and it essentially means that the fluid wets the surface. This is an experimentally observed phenomenon and applies in most situations. For our flow, then, the fluid at the bottom surface has zero velocity whereas the fluid at the top surface takes on the velocity of the plate, v_0.

We can imagine the flow in Fig. 12.4 as comprising several fluid layers all traveling to the right. If all the fluid layers were moving to the right with the same speed, no forces would exist between them. However, since each fluid layer has a different velocity than the layer immediately above or below it, a friction-like force develops between them. This happens in much the same way that friction develops when you rub your hands together.

Let us examine this force in more detail. Figure 12.5 shows a small portion of the flowing fluid. Also shown in the figure is the force per unit area, i.e., the stress that develops between fluid layers. Since this force is tangent to the surface of interest, it is a shear stress.

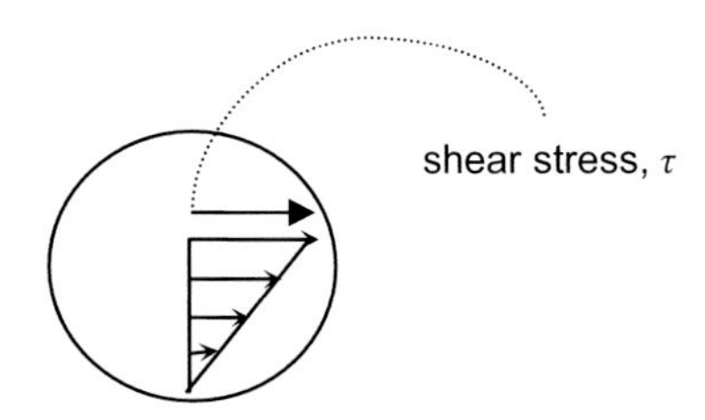

Fig.12.5. Shear stress in a moving fluid

For a **Newtonian fluid**[2], the shear stress is linearly related to the velocity gradient. For our example,

$$\tau \propto \frac{dv}{dy}$$

$$\tau = \eta \frac{dv}{dy} \tag{12.3}$$

where τ is the shear stress and V is the local fluid velocity. This relationship is very similar to Eq. (5.11) for shear stress and strain in an elastic solid with dv/dy in place of the strain.

For flows that are not one dimensional the relationship between viscous stress and velocity gradient is more complicated. This is primarily due to velocity being a vector quantity rather than a scalar. As such, we must relate the stress tensor to the rate of strain tensor, again much in the same way as in Chapter 5 for stress and strain. Sparing the details of the derivation, the result for an incompressible fluid looks like this:

$$\tau_{xx} = 2\eta \frac{\partial u}{\partial x} \tag{12.4}$$

$$\tau_{yy} = 2\eta \frac{\partial v}{\partial y} \tag{12.5}$$

$$\tau_{zz} = 2\eta \frac{\partial w}{\partial z} \tag{12.6}$$

$$\tau_{xy} = \tau_{yx} = \eta \left(\frac{\partial u}{\partial y} + \frac{\partial v}{\partial x} \right) \tag{12.7}$$

$$\tau_{yz} = \tau_{zy} = \eta \left(\frac{\partial v}{\partial z} + \frac{\partial w}{\partial y} \right) \tag{12.8}$$

$$\tau_{zx} = \tau_{xz} = \eta \left(\frac{\partial w}{\partial x} + \frac{\partial u}{\partial z} \right) \tag{12.9}$$

---

[2] This is a *constitutive relation*, not a universal law. Not all fluids are Newtonian! However, many fluids, including air and water, are well-modeled as Newtonian.

As outlined in Chapter 5, the notation τ_{xy} implies the stress on the $\underline{x}$ surface in the $\underline{y}$ direction. The symbols u, v and w are used for the x, y and z components of velocity, respectively. Also note that there are *normal* viscous stress components as well as shear stress components; e.g., τ_{xx}. What do you think these represent? Can you think of a common fluid situation in which normal viscous stress components are important?

12.2.2 Entrance lengths

Consider a large tank that drains into a long pipe as in Fig. 12.6. Right at the entrance of the pipe, all the fluid will have only one velocity, U. Because of the no slip boundary condition, however, we know that the fluid touching the pipe walls has zero velocity. This is true no matter the axial location of the fluid. However, as the fluid travels down the length of the pipe, more and more fluid near the center of the pipe feels the effect of the wall slowing it down. This region near the entrance of the flow channel is known as the **entrance length**, and is characterized by a velocity distribution which changes in the flow direction.

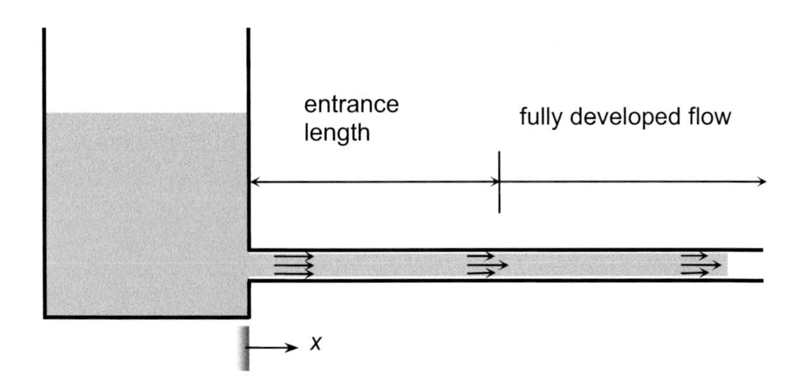

Fig. 12.6. Entrance length

Eventually the fluid reaches a location where the wall has made its presence known to all the fluid across the channel cross section. From this point downstream, the flow is called **fully developed**. Once a flow is fully developed, the *velocity distribution does not change in the flow direction.* Hence, fully developed flow is one dimensional. Mathematically, this translates into

$$\frac{\partial u}{\partial x} = 0 . \tag{12.10}$$

The entrance length is a function of the flow regime as well as the Reynolds number. For laminar flow, entrance lengths are approximately $0.05 Re \cdot D$. Since microfluidic devices are characterized by both small dimensions and small Reynolds numbers, the entrance region is often ignored and entire flows are modeled as fully developed.

12.3 Basic equations of fluid mechanics

12.3.1 Conservation of mass

The law of conservation of mass states that *mass can neither be created nor destroyed.* The universe has all the mass it ever will. (Actually, this is not quite true, as Einstein showed that mass is actually a form of energy and strictly speaking it is only energy that is always conserved. But as long as we stay away from speeds close to the speed of light, which is not a problem in microfluidic devices, mass will be conserved.) For a **closed system**, or a system that consists of a fixed mass, conservation of mass reduces to

$$m = \text{constant}. \tag{12.11}$$

Closed systems are useful tools for analysis in dynamics and solid mechanics. For fluid mechanics, however, it is much more useful to set aside a region of space for analysis, one for which we allow mass to enter or leave. Such a system is called an **open system**. Figure 12.7 illustrates an open system.

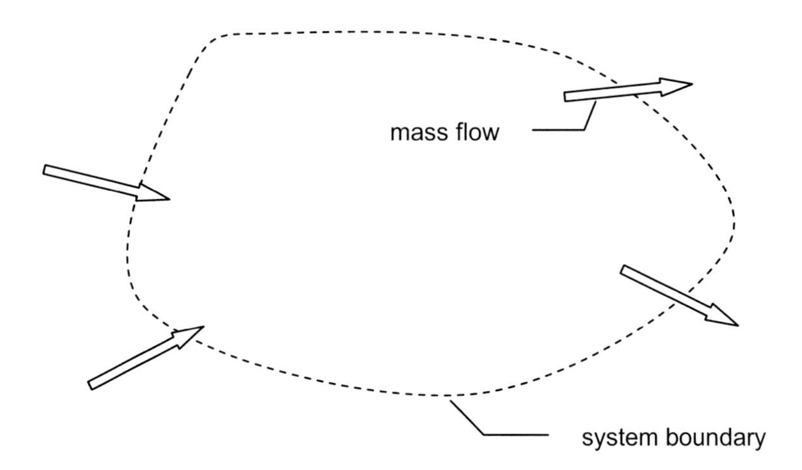

Fig. 12.7. Conservation of mass for an open system

In words the conservation of mass is represented by the equation

$$\begin{bmatrix}\text{Rate of accumulation}\\ \text{of mass in the system}\end{bmatrix} = \begin{bmatrix}\text{Rate of mass}\\ \text{flowing in}\end{bmatrix} - \begin{bmatrix}\text{Rate of mass}\\ \text{flowing out}\end{bmatrix}$$

In symbols this equation becomes

$$\frac{dm_{sys}}{dt} = \sum \dot{m}_{in} - \sum \dot{m}_{out} \tag{12.12}$$

Here $\dot{m}$ is the **mass flow rate** and is given by

$$\dot{m} = \rho A v \,.\text{[3]} \tag{12.13}$$

where A is the cross sectional area where mass crosses the boundary of the system. It is important to note that here $\dot{m}$ does not represent a time derivative, but rather the *rate at which mass is crossing the boundary of the system.* Eq. (12.13) gives us a convenient way to calculate this rate.

12.3.2 Conservation of linear momentum

Just as mass is a conserved quantity, so is linear momentum. That is, *linear momentum can neither be created nor destroyed.* The working equations

[3] This equation assumes that v is constant across A. What would this expression look like if v were not constant across A?

are a bit more complicated than those for conservation of mass, however. Therefore let's start with a more familiar equation which is actually a specialized form of conservation of momentum.

Recall Newton's second law for a particle from basic mechanics:

$$m\vec{a} = \sum \vec{F} \tag{12.14}$$

One interpretation of Newton's second law is that you have to apply a force (a push or pull) to a particle in order to change its state of motion (to accelerate it). With a bit of rearrangement, we can give Newton's second law another interpretation.

$$m\frac{d\vec{v}}{dt} = \sum \vec{F} \tag{12.15}$$

$$\frac{d(m\vec{v})}{dt} = \sum \vec{F} \tag{12.16}$$

Cast in this form we recognize *mv* as the linear momentum of the particle. Now we can say that in order to change the momentum of a particle, we must apply some external force to it. Furthermore, since we know that momentum is conserved, we can interpret force as *rate of momentum transport into the particle*. In words this equation is

$$\begin{bmatrix}\text{Time rate of change}\\ \text{of a particle's momentum}\end{bmatrix} = \begin{bmatrix}\text{Rate of momentum transport}\\ \text{into particle via external forces}\end{bmatrix}$$

In the event that our closed system is not a particle, we can replace particle velocity with the center of mass velocity and everything works fine. This is the realm of rigid body mechanics.

As we saw earlier, closed systems are not particularly useful for fluid mechanics, as it is most difficult to keep track of a fixed mass in a flowing fluid. Even worse, fluids are not rigid bodies – they can deform. We therefore need to expand our concept of conservation of momentum to open systems. Consider the open system shown in Fig. 12.8 In addition to momentum being transported in and out of the system via forces, it can now also be transported via *mass flow*.

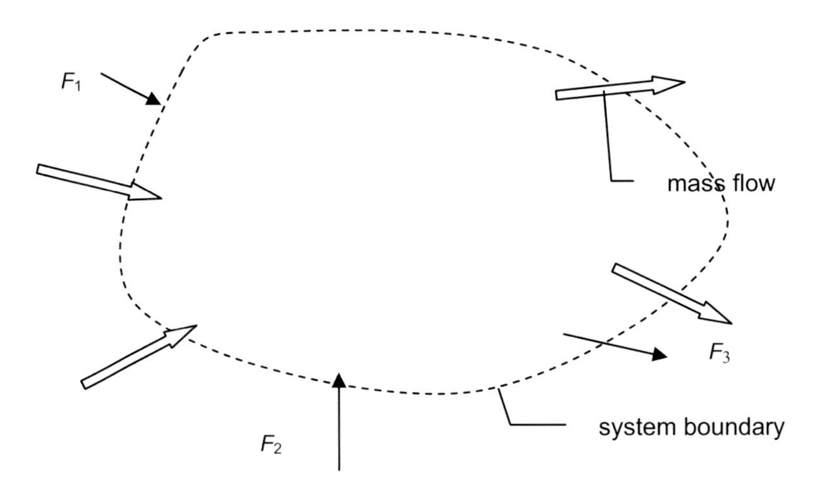

Fig. 12.8. Conservation of linear momentum for an open system

In words, the new improved conservation of momentum looks like

$$\begin{bmatrix}\text{Time rate of change}\\ \text{of a system's}\\ \text{momentum}\end{bmatrix}$$

$$=\begin{bmatrix}\text{Rate of momentum}\\ \text{transport into system}\\ \text{via external forces}\end{bmatrix}+\begin{bmatrix}\text{Rate of momentum}\\ \text{transport into system}\\ \textit{via mass flow}\end{bmatrix}-\begin{bmatrix}\text{Rate of momentum}\\ \text{transport out of system}\\ \textit{via mass flow}\end{bmatrix}$$

In symbols the equation becomes

$$\frac{d(m\vec{v})}{dt}=\sum\vec{F}+\sum_{in}\dot{m}\vec{v}-\sum_{out}\dot{m}\vec{v} \tag{12.17}$$

Our new improved form of conservation of momentum allows for the transport of momentum with mass flow as well as through forces. Note that for open systems, we cannot categorically replace the left hand side of the equation with ***ma*** because *both the velocity and mass of an open system can change.*

The previously derived equations are usually referred to as **macroscopic conservation equations**. They are fairly general and can be applied to most systems. However, they are most useful when a system is of finite size (hence the term macroscopic) where we don't care much about the details going on inside the system itself. Indeed, when using macroscopic conservation equations we speak of the *momentum of the system* or the *velocity of the system* as if there were only one value of those quantities at any given time.

Since the no-slip boundary condition requires the fluid touching a solid surface to have zero relative velocity, however, it is easy to see that fluid particles have different velocities at different locations within a system. Therefore, if we define a system boundary to coincide with the entire fluid inside a pipe and then apply the macroscopic conservation equations, it is impossible to extract information about the spatial variation of fluid velocity. To complete this job we need to derive specialized forms of the conservation equations which are applicable at any point within a fluid.

12.3.3 Conservation equations at a point: Continuity and Navier-Stokes equations

We will derive two equations (four, really – one is a vector equation) which represent conservation of mass and linear momentum applicable to any point within a flowing fluid. We will limit our discussion to incompressible Newtonian fluids only. The derivations won't be particularly rigorous, though hopefully they will have heuristic value.

Conservation of Mass: Continuity:

For illustrative purposes let us begin by considering two-dimensional flow. Figure 12. 9 shows fluid flow through a square of dimensions Δx and Δy. (For simplicity let the dimension into the page measure one unit.) The x direction velocity is denoted by u and the y direction velocity is denoted by v. Δx and Δy are small enough to assume that u and y have only one value on any given face.

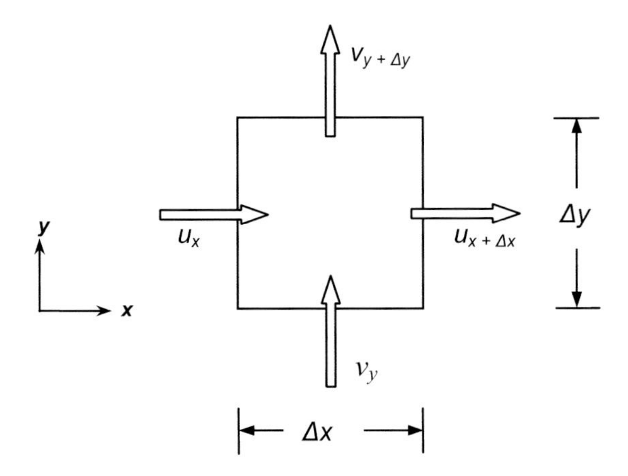

Fig. 12.9. Fluid flowing through a small square

The macroscopic conservation of mass is

$$\frac{dm_{sys}}{dt} = \sum \dot{m}_{in} - \sum \dot{m}_{out} \tag{12.18}$$

For our system this becomes

$$\frac{\partial}{\partial t}(\rho \cdot \Delta x \cdot \Delta y) = \rho \cdot \Delta y \cdot u_x + \rho \cdot \Delta x \cdot v_y - \rho \cdot \Delta y \cdot u_{x+\Delta x} - \rho \cdot \Delta x \cdot v_{y+\Delta y} \tag{12.19}$$

Rearranging,

$$\frac{\partial \rho}{\partial t} + \frac{(\rho u)_{x+\Delta x} - (\rho u)_x}{\Delta x} + \frac{(\rho v)_{y+\Delta y} - (\rho v)_y}{\Delta y} = 0\,. \tag{12.20}$$

In the limit as Δx and $\Delta y \rightarrow 0$,

$$\frac{\partial \rho}{\partial t} + \frac{\partial(\rho u)}{\partial x} + \frac{\partial(\rho v)}{\partial y} = 0\,. \tag{12.21}$$

This is the **continuity equation**, so named because it assumes that ρ has a value at a point, which is to say that it is continuous. It essentially represents conservation of mass applied to a point. If we had 3-D flow it would look like this (w is the z direction velocity):

$$\frac{\partial \rho}{\partial t} + \frac{\partial}{\partial x}(\rho u) + \frac{\partial}{\partial y}(\rho v) + \frac{\partial}{\partial z}(\rho w) = 0\,. \tag{12.22}$$

A more compact way to write this is

$$\frac{\partial \rho}{\partial t} + \vec{\nabla} \bullet (\rho \vec{V}) = 0 \tag{12.23}$$

where $\vec{\nabla}$ is the "del" operator and in rectangular coordinates is given by

$$\vec{\nabla} = \frac{\partial}{\partial x}\hat{i} + \frac{\partial}{\partial y}\hat{j} + \frac{\partial}{\partial z}\hat{k}\,. \tag{12.24}$$

The operation $\vec{\nabla} \bullet (\,)$ is known as taking the **divergence** of the quantity ( ). The convenience of this notation is that we can actually talk about divergence without reference to a particular coordinate system.

Most flows of interest to us can be modeled as incompressible. For such flows, the density of the fluid has a constant value. In this case, Eq. (12.21) reduces to

$$\frac{\partial u}{\partial x} + \frac{\partial v}{\partial y} = 0 \tag{12.25}$$

or

$$\vec{\nabla} \bullet (\vec{V}) = 0\,. \tag{12.26}$$

Conservation of Linear Momentum: The Navier-Stokes Equations

Let us now turn our attention to the conservation of linear momentum. We will consider only the x-direction linear momentum in our derivation.

Figure 12.10 shows the needed velocities and stresses. In addition to the forces caused by the pressure and stresses, a **body force** is also shown in the figure. Forces due to stress are known as **surface forces**, as they act on a surface. A body force, on the other hand, does not have any specific point of application. A body force arises due to the attraction or repulsion of something located some distance away from the system at hand. For this reason, body forces are often referred to as "forces at a distance." Gravity is an example of a body force. (You don't need to touch the earth to feel the effects of its pull on you.) Forces on charged particles within an electric field are also body forces. In the figure b_x represents some generic body force *per unit volume* applied in the x-direction.

As before, we consider two dimensional flow with the dimension into the page measuring one unit, and that Δx and Δy are small enough to assume that u and v are uniform across any surface of the square. We will only consider incompressible fluids (ρ = constant).

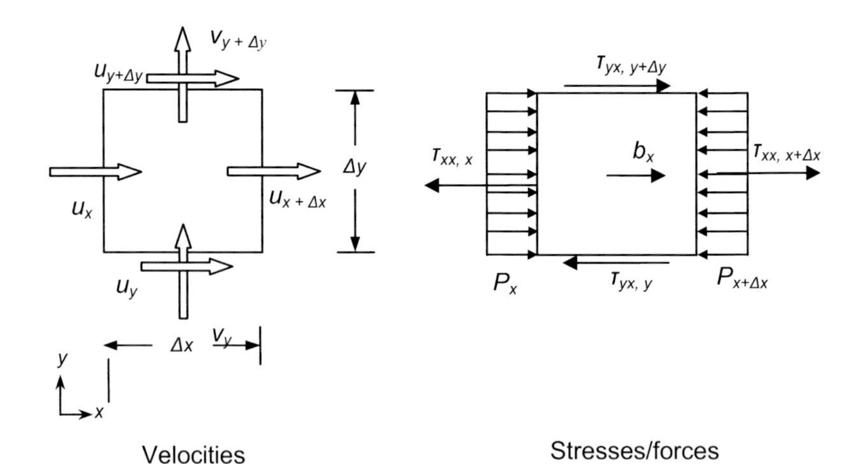

Fig. 12.10. Relevant velocities and stresses for x-direction linear momentum

The conservation of linear momentum is given by Eq. (12.17).

$$\frac{d(m\vec{V})}{dt} = \sum \vec{F} + \sum_{in} \dot{m}\vec{V} - \sum_{out} \dot{m}\vec{V} \tag{12.17}$$

The x-direction component is

$$\frac{d(mu)}{dt} = \sum F_x + \sum_{in} \dot{m}u - \sum_{out} \dot{m}u \tag{12.27}$$

In terms of density and the quantities given in the Fig. 12.10 this becomes

$$\begin{aligned} &\frac{d}{dt}(\rho \cdot \Delta x \cdot \Delta y \cdot u) = (P_x - P_{x+\Delta x})\Delta y - (\tau_{xx,x} + \tau_{xx,x+\Delta x})\Delta y \\ &-(\tau_{yx,y} + \tau_{yx,y+\Delta y})\Delta x + b_x \cdot \Delta x \cdot \Delta y + (\rho \cdot \Delta y \cdot u_x)u_x \\ &+(\rho \cdot \Delta x \cdot v_y)u_y - (\rho \cdot \Delta y \cdot u_{x+\Delta x})u_{x+\Delta x} - (\rho \cdot \Delta x \cdot v_{y+\Delta y})u_{y+\Delta y} \end{aligned} \tag{12.28}$$

Gathering terms and dividing through by $\Delta x\ \Delta y$,

$$\begin{aligned} &\rho\frac{\partial u}{\partial t} = -\frac{P_{x+\Delta x} - P_x}{\Delta x} + \frac{\tau_{xx,x+\Delta x} - \tau_{xx,x}}{\Delta x} + \frac{\tau_{yx,x+\Delta y} - \tau_{yx,y}}{\Delta y} + b_x \\ &-\rho\frac{(u_{x+\Delta x})^2 - u_x^{\ 2}}{\Delta x} - \rho\frac{(uv)_{y+\Delta y} - (uv)_y}{\Delta y} \end{aligned} \tag{12.29}$$

In the limit as Δx and $\Delta y \rightarrow 0$,

$$\rho\frac{\partial u}{\partial t}=-\frac{\partial P}{\partial x}+\frac{\partial \tau_{xx}}{\partial x}+\frac{\partial \tau_{yx}}{\partial y}+b_x-\rho\frac{\partial(u^2)}{\partial x}-\rho\frac{\partial(uv)}{\partial y} \tag{12.30}$$

With the aid of continuity, we can simplify this expression further. Expanding the last two derivatives on the right hand side of Eq. (12.30),

$$\rho\frac{\partial u}{\partial t}=-\frac{\partial P}{\partial x}+\frac{\partial \tau_{xx}}{\partial x}+\frac{\partial \tau_{yx}}{\partial y}+b_x-\rho\left[2u\frac{\partial u}{\partial x}+u\frac{\partial v}{\partial y}+v\frac{\partial u}{\partial y}\right] \tag{12.31}$$

From continuity for incompressible fluids we know

$$\frac{\partial u}{\partial x}+\frac{\partial v}{\partial y}=0 \tag{12.25}$$

Therefore

$$\rho\frac{\partial u}{\partial t}=-\frac{\partial P}{\partial x}+\frac{\partial \tau_{xx}}{\partial x}+\frac{\partial \tau_{yx}}{\partial y}+b_x-\rho\left[u\frac{\partial u}{\partial x}+v\frac{\partial u}{\partial y}\right] \tag{12.32}$$

Rearranging a bit

$$\rho\left[\frac{\partial u}{\partial t}+u\frac{\partial u}{\partial x}+v\frac{\partial u}{\partial y}\right]=-\frac{\partial P}{\partial x}+\frac{\partial \tau_{xx}}{\partial x}+\frac{\partial \tau_{yx}}{\partial y}+b_x \tag{12.33}$$

Some texts define the operator

$$\frac{D}{Dt}()\equiv\frac{\partial()}{\partial t}+u\frac{\partial()}{\partial x}+v\frac{\partial()}{\partial y} \tag{12.34}$$

as the **material derivative**. One can interpret the material derivative as the derivative of a quantity as *a fixed mass is followed* within a flowing fluid. In terms of the material derivative, Eq. (12.34) becomes

$$\rho\frac{Du}{Dt}=-\frac{\partial P}{\partial x}+\frac{\partial \tau_{xx}}{\partial x}+\frac{\partial \tau_{yx}}{\partial y}+b_x \tag{12.35}$$

This equation has a $m\cdot du/dt=\sum F_x$ flavor to it, but on a per unit volume basis. This makes sense when you consider that the idea of a material derivative is to follow a fixed mass of fluid. The system is then a *closed* system and you don't have the momentum transport by mass flow terms appearing explicitly. Of course the downside is that now you need a special form of your time derivative (Dv/Dt instead of dv/dt).

And so now that we have an equation describing conservation of linear momentum at a point in an incompressible fluid, the last step is to plug in something for the stresses. If our fluid is Newtonian, then Eqs. (12.4) to (12.9) apply. For the x-direction linear momentum equation, we need

$$\tau_{xx} = 2\eta \frac{\partial u}{\partial x} \tag{12.4}$$

and

$$\tau_{xy} = \tau_{yx} = \eta\left(\frac{\partial u}{\partial y} + \frac{\partial v}{\partial x}\right) \tag{12.7}$$

Equation (12.35) becomes

$$\begin{aligned} &\rho\left[\frac{\partial u}{\partial t} + u\frac{\partial u}{\partial x} + v\frac{\partial u}{\partial y}\right] \\ &= -\frac{\partial P}{\partial x} + \frac{\partial}{\partial x}\left(2\eta\frac{\partial u}{\partial x}\right) + \frac{\partial}{\partial y}\left(\eta\left(\frac{\partial u}{\partial y} + \frac{\partial v}{\partial x}\right)\right) + b_x \end{aligned} \tag{12.36}$$

If we assume that viscosity is constant and make use of the fact that the order of differentiation is arbitrary,

$$\begin{aligned} &\rho\left[\frac{\partial u}{\partial t} + u\frac{\partial u}{\partial x} + v\frac{\partial u}{\partial y}\right] \\ &= -\frac{\partial P}{\partial x} + \eta\left(\frac{\partial^2 u}{\partial x^2} + \frac{\partial^2 u}{\partial y^2}\right) + \eta\frac{\partial}{\partial x}\left(\frac{\partial u}{\partial x} + \frac{\partial v}{\partial y}\right) + b_x \end{aligned} \tag{12.37}$$

We can again use continuity to simplify

$$\rho\left[\frac{\partial u}{\partial t} + u\frac{\partial u}{\partial x} + v\frac{\partial u}{\partial y}\right] = -\frac{\partial P}{\partial x} + \eta\left(\frac{\partial^2 u}{\partial x^2} + \frac{\partial^2 u}{\partial y^2}\right) + b_x \tag{12.38}$$

Equation (12.38) represents the x-direction conservation of linear momentum at a point for a Newtonian fluid for two dimensional flow. If we had derived this equation for three dimensional flow instead, then the result would also contains the z-direction derivatives for velocity.

$$\rho\left[\frac{\partial u}{\partial t}+u\frac{\partial u}{\partial x}+v\frac{\partial u}{\partial y}+w\frac{\partial u}{\partial z}\right]$$
$$=-\frac{\partial P}{\partial x}+\eta\left(\frac{\partial^2 u}{\partial x^2}+\frac{\partial^2 u}{\partial y^2}+\frac{\partial^2 u}{\partial z^2}\right)+b_x \tag{12.39}$$

The y-direction and z-direction equations look very similar and are presented below without derivation, respectively.

$$\rho\left[\frac{\partial v}{\partial t}+u\frac{\partial v}{\partial x}+v\frac{\partial v}{\partial y}+w\frac{\partial v}{\partial z}\right]$$
$$=-\frac{\partial P}{\partial y}+\eta\left(\frac{\partial^2 v}{\partial x^2}+\frac{\partial^2 v}{\partial y^2}+\frac{\partial^2 v}{\partial z^2}\right)+b_y \tag{12.40}$$

$$\rho\left[\frac{\partial w}{\partial t}+u\frac{\partial w}{\partial x}+v\frac{\partial w}{\partial y}+w\frac{\partial w}{\partial z}\right]$$
$$=-\frac{\partial P}{\partial z}+\eta\left(\frac{\partial^2 w}{\partial x^2}+\frac{\partial^2 w}{\partial y^2}+\frac{\partial^2 w}{\partial z^2}\right)+b_z \tag{12.41}$$

Equations (12.39) to (12.40) are collectively known as the **Navier-Stokes equations** for an incompressible fluid. Though the derivation was painful, the equations themselves represent a simple idea – that linear momentum is conserved.

The three Navier-Stokes equations are actually the three components of a vector equation given by

$$\rho\left[\frac{\partial \vec{V}}{\partial t}+u\frac{\partial \vec{V}}{\partial x}+v\frac{\partial \vec{V}}{\partial y}+w\frac{\partial \vec{V}}{\partial z}\right]=-\vec{\nabla}P+\eta\nabla^2\vec{V}+\vec{b} \tag{12.42}$$

where ∇^2 is the **Laplacian** operator, which in rectangular coordinates is given by

$$\nabla^2 \equiv \frac{\partial^2}{\partial x^2}+\frac{\partial^2}{\partial y^2}+\frac{\partial^2}{\partial z^2} \tag{12.43}$$

If we use the material derivative notation, the Navier–Stokes equations take on a much less intimidating form that lends itself well to interpretation:

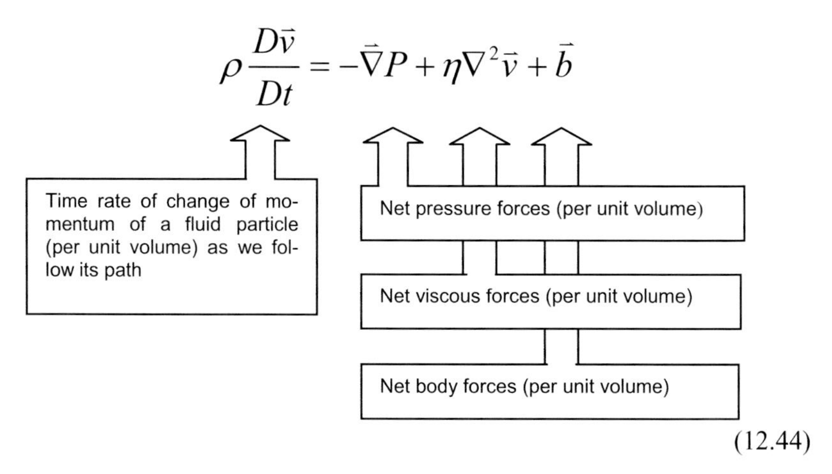

(12.44)

Again, this has a $m{\cdot}d\mathbf{v}/dt = \sum \boldsymbol{F}$ feel to it, but on a per unit volume basis.

It is helpful at this point to review a few points about the Navier-Stokes equations:

1. The form of the Navier-Stokes equations given here is applicable to *incompressible fluids*.
2. The Navier-Stokes equations are applicable only to *Newtonian fluids*; that is, fluids for which the viscous stress/strain relationship is given by relationships such as in Eqs. (12.4) to (12.9).
3. The Navier-Stokes equations represent *conservation of linear momentum* and *apply at a point* within a fluid.

The third point shows us that if we wish to solve the Navier-Stokes equations, we need to do so over a region of space for which we know the appropriate boundary conditions[4]. This will give us information as to how the velocity changes over that region, which is exactly why we wanted to develop these equations in the first place! In the unsteady case, we also require the appropriate initial condition. Solving the Navier-Stoke equations is no easy task, however, as we see that they are non-linear partial differential equations for which there is no general solution method.[5]

[4] The boundary condition typically is no slip. However, there are several cases encountered in microfluidics in which the no slip boundary condition must be abandoned because of the length scale involved. We will not concern ourselves with such cases in this chapter.

[5] And if our flows are turbulent, the situation is much worse. In the turbulent case, which is intrinsically *unsteady*, we must time-average the equations. This results in an additional unknown quantity known as the turbulent stress for which there

12.4 Some solutions to the Navier-Stokes equations

12.4.1 Couette flow

Let us reconsider the simple case of steady flow between two infinite plates as shown in Fig. 12.11. The top plate moves to the right with a constant speed U_0 whereas the bottom plate is stationary.

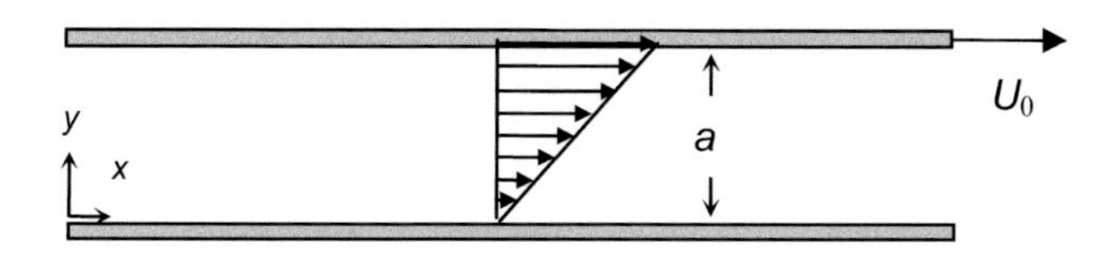

Fig. 12.11. Couette flow

By considering the no slip boundary condition and the idea of viscous drag, we already intuited that the velocity profile should be linear as shown in Fig. 12.4. Let us now prove this by solving the Navier-Stokes for this case.

First let's write continuity. Since we are considering incompressible flow, we can use Eq. (12.25).

$$\frac{\partial u}{\partial x}+\frac{\partial v}{\partial y}=0 \tag{12.25}$$

Since we are also considering one-dimensional flow, $v=0$ and thus,

$$\frac{\partial u}{\partial x}=0\,. \tag{12.45}$$

Hence, u is a function of y only.

Next, let's apply the x component of Navier-Stokes.

$$\rho\left[\frac{\partial u}{\partial t}+u\frac{\partial u}{\partial x}+v\frac{\partial u}{\partial y}\right]=-\frac{\partial P}{\partial x}+\eta\left(\frac{\partial^2 u}{\partial x^2}+\frac{\partial^2 u}{\partial y^2}\right)+b_x\,. \tag{12.38}$$

---

is no universal constitutive relation. This is the fundamental issue in turbulence and is known as the *closure problem*. Luckily for us, however, most microfluidic flows are laminar, not turbulent. Why is that?

From continuity we know that u depends on y only. Furthermore, we are considering steady-state flow with no applied pressure gradient or body forces; we are only concerned with the simple case of the top plate moving to the right with a constant velocity U_0. This greatly simplifies Eq. (12.38):

$$\rho\left[\frac{\partial u}{\partial t}+u\frac{\partial u}{\partial x}+v\frac{\partial u}{\partial y}\right]=-\frac{\partial P}{\partial x}+\eta\left(\frac{\partial^2 u}{\partial x^2}+\frac{\partial^2 u}{\partial y^2}\right)+b_x. \tag{12.46}$$

Steady-state

$u \neq f(x)$

1-D flow

No applied P

$u \neq f(x)$

No applied body forces

We are left with a simple, linear, ordinary differential equation that can be solved by direct integration.

$$\frac{d^2u}{dy^2}=\frac{d}{dy}\left(\frac{du}{dy}\right)=0 \tag{12.47}$$

Integrating twice gives

$$u=c_1 y+c_2\,. \tag{12.48}$$

The two constants of integration are determined by applying the no-slip boundary condition at $y = 0$ and $y = a$. Mathematically,

$$u(y=0)=0 \text{ and } u(y=a)=U_0 \qquad (12.49), (12.50)$$

Equation (12.48) yields

$$0=c_1\cdot 0+c_2 \text{ and } U_0=c_1\cdot a+c_2 \qquad (12.51), (12.52)$$

from which we determine that $c_1 = U_0/a$ and $c_2 = 0$. Finally,

$$u=\frac{U_0}{a}y \tag{12.53}$$

the result we expected.

12.4.2 Poiseuille flow

In the case of Couette flow, we move a fluid by moving one of the surfaces with which the fluid is in contact. A more common situation is one in which the walls of a flow channel are stationary and the fluid is moved by some other means. Flow within a pipe is a good example. In the case of Poiseuille flow, the application of a pressure gradient in the flow direction creates the flow.

Let us now consider a flow channel consisting of two infinite, stationary plates separated by a distance of a as shown in Fig. 12.12. A constant pressure gradient of $K = -dP/dx$ is applied in the flow direction. (It is convenient to use the negative sign in the definition of K since a negative pressure gradient will produce flow in the positive x direction.) For fully developed, steady, one-dimensional laminar flow, the resulting velocity distribution is parabolic.

$$u = \left(\frac{Kay}{2\eta}\right)\left(1 - \frac{y}{a}\right) \tag{12.54}$$

It is left as an exercise for you to derive this result.

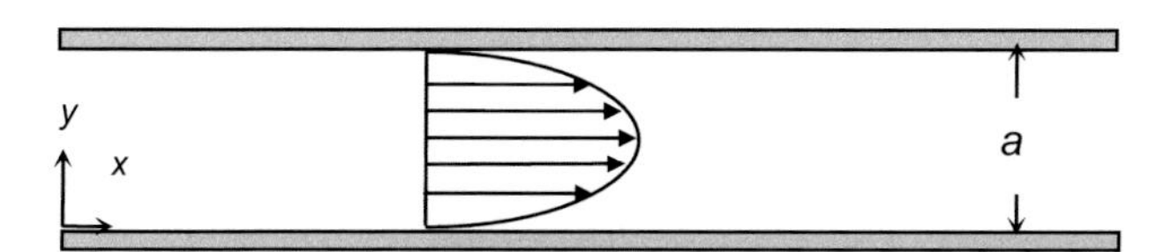

Fig. 12.12. Pressure-driven Poiseuille flow results in a parabolic velocity profile.

12.5 Electro-osmotic flow

One of the most common methods of creating fluid motion in microfluidic devices is by applying an electric field in the flow direction. The application of an electric field causes a body force on charged particles within the fluid which in turn viscously drag the surrounding fluid along with them. Such a flow is known as **electro-osmotic (EO) flow**. Unlike pressure driven flows such as Poiseuille flow, EO flows result in an almost perfectly *flat* velocity profile. Such a profile has particular advantages in chemical detection devices, which are among the most common microfluidic devices.

As electric fields play such a prominent role in EO flows, it behooves us to review some of the fundamentals of electrostatics encountered in freshmen-level physics courses.

12.5.1 Electrostatics

Coulomb's Law

The starting place for electrostatics is **Coulomb's law** which states that the force that one electrically charged particle of electric charge q_1 exerts on another electrically charged particle of charge q_2 is directly proportional to the charges and inversely proportional to the square of the distance between them. Mathematically, Coulomb's Law is given by

$$\vec{F} = \frac{kq_1q_2}{r^2}\hat{e}_r = \frac{1}{4\pi\varepsilon}\frac{q_1q_2}{r^2}\hat{e}_r \tag{12.55}$$

where r is the distance between the particles, $\hat{e}_r$ is the unit vector in the direction of r from q_1 to q_2, and ε is the electrical permittivity. (You may be more familiar with the form of this law which includes Coulombs' constant, k. Don't worry. This is the same law and k is equal to $1/4\pi\varepsilon$. We prefer ε here because later on a 4π term shows up to cancel the one in Eq. (12.55). It helps to know what your results will be.) We can see from Eq. (12.55) that the force between the charged particles is repulsive if the two charges have the same sign (both positive or both negative) and attractive if the two charges have opposite sign (one positive and one negative.) We also see that Coulomb's law has the same form as Newton's law of gravity. In fact, many helpful analogies can be drawn between gravitational forces and electrostatic forces.

Electric fields

Equation (12.55) gives the force between only two electrically charged particles. In the case where there are large numbers of electrically charged particles, each one of them exerts a force on every other one, and the resulting distribution of forces is potentially quite complex. A useful tool to help us find electrical forces in such a case is that of the **electric field**. In its simplest form, an electric field gives the value of the net electric force exerted on a particle q at any point in space, but *per unit charge*. In other words, when we would like to know the value of the force exerted on a particle of arbitrary electrical charge q, we simply multiply q by the electric field at that point in space, as given in Eq. (12.56).

$$\vec{F} = q\vec{E} \tag{12.56}$$

The gravitational analogue of an electric field is the gravitational field, which gives the value of the gravitational force exerted on a mass per unit mass. Near its surface, the earth's gravitational field has a constant value given by $g = 9.81$ N/kg in a direction toward the earth's surface. The resulting equation is the familiar expression relating an object's mass to its weight:

$$\vec{F} = m\vec{g}\,. \tag{12.57}$$

The simplicity of Eq. (12.57) is due to the fact that the earth is very large compared to objects on its surface, resulting in a constant value for g. The gravitational forces exerted between astronomical bodies on one another, however, depend heavily on position, making the use of the gravitational field idea indispensable. This is often the case with electrical fields too.

Electric potential

The most common intuitive definition of work is given as "a force through a distance." For example, if you want to lift a 1000 lb ball off of the ground, you have to perform work on it, the magnitude of which is equal to the ball's weight multiplied by the vertical distance which you raise it. If you wish to roll it along the ground, however, this requires no work against gravity. (You must still overcome friction, of course.) Another way to look at this is that one must perform work on a mass in order to move it around in a gravitational field.

The same is true about moving charges around electric fields. For a charged particle of charge q, the work needed to move it from location 1 to location 2 in a *constant* electric field, then, is given by

$$W_{1\to 2} = -\vec{F} \bullet \vec{r}_{1\to 2} = -q\vec{E} \bullet \vec{r}_{1\to 2} \tag{12.58}$$

The negative sign is needed since the force that must be exerted on the charge to move it is equal and opposite to the field force which must be overcome. The dot product reflects the fact that only displacements parallel to the direction of force contribute to the work.

By performing work on the charge, its energy has increased. Energy, like mass and momentum, is also a conserved quantity. In this case, this increase in energy is stored as electrical potential energy, $\mathcal{U}$. (We use $\mathcal{U}$ rather than E for energy so as not to confuse it with electric field.) Thus,

$$W_{1\to 2} = -q\vec{E} \bullet \vec{r}_{1\to 2} = \mathcal{U}_2 - \mathcal{U}_1 . \tag{12.59}$$

We define **electric potential**, φ, as this electric potential energy *per unit charge*. For our example, then,

$$\varphi_2 - \varphi_1 = -\vec{E} \bullet \vec{r}_{1\to 2} . \tag{12.60}$$

In other words, the work *per unit charge* required to move a charge around in an electric field is equal to its change in electric potential. If the electric field is not constant, then we get the more general relationship

$$\varphi_2 - \varphi_1 \equiv -\int_1^2 \vec{E} \bullet d\vec{r} . \tag{12.61}$$

The utility of the electric potential is that it manifests itself most often as a voltage, an easily measured quantity and one for which we have some physical feel.

Now there is a handy theorem in vector calculus that tells us that if the integral of a vector dotted with a differential position vector (a so-called path integral) is equal to the difference of a scalar quantity, then the vector must be the gradient of the scalar. Eq. (12.61) has this form, and so

$$\vec{E} = -\vec{\nabla}\varphi . \tag{12.62}$$

Intuitively this makes sense if we consider the work per unit charge required to move a charge *parallel* to an electric field. In this case, the force and displacement are in the same direction and we can drop the dot in Eq. (12.61) to get

$$\varphi_2 - \varphi_1 \equiv -\int_1^2 E dr \tag{12.63}$$

which implies

$$\frac{d\varphi}{dr} = -E . \tag{12.64}$$

Of course this is a special form of the more general Eq. (12.62), but at least it shows you that the gradient in Eq. (12.62) makes sense.

Gauss's Law

Gauss's law represents a convenient relationship between some of the electrostatic properties we have just reviewed. We can develop Gauss's law by considering a point charge of charge q_0 at the center of a spherical surface of radius R as shown in Fig. 12.13.

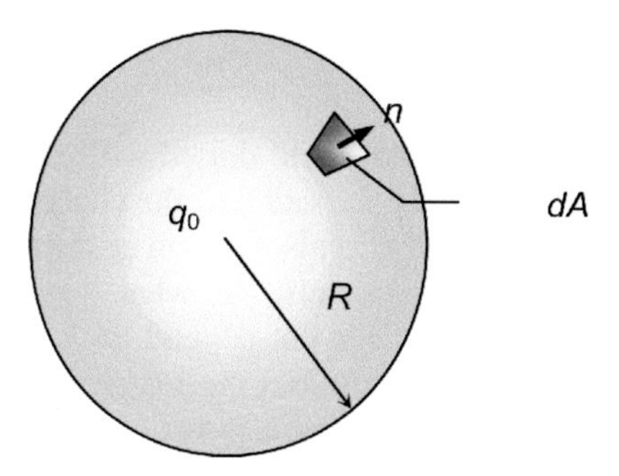

Fig. 12.13. Derivation of Gauss's Law

To find the value of the electrostatic force q_0 exerts on an arbitrary point charge q located on the spherical surface, we use Eq. (12.55) to get

$$\vec{F} = \frac{1}{4\pi\varepsilon}\frac{q_0 q}{R^2}\hat{e}_R . \tag{12.65}$$

To get the electric field on the surface (the force per unit charge) we need only divide by q:

$$\vec{E} = \frac{1}{4\pi\varepsilon}\frac{q_0}{R^2}\hat{e}_R . \tag{12.66}$$

And so, a point charge q_0 creates an electric field in space at distance R away as given by Eq. (12.66).

For reasons that will become apparent shortly, we would also like to know the value of the integral

$$\oint_{A_{sp}} \vec{E} \bullet \hat{n} \cdot dA$$

evaluated on the surface area of the sphere, A_{sp}. Here, $\boldsymbol{n}$ is an outward pointing normal to the surface and therefore coincides with $\boldsymbol{e_R}$. As long as we stay on the surface of the sphere, R has a constant value, and the integral takes on a simple form:

$$\oint_{A_{sp}} \vec{E} \bullet \hat{n} \cdot dA = \oint_{A_{sp}} \frac{1}{4\pi\varepsilon} \frac{q_0}{R^2} \hat{e}_R \bullet \hat{n} \cdot dA \tag{12.67}$$

$$= \frac{1}{4\pi\varepsilon} \frac{q_0}{R^2} \oint_{A_{sp}} dA = \frac{1}{4\pi\varepsilon} \frac{q_0}{R^2} A_{sp} = \frac{1}{4\pi\varepsilon} \frac{q_0}{R^2} \cdot 4\pi R^2$$

$$= \frac{q_0}{\varepsilon}$$

Therefore the value of an electric field dotted with an outward pointing area element of a sphere is simply the value of the charge *inside* the sphere divided by ε. The area integral of $\boldsymbol{E}$ is also known as the **electric flux**.

Now you may be thinking to yourself, "So what?" Well, the interesting thing about Eq. (12.67) is that it does not depend on R. You can make the sphere as big or as little as you want and the integral is still q_0/ε. In fact, since there is no dependence on R, there is really no dependence on the *shape* of the enclosing surface where the integral is evaluated, or the nature or location of q within in it either. This is quite a useful result, and this is what is known as Gauss's law:

$$\oint_{A_{surface}} \vec{E} \bullet \hat{n} \cdot dA = \frac{q_{inside}}{\varepsilon}. \tag{12.68}$$

Most of the time we use Gauss's Law to determine $\boldsymbol{E}$ from knowledge of q_{inside}. This is most easily done when a high degree of symmetry exists in the problem and we can therefore choose a simple enclosing shape.

Charge density and Gauss's Law at a point

Thus far we have talked about electrostatics in terms of point charges – discreet little points that have a fixed charge. Much of the discussion so far may in fact be a review of material with which you are already familiar. However, just as the mechanics of particles is often inadequate to describe the mechanics of fluids, so is the idea of point charges in describing the electrostatics of fluids. We therefore need to introduce some more advanced ideas, such as that of **charge density**, which plays the same role for electric phenomena as mass density does for mechanical phenomena. Mathematically,

$$\rho_e \equiv \lim_{V\to 0}\left(\frac{dq}{dV}\right) \tag{12.69}$$

Thus, charge density is nothing more than the net electric charge per unit volume at a point within a substance. As with mass density, we must assume that our fluid is a continuum in order to use this idea.

Since we said that Gauss's law doesn't care about the nature of the charge inside the enclosing surface, we can just as easily apply it to a fluid with a continuously varying charge distribution as to point charges. For the former case Gauss's law becomes

$$\oint_{A_{surface}} \vec{E} \bullet \hat{n} \cdot dA = \frac{q_{inside}}{\varepsilon} = \frac{\oint_V \rho_e \cdot dV}{\varepsilon}. \quad (12.70)$$

We now make use of another handy theorem from vector calculus which states that the integral of a vector dotted with an outward normal area of an enclosing surface is equal to the integral of the divergence of the vector within the *volume* of the surface. In other words, if you take the divergence of something, it can turn a surface integral into a volume integral. This is called, not surprisingly, the **divergence theorem**. Using the divergence theorem on Eq. (12.70) yields

$$\oint_{A_{surface}} \vec{E} \bullet \hat{n} \cdot dA = \oint_V \vec{\nabla} \bullet \vec{E} \cdot dV = \frac{\oint_V \rho_e \cdot dV}{\varepsilon}. \quad (12.71)$$

Since the volume over which we are evaluating these integrals is arbitrary, we don't have to do the integral at all. We can just state that the value of what's in the integral on the left hand side is equal to what's in the integral on the right hand side.

$$\vec{\nabla} \bullet \vec{E} = \frac{\rho_e}{\varepsilon}. \quad (12.72)$$

The divergence theorem has just let us express Gauss's law *at a point*, much in the same way that Navier-Stokes expresses the conservation of momentum at a point.

We can take this one step further by making use of Eq. (12.62):

$$\vec{\nabla} \bullet E = \vec{\nabla} \bullet \left(-\vec{\nabla}\varphi\right) = -\vec{\nabla}^2 \varphi = \frac{\rho_e}{\varepsilon} \quad (12.73)$$

It's left as an exercise to show that $\vec{\nabla} \bullet (\vec{\nabla}\varphi) = \vec{\nabla}^2 \varphi$.[6]

[6] It's actually pretty easy if you just use rectangular coordinates. If you want to know more about Laplacians, gradients, divergence and all those handy integral

The expression of Gauss's Law given by Eq. (12.73) is the most useful formulation for fluids applications, as it is valid at a point within a continuum. Furthermore, it is expressed in terms of the often more convenient (that is, measurable) quantity of electric potential rather than electric field.

If your background hasn't permitted you to work with electrostatic phenomena very often, and the previous equations are therefore not particularly familiar, just keep in mind that *all* of them are really only a restatement of Coulomb's law, Eq. (12.55). We have made use of some mathematical gymnastics, but all of this is only for the sake of convenience. Simply stated, all these equations say, in one form or another, that big charges attract or repulse each other more than small charges do, and that the closer those charges are to each other, the stronger that attraction or repulsion is.

12.5.2 Ionic double layers

In order for EO flow to work, the fluid itself must be an **electrolyte**, or a fluid capable of conducting electricity. The mode of electric conduction in an electrolyte is via the movement of ions within it. Now usually an electrolyte is going to be neutral in bulk; that is, for most of the fluid there are just as many negative ions capable of conducting electricity floating around as there are positive ions, resulting in $\rho_e = 0$. When an electrolyte is brought in contact with a solid surface, however, complicated interactions will leave the surface with a net charge. Therefore, the fluid near surface will also have a net charge associated with it. The thin layer next to the solid surface where all this occurs is known as the **ionic double layer**. It's where all the action takes place in EO flow.

The nature of the ionic double layer, including how thick or thin it is, what the surface charge density of the wall is, etc. depends heavily on the fluid/surface combination. For water next to a glass surface, the mechanism for the creating of the ionic double layer appears to be the deprotonation of silanol groups within the glass. (See Fig. 12.14.) This is just a fancy way of saying that SiOH in the glass near the solid/liquid surface gives up a proton to the water. This proton, at least within the double layer, becomes the source of local non-zero charge density within the fluid.

---

theorems, you should take a look at a text on vector calculus; even better, take a course on it. Familiarity with vector calculus adds tremendous insight into many physical phenomena even when the resulting equations are ones that are not routinely solved.

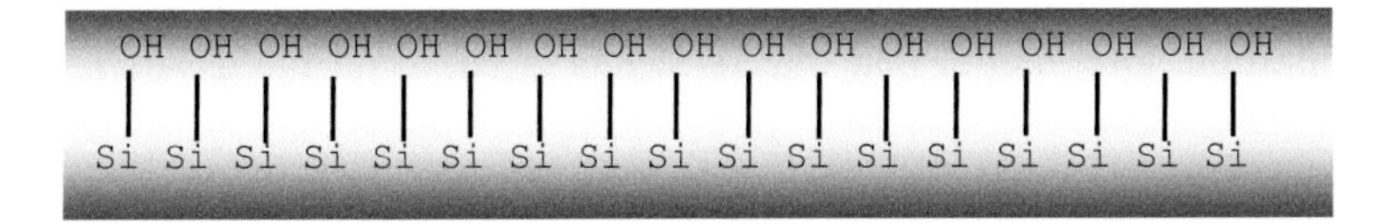

Dry glass

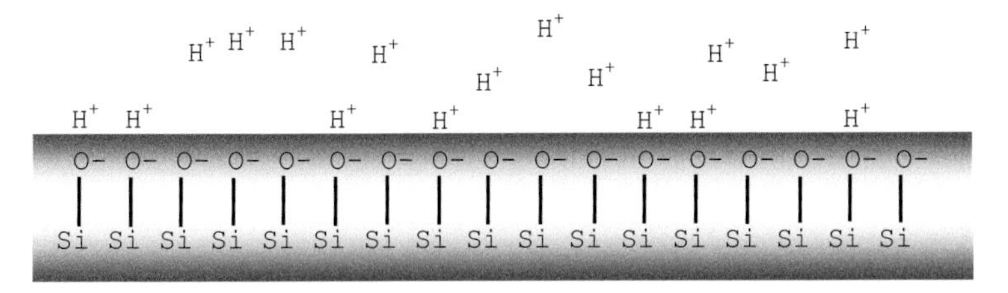

glass surface wetted by water

Fig. 12.14. Ionic double layer formed by water on glass

The name "ionic double layer" comes from modeling the region of non-zero charge density as being made up of two layers. The inner layer, sometimes called the Stern layer, is made up of immobile ions on the solid surface. The *diffuse layer*, also called the Gouy-Chapman layer, extends to the point where all the other ions meet the rest of the fluid. It is in the diffuse layer that externally applied electric fields can affect the fluid's state of motion, since it contains mobile ions. The plane separating the two layers two is called the shear surface, as relative motion can occur across it.

The DeBeye length

If we are to quantitatively describe EO flow, we must quantitatively describe the distribution of charge within the ionic double layer. Specifically, we need to know the value of charge density ρ_e and its spatial variation within the double layer. The net charge density depends on the charges of all ions within the electrolyte and their respective concentrations. Hence,

$$\rho_e = \sum_i n_i z_i q_e \,. \tag{12.74}$$

where n_i is the concentration of species i per unit volume and z_i is the corresponding charge of that species in units of electronic charge q_e.[7]

The key, then, is to find the spatial variation of the each ionic species' concentration. One model that that does this is based on the assumption that the two main things which tug on the ions in the double layer, **electromigration** and **mass diffusion**, perfectly balance each other. This is equivalent to saying that within the double layer, we are always fairly close to equilibrium.

Electromigration refers to nothing other than the movement of charge due to a potential difference. The resulting charge flow per unit time is the familiar electric current, i. You may recall that for simple DC circuits this current is opposed by the internal resistance of the conductor. For a material which obeys Ohm's law, this relationship is

$$\varphi_1 - \varphi_2 = iR. \tag{12.75}$$

This is a macroscopic-type relation. Again, we must concern ourselves with how quantities vary spatially within the flow field, and we therefore need the version of Ohm's law which applies at a point in the fluid. This is given by

$$\vec{E} = \vec{J}\rho, \tag{12.76}$$

where $\boldsymbol{J}$ is the **current density**, or current per unit area normal to the current direction, and ρ is the electrical resistivity. (The symbol ρ is also used for mass density, but you'll be able to figure out which one is which from context.) In terms of the electrical conductivity of the material, $\sigma = 1/\rho$, Ohm's law is

$$\vec{J} = \sigma\vec{E}. \tag{12.77}$$

Electrical conductivity is a measure of how well electricity moves through the material. It is a property with which you are most likely already familiar. When we deal with electrolytic solutions, however, the conductivity can change based on the concentrations of the various ions, what those ions are and so on. We therefore need a more detailed version of electrical conductivity which includes these effects.

---

[7] We are in fact referring to z_i when we say that the charge of an ion is "+1"or "-2," but this usage carries the danger of our forgetting that we are talking about a physically measurable quantity with both *dimensions* and *units*. z_i has the dimension of electric charge and units of the charge of an electron, or 1.602×10^{-19} coulombs. In light of this fact, q_e as it appears in Eq. (12.74) is actually a conversion factor, not a variable.

Electrical conductivity depends in part on how many charges there are which can freely move about the material at large. These mobile charges are called **charge carriers**. Unbound electrons serve as the charge carriers in metals. In electrolytes, each ionic species *i* performs this function. Conductivity also depends on how likely individual charge carriers are to move around, the measure of which is the **electric mobility**. In terms of these quantities, the electrical conductivity is

$$\sigma = \sum n_i z_i q_e \mu_i , \tag{12.78}$$

where μ_i is the mobility of the charge carrier.

When an ionic double layer forms, the resulting distribution of charges creates an electric field. For an electric field $\boldsymbol{E}$ within the double layer, the current density for ionic species *i* is found from combining Eqs. (12.77) and (12.78)

$$\vec{J}_i = n_i z_i q_e \mu_i \vec{E} . \tag{12.79}$$

Eliminating $\boldsymbol{E}$ in favor of potential (Remember, $\vec{E} = -\vec{\nabla}\varphi$),

$$\vec{J}_i = -n_i z_i q_e \mu_i \vec{\nabla}\varphi . \tag{12.80}$$

Keep in mind that this equation really represents Ohm's law at a point within the double layer. Figure 12.15 illustrates this. In Fig. 12.15, we have assumed that the solid surface is negatively charged and the resulting field points toward the surface, but Eq. (12.80) will work for the general case.

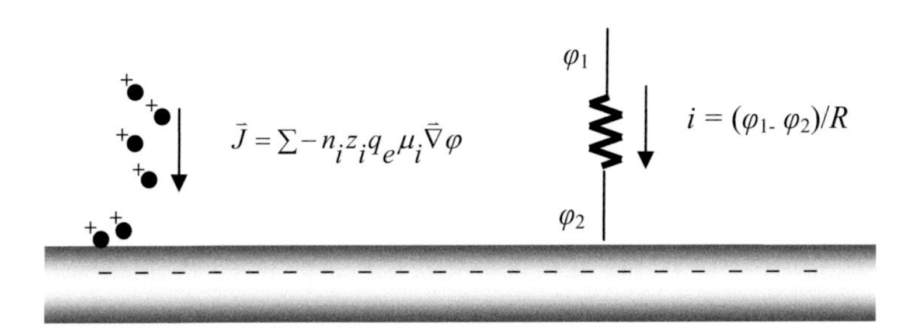

Fig. 12.15. Electromigration of ions within the double layer resembles the flow of electric current through a resistor.

Balancing electromigration within the double layer is **mass diffusion**. Substances tend to move in the direction of decreasing concentration. You have no doubt experienced this effect when using a paper towel to pick up a liquid spill. When the spill is initially touched by the towel, the liquid quickly moves into the dry regions of the towel. If the spill is large enough, the liquid will eventually saturate the entire towel. Ions within the double layer are no different. If there is a large concentration of positive ions near the wall, they will want to move out into the fluid where a smaller concentration of positive ions exists.

The mathematical relationship describing mass transfer due to a concentration gradient is quite simple and is given by

$$\vec{j}_i = -D_i \vec{\nabla} n_i , \tag{12.81}$$

where j_i is the **mass flux**, or the mass flow per unit area normal to the flow direction, and D_i is the **mass diffusivity** of species i within a mixture. The negative sign indicates that mass travels in the direction of decreasing concentration.

For the most part mass diffusivity is just another material property like thermal conductivity or viscosity.[8] One difference, however, is that D_i not only depends on the species i, but also the substance through which species i is diffusing. Indeed, paper towel manufacturers would all have you believe that their particular brand effects a larger value of D_i than do their competitors'.

Though we use the term "mass transfer," we often don't use dimensions of mass for it. In fact, the form of Eq. (12.81) will yield j_i in dimensions of [amount of substance]/[time][area], a typical unit for which would be moles/s-m^2. To actually get mass flux in dimensions of [mass]/[time][area], we would simply use the density gradient of species i rather than concentration gradient. In our case, we really don't want either, but the current density, $\boldsymbol{J}_i$ of species i due to mass diffusion instead. We need only multiply Eq. (12.81) by the charge of species i to achieve this:

$$\vec{J}_i = -z_i q_e D_i \vec{\nabla} n_i \tag{12.82}$$

We are now in a position to determine what the concentration of ions looks like within the double layer. Our model simply says that the effect of electromigration and mass diffusion balance each other. In other words, the current density due to electromigration and the current density due to mass diffusion for any species i must add to zero.

[8] Equation ((12.81) is the same as Eq. (2.2) given in Chapter 2. Comparing the form of Eq. ((12.81) with Eqs. (11.4) and (12.3) we are again reminded that the one dimensional fluxes of mass, heat flux and viscous stress all follow Fick's law of diffusion.

$$\vec{J}_{total,i} = \vec{J}_{electromigration,i} + \vec{J}_{diffusion,i} = 0 \tag{12.83}$$

$$\frac{1}{n_i}\vec{\nabla} n = -\frac{\mu_i}{D_i}\vec{\nabla}\varphi \tag{12.84}$$

Thus,

In our study, we concern ourselves only with variations in the direction normal to the solid surface, as illustrated in Fig. 12.16.

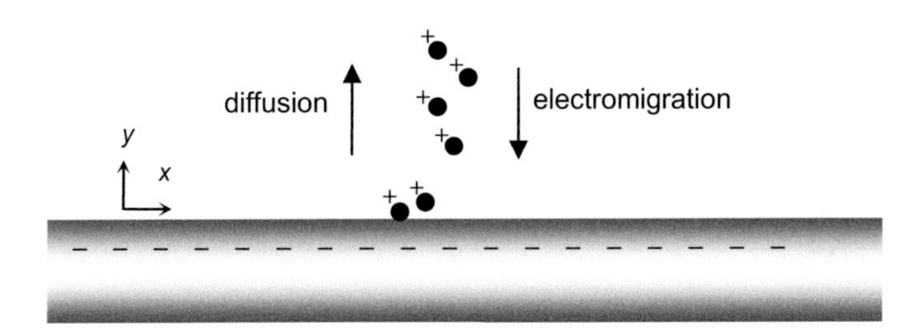

Fig. 12.16. Mass transfer by diffusion is balanced by electromigration within the double layer.

Equation (12.84) therefore reduces to

$$\frac{1}{n_i}\frac{dn_i}{dy} = -\frac{\mu_i}{D_i}\frac{d\varphi}{dy}$$

or

$$\frac{1}{n_i}dn_i = -\frac{\mu_i}{D_i}d\varphi\,. \tag{12.85}$$

Equation (12.85) can be integrated directly to obtain the relationship between concentration and potential. We will set the reference for potential equal to zero far from the wall where the concentration of species i is its bulk concentration, $n_{i,0}$. Integrating Eq. (12.85), then,

$$\int_{n_{i,0}}^{n_i} \frac{1}{n_i}dn_i = \int_0^{\varphi} -\frac{\mu_i}{D_i}d\varphi$$

$$\ln\left(\frac{n_i}{n_{i,0}}\right) = -\frac{\mu_i}{D_i}\varphi$$

$$n_i = n_{i,0}\exp\left(-\frac{\mu_i}{D_i}\varphi\right). \tag{12.86}$$

Albert Einstein realized that the mass diffusivities of charge carriers and their electric mobilities must somehow be related. This relation is given by

$$\frac{\mu_i}{D_i} = \frac{z_i q_e}{k_B T}, \tag{12.87}$$

which is appropriately referred to as the **Einstein relation**.[9] Here k_B is Boltzman's constant, the equivalent of the universal gas constant per molecule, and T is absolute temperature. Making use of the Einstein relation in Eq. (12.86) yields

$$n_i = n_{i,0}\exp\left(-\frac{z_i q_e}{k_B T}\varphi\right) \tag{12.88}$$

We are almost there. Eq. (12.88) gives us the concentration of each ionic species which we can now plug into Eq. (12.74) to get ρ_e.

$$\rho_e = \sum_i n_i z_i q_e = \sum_i n_{i,0}\exp\left(-\frac{z_i q_e}{k_B T}\varphi\right) z_i q_e \tag{12.89}$$

What Eq. (12.89) lacks, however, is the *spatial* variation of ρ_e. To get that, we make use of Gauss's law at a point. (We didn't look at electrostatics for nothing.) And so, let's "simplify" Eq. (12.89) by plugging it into Gauss's law at a point, Eq. (12.73).

$$\vec{\nabla}^2\varphi = -\frac{\rho_e}{\varepsilon} = -\frac{\sum_i n_{i,0}\exp\left(-\frac{z_i q_e}{k_B T}\varphi\right) z_i q_e}{\varepsilon} \tag{12.90}$$

As stated earlier, we concern ourselves only with variations normal to the solid surface. Eq. (12.90) then becomes

[9] The derivation of the Einstein relation is beyond the scope of this text. Suffice it to say that the Einstein relation is a sometimes useful relationship between diffusivity and mobility much in the same way that the ideal gas law is a sometimes useful relationship between pressure, temperature and volume.

$$\frac{d^2\varphi}{dy^2} = -\frac{\sum_i n_{i,0} \exp(-\frac{z_i q_e}{k_B T}\varphi) z_i q_e}{\varepsilon} \tag{12.91}$$

We might be able to solve this equation if only there weren't an exponential term in it. Lucky for us, Einstein wasn't the only clever fellow with short-cuts to contribute to our journey. Another clever fellow named Debeye figured out a way around this problem. Debeye realized that for small $z_i q_e \varphi / k_B T$ we can do a Taylor expansion on the exponential term to get

$$\exp(-\frac{z_i q_e}{k_B T}\varphi) \approx 1 - \frac{z_i q_e}{k_B T}\varphi + \text{terms of higher order}$$

This turns Eq. (12.91) into

$$\frac{d^2\varphi}{dy^2} = -\frac{\sum_i n_{i,0}(1 - \frac{z_i q_e}{k_B T}\varphi) z_i q_e}{\varepsilon} \tag{12.92}$$

$$= -\frac{\sum_i n_{i,0} z_i q_e}{\varepsilon} + \frac{\sum_i n_{i,0} \frac{(z_i q_e)^2}{k_B T}\varphi}{\varepsilon}$$

Remember how we said that in the bulk our electrolytes are neutral, meaning that there are just as many positive ions floating around as negative ions with the opposite charge? For this reason, we see that the first term on the right hand side of Eq. (12.92) must be zero. The second term, however, is non-zero since $z_i q_e$ is squared. This allows us to write Eq. (12.92) as

$$\frac{d^2\varphi}{dy^2} = \frac{1}{\lambda_D^2}\varphi, \tag{12.93}$$

where we have taken the liberty of replacing all that stuff in the summation with $1/\lambda_D^2$. λ_D is called the **Debeye length**. It is given by

$$\lambda_D = \left(\frac{1}{\varepsilon k_B T}\sum_i n_{i,0}(z_i q_e)^2\right)^{-1/2} \tag{12.94}$$

Eq. (12.93) is a homogenous linear first order differential equation with constant coefficients, and one of the more readily solvable types. The general solution for such an equation is

$$\varphi = C_1 e^{(y/\lambda_D)} + C_2 e^{-(y/\lambda_D)} \tag{12.95}$$

To obtain the coefficients C_1 and C_2 we apply the appropriate boundary conditions. We have already stated that our reference for φ is zero far from the wall, making $C_1 = 0$. At the wall itself, the potential has some finite value that we will call the wall potential, φ_w.[10] C_2, then, must be φ_w and thus

$$\varphi = \varphi_w e^{-(y/\lambda_D)} \tag{12.96}$$

Eq. (12.96) gives us the interpretation of the Debeye length as the exponential decay distance for the wall potential. In three or four Debeye lengths, the wall potential fades to zero and we have left the ionic double layer. Debeye lengths vary from a fraction of a nanometer to a fraction of a micron. Typical microfluidic channels have widths on the order of microns to a millimeter; thus, ionic double layers are indeed thin compared to channel widths.

Let's not loose sight of where we are headed. Remember that we wanted an expression for the *spatial* variation of charge density. The simple form of Eq. (12.96) let's us do exactly that. We need only invoke Gauss's law at a point one more time while incorporating Eq. (12.96) for the potential.

$$\vec{\nabla}^2 \varphi = \frac{d^2\varphi}{dy^2} = -\frac{\rho_e}{\varepsilon} \tag{12.97}$$

Solving for ρ_e and using Eq. (12.96),

$$\rho_e = -\varepsilon \frac{d^2\varphi}{dy^2} = -\varepsilon \frac{\varphi_w}{\lambda_D^2} e^{-(y/\lambda_D)}. \tag{12.98}$$

We have achieved our goal. Eq. (12.98) gives the spatial variation of charge density within the ionic double layer. We see that close to the surface ρ_e has a non-zero value but quickly diminishes to its bulk value of zero within three to four Debeye lengths.

It is important to keep in mind that this section presented only one model for approximating the distribution of charge within the ionic double layer. Though the model presented here is the most common, it is not always appropriate and other models are out there. Indeed, understanding

[10] The wall potential φ_w is often referred to as the **zeta potential** ζ in the EO literature. Manipulation and control of the zeta potential, of course, is a big deal and a most active research area.

and characterizing ionic double layers represents one of the most important issues in microfluidics.

12.5.3 Navier-Stokes with a constant electric field

We are now in a position to put our knowledge of fluid mechanics together with our knowledge of ionic double layers to describe EO flow. Consider a fluid bounded by a flat surface, the combination of which produces an ionic double layer as shown in Fig. 12.17. By applying a constant electric field E_x across the fluid *parallel to the surface* we should be able to get the fluid in the double layer, fluid with nonzero charge density, to move. We will assume here that the applied field in the flow direction is small enough compared to the field perpendicular to the double layer so that our previous characterization is valid. This condition is satisfied in most microfluidics devices.

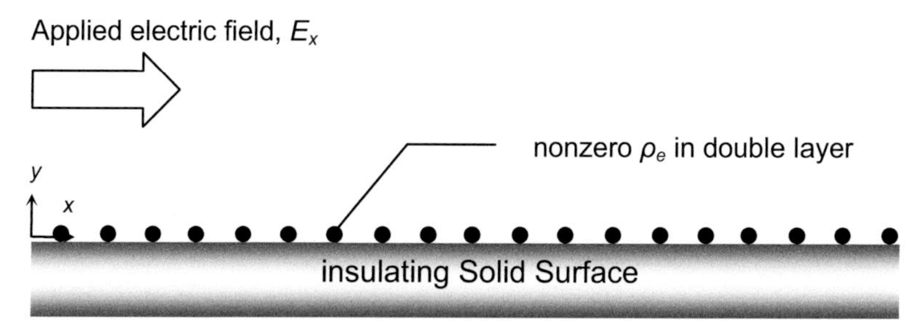

Fig. 12.17. Electroosmotic flow

To describe the variation of fluid velocity with position, we need the x-direction Navier Stokes equation, Eq. (12.38).

$$\rho\left[\frac{\partial u}{\partial t}+u\frac{\partial u}{\partial x}+v\frac{\partial u}{\partial y}\right]=-\frac{\partial P}{\partial x}+\eta\left(\frac{\partial^2 u}{\partial x^2}+\frac{\partial^2 u}{\partial y^2}\right)+b_x \qquad ((12.38)$$

Let's assume we have steady, one-dimensional flow with no applied pressure gradient. Eq. (12.38) reduces to

$$0=\eta\left(\frac{\partial^2 u}{\partial y^2}\right)+b_x \qquad (12.99)$$

Notice that the body force hasn't dropped out this time. Indeed, there is an *electrical* body force applied in the flow direction, which in fact drives the flow. Now the force on a charge of q due to the applied field E_x would be given by

$$F_x = qE_x.$$

We need the body force per unit volume, though, and so we see that

$$b_x = \rho_e E_x,$$

which turns Eq. (12.99) into

$$0 = \eta\left(\frac{d^2u}{dy^2}\right) + \rho_e E_x \tag{12.100}$$

Using Eq. (12.98) for ρ_e we get

$$0 = \eta\left(\frac{d^2u}{dy^2}\right) - \varepsilon\frac{\varphi_w}{\lambda_D^2}e^{-(y/\lambda_D)} \cdot E_x \tag{12.101}$$

Rearranging

$$\frac{d}{dy}\left(\frac{du}{dy}\right) = \frac{\varepsilon}{\eta}\frac{\varphi_w}{\lambda_D^2}e^{-(y/\lambda_D)} \cdot E_x\,. \tag{12.102}$$

Eq. (12.102) is a linear ordinary differential equation which can be solved by direct integration. Integrating twice results in

$$u = \frac{\varepsilon\varphi_w E_x}{\eta} \cdot e^{-(y/\lambda_D)} + C_1 y + C_2\,. \tag{12.103}$$

The constants of integration C_1 and C_2 can be found by applying the appropriate boundary conditions. The no slip boundary condition requires that $u = 0$ at $y = 0$, giving

$$C_2 = -\frac{\varepsilon\varphi_w E_x}{\eta}\,. \tag{12.104}$$

Far from the wall, as $y \rightarrow \infty$, u must have a finite value, necessitating that $C_1 = 0$. Finally we have our velocity distribution for EO flow

$$u = \frac{\varepsilon\varphi_w E_x}{\eta}\left(e^{-(y/\lambda_D)} - 1\right). \tag{12.105}$$

As we have seen, λ_D is usually very small, on the order of nanometers. In three or four Debeye lengths from the solid surface, the exponential term in Eq. (12.105) goes away and u becomes a constant value.

$$u \approx U_{avg} = -\frac{\varepsilon \varphi_w E_x}{\eta}. \tag{12.106}$$

In most microfluidic devices, the width of the flow channel is large compared to the Debeye length and thus, except for very close to the channel walls, the resulting velocity profile is flat. As we have already seen, the profile for *pressure-driven* flow is parabolic. (See Fig. 12.18.) Flat profiles have tremendous advantages in chemical analysis systems, as will be shown in the next section.

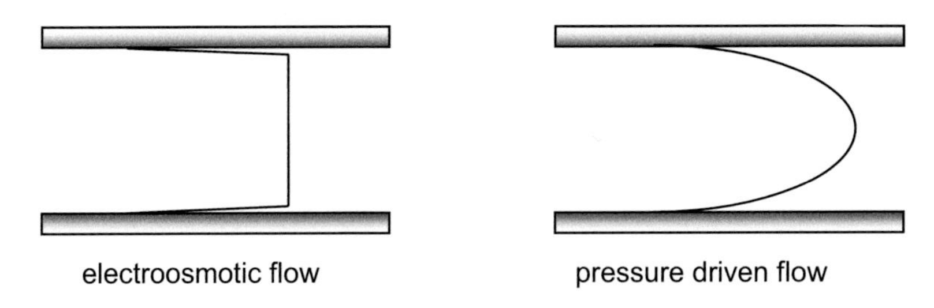

Fig. 12.18. Comparison of velocity profiles for electroosmotic and pressure driven flows

In addition to its flat velocity profile, EO flow differs from pressure driven flow in that the resulting fluid velocities are independent of channel geometry, namely channel width. This is not the case at all with pressure driven flows. In fact, the smaller the channel dimensions, the larger the required *dP*/*dx* to achieve the same average velocity. (If you have ever tried to use one of those little plastic coffee stirrers as a straw, you know what I mean. A big straw lets you suck up your beverage without any problem, but you have to suck really hard on the coffee stirrer to get just a trickle of your drink.) Not true with EO flow, however. Eq. (12.106) shows us that for a given E_x and zeta potential (φ_w) you get a certain velocity. This is another reason why EO flow is so popular in microfluidics.

12.6 Electrophoretic separation

In EO flow we can visualize the applied electric field as pulling very thin "plates of charge" next to the walls along the microchannel. These “charge plates” are made up of the ions within the diffuse layer of the ionic double layer, and viscously drag the bulk fluid in the rest of the channel with them, creating the characteristic *plug flow* seen in EO. (Fig. 12.19) This scheme works because the source of the charge density in the fluid is the electrolyte's ions themselves, which exist in sufficient concentrations to be considered in the background. In other words, there are enough charges in the fluid so that it makes sense to talk about the charge density ρ_e, or "charge at a point."

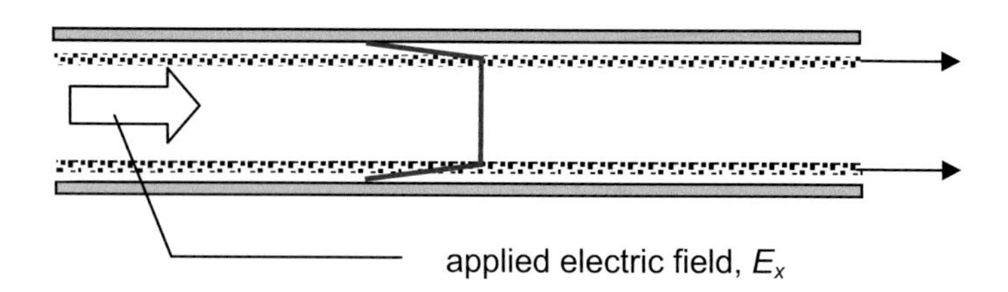

Fig. 12.19. "Charge plates" in electroosmotic flow

Imagine now that in addition to these background ions, there are also some other ions of different chemical species in the fluid. These ions exist in small enough amounts so that they don't mess up the background charge density we've already calculated. Now electric fields affect these new ions differently than the background ions, allowing them to move relative to the bulk fluid. Manipulating these ions and their relative motions makes up the area of **electrophoresis**, the backbone of chemical analysis in microfluidic devices.

An analogy may be helpful here. Visualize our fluid as vegetable soup made up of broth and small chunks of vegetables. Now broth is mostly water, but unlike water, it's not colorless (unless you live in Terre Haute, that is). The stuff that gives the broth its color is the background ion concentration. The chunks of vegetables are the new ions we've just thrown in. The addition of a few carrots or mushrooms does not affect the nature of the broth, just as the presence of small concentrations of ions does not affect the background ions in an electrolyte. (Fig. 12.20.)

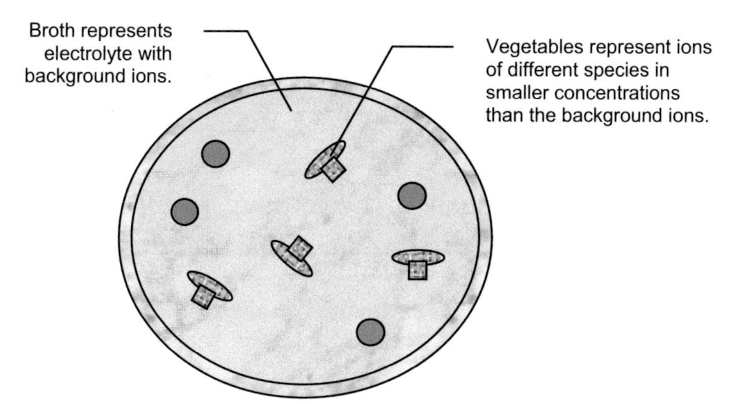

Fig. 12.20. Chemical sample as vegetable soup

Furthermore, when you spoon a ladle full of vegetable soup, you observe that the carrots and mushrooms move around relative to the broth. This relative motion of veggies to broth is the analogue of electrophoresis. Such a relative motion is frequently referred to as **drift**.

The effect of an applied electric field on our ions of low concentration is captured in a property known as the **electrophoretic mobility**, given by

$$u_{ep} = \mu_{ep} E_{x,} \tag{12.107}$$

where u_{ep} is velocity of the ion relative to the bulk fluid (the **drift velocity**) and μ_{ep} is the electrophoretic mobility.

Figure 12.21 shows a typical microfluidic device used in chemical assays. The device consists of a separation column, a junction, and entry ports by which fluid is introduced to the device. Typical channels measure 80-100 μm wide by 10-30 μm deep and are usually etched in a flat piece of glass. A flat glass cover plate is bonded to the channel plate and holes are drilled in it to form the ports. The length of the separator column itself varies and serves as a design parameter for the device.

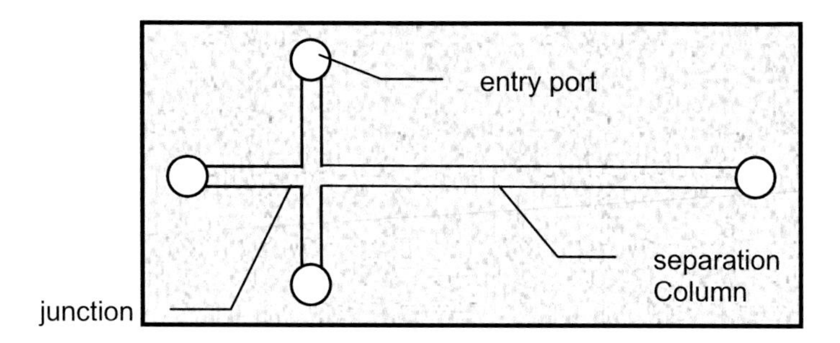

Fig. 12.21. Typical microfluidic device used for electrophoretic separation

Figure 12.22 shows the use of such a device as a chemical analysis system. First the channels are filled with a buffered electrolyte. (Buffer solutions are ones for which the pH is stable.) A plug of sample is then introduced into one of the entry ports of the shorter channel. A voltage drop applied across the short channel results in EO flow; the sample therefore fills the short channel. A voltage drop is then applied across the separator column resulting in EO flow in that direction. As a slug of the sample is carried through the separator channel, the different electophoretic mobilities of the sample's constituents results in their traveling at different speeds relative to the bulk fluid motion. As the plug flows down the separation column, the different components will therefore form separate, distinct bands. These bands can then be detected via an optical technique such as fluorescence. The flat velocity profiles in EO flow greatly enhance the effectiveness of these optical detection techniques.

Returning to our analogy, vegetable soup is introduced into the separator column. The soup is carried down the column by EO flow. Due to the difference in electrophoretic mobilities of the vegetables, however, separate bands of carrots and mushrooms form in the flow, allowing them to be detected.

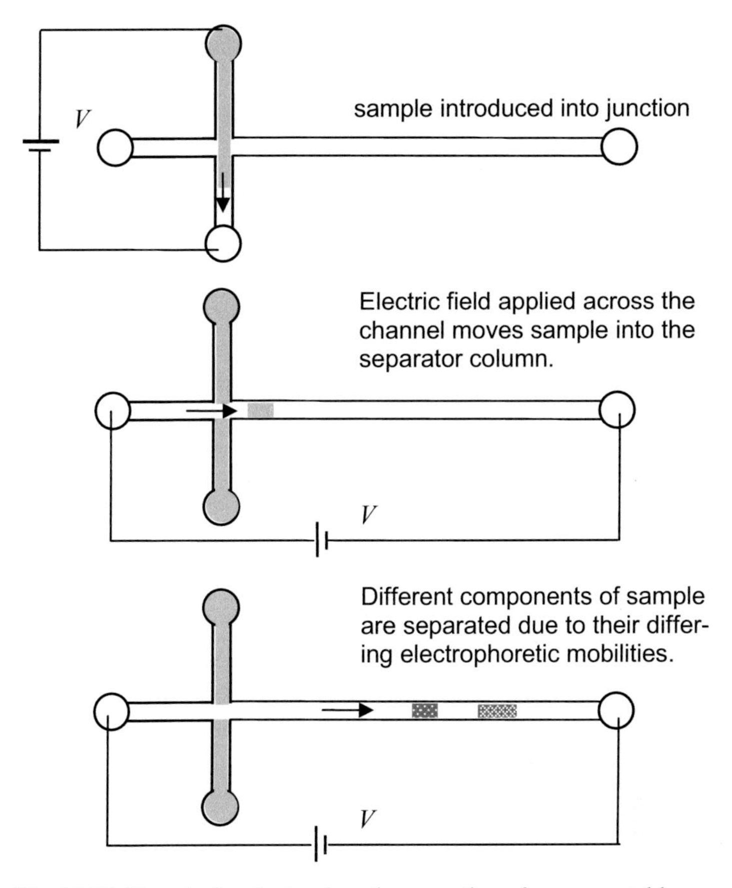

Fig. 12.22. Steps in the electrophoretic separation of, say, vegetable soup

For negative zeta potentials (φ_w) the bulk flow will be in the direction of the applied electric field, which is to the right in Fig. 12.22. Ions of positive charge will travel faster than the bulk fluid, and ions of negative charge will travel slower than the bulk fluid. Using our analogy one last time, if carrots are positively charged, they will travel faster than the surrounding soup. If mushrooms are negatively charged, they will travel slower than the surrounding soup.

Now in addition to this electrophoretic drift, the ions in the bulk liquid will also diffuse into regions of lower concentration. This has the potential to mess up the analysis we have performed, and worse yet, the potential to mess up the sharp-edged bands of chemical species which are so readily detected with optical methods. However, the physical size of a microchannel allows us to minimize these diffusion effects. Furthermore, this has the added benefit of minimizing the time required to do a chemical separation. Separations on the order of seconds are typical in microchannels as compared to separations on the order of minutes in larger channels.

Essay: Detection Schemes Employed in Microfluidic Devices for Chemical Analysis

Daniel J. Morris, Jr.
Department of Chemistry, Rose-Hulman Institute of Technology

Microfluidic systems for chemical analysis are referred to as "lab-on-a-chip" devices or miniaturized total analysis systems (μ-TAS). The devices are small, portable and self-contained and provide a single platform on which all aspects of analysis from sample prep (pre-concentration or dilution, sample clean-up, etc.) to detection are performed. Detection of the species of interest (analyte) is one of the very challenging aspects of chemical analysis using microfluidic devices. While the ability to handle small sample sizes are a distinct advantage of such devices, this makes the overall quantity of analyte to be detected very small. Therefore, sensitive detection methods are required. In addition, the ideal detection scheme should be universal in its ability to respond to analytes and be easily adaptable to the various materials used in microfluidic devices (glass, polymers, silicon, etc.).

Optical detection is currently the most common and widely applied detection method in lab-on-a-chip applications. Interfacing microfluidic devices with optical detection schemes is straightforward, and the relatively low-cost detection elements can be integrated within the actual device. Lased-induced fluorescence (LIF) detection is most commonly applied because of its high sensitivity. However, very few compounds exhibit fluorescent behavior, and non-fluorescent compounds are often "tagged" with a fluorescent entity using an additional chemical reaction to make them detectable. An additional limitation of fluorescence is that it provides virtually no structural information about the molecule.

An optical detection technique gaining acceptance in limited applications is that of Raman spectroscopy. Raman spectroscopy is based on the inelastic scattering of light. When Raman scattering occurs from a sample, the light scattered from a source is shifted in wavelength according to the energies of vibrational transitions within the molecules. It suffers from being an inherently weak signal as 99.999% of light is scattered elastically (i.e. reflected) with no shift in wavelength. However, the advantage of Raman spectroscopy is its ability to provide structural information about a molecule based on its vibrational motions. Low-cost, high-power lasers and highly sensitive detectors (CCD's) are making Raman spectroscopy become a more common detection technique. The low intensities of Raman signals can be enhanced by immobilizing the compound of interest on a metal substrate (surface-enhanced Raman scattering) or pre-concentrating analytes within a microfluidic device prior to detection.

Optical detection on microfluidic devices can also be accomplished using refractive index changes. When a laser beam is reflected from a microchannel a diffraction pattern results, similar to shining a laser beam through a pinhole aperture. The diffraction pattern shifts spatially based on the refractive index of the solution filling the channel. When the laser spot size is small enough, the pattern exhibits a shift when an analyte plug (sandwiched between the bulk solution filling the channel) passes through the laser beam. The intensity of one of the fringes is monitored using a small area detector, and when the fringe pattern shifts the signal output changes, effectively recording when the analyte plug passes through the channel. This technique is called interferometric backscattering detection (IBSD) and is truly a universal detection technique in that it will respond to all analytes. Its ability to detect analytes at low concentration is reported to rival that of LIF.

Electrochemical detection is well established as an analysis tool capable of measuring low levels of analytes. Its adoption as a detection technique for microfluidic devices has not been as widespread as optical methods due to 1) the additional fabrication step of integrating electrodes in a microfluidic device and 2) the difficulty of decoupling the electronics associated with electrochemical detection from the voltages applied for electrophoretic separations. Recently, several workers have developed very effective means of decoupling the detection and separation voltages, and it is becoming more feasible to take advantage of the low detection limits and universal response of electrochemical detection methods for lab-on-a-chip technology.

Perhaps the highest impact detection method to be applied in lab-on-a-chip technology is mass spectrometry (MS). A mass spectrometer is a very sensitive detector that provides highly accurate mass values and structural information. Samples are converted into ion fragments, detected and characterized based on their mass/charge (m/z) ratios. Knowing the ion fragments generated from a sample, the original structure from which they were formed can be ascertained. Electrophoretic microfluidic devices for chemical analysis are very compatible with MS detection via electrospray ionization (ESI) because the ESI-MS interface is merely an extension of the microfluidic channel. EOF on a microfluidic device is driven by an applied voltage and the ESI interface for the mass spectrometer operates under an applied voltage at atmospheric pressure and ambient temperature. Required flow rates for the ESI-MS interface are on the order of μL/min, flow rates that can be achieved easily using EOF.

A microfabrication-based detection technology experiencing significant advances in development and application is microcantilever (MC)-based sensors. Analytes of interest adsorb to the surface of a cantilever beam, resulting in beam deflection. The presence of a deflection indicates the presence of a particular analyte, and the magnitude of deflection can be measured and related to changes in surface stress. The selectively of analyte adsorption can be controlled based on functionalization of the cantilever surface to provide specific receptor sites or even by applied potentials.

Significant opportunities exist for the development of new and improved detectors for microfluidic applications. The above list should not be considered comprehensive or exhaustive. It will indeed be very exciting to watch future developments.

References and suggested reading

Angrist SW (1982) Direct Energy Conversion, 4th edn. Allyn and Bacon, Boston

Bird, R. B., Stewart, W. E., and Lightfoot, E. N, (2001) Transport Phenomena, 2nd edn. Wiley, New York

Davis HF, Snider AD (1988) Introduction to Vector Calculus, 5th edn. Brown, Dubuque, IA

Devasenathipathy S (2003) "Liquid Flow in Microdevices," presented at Third Annual MEMS Technology Seminar, Los Angeles, CA, May 19-21, 2003

Madou MJ (2002) Fundamentals of Microfabrication – The Science of Miniaturization, 2nd edn. CRC Press, New York

Munson BR, Young DF, Okiishi TH, (1998) Fundamentals of Fluid Mechnics, 3rd edn. Wiley, New York

Nagle RK Saff EB (1986) Fundamentals of Differential Equations. Benjamin/Cummings, Menlo Park, CA

Senturia SD (2001) Microsystem Design. Kluwer Academic, Boston, 2001.

Thomas GB Finney RL (1984) Calculus and Analytic Geometry, 6th edn. Addison Wesley, Reading, MA

Tipler PA. (1986) Physics, 2nd edn. Worth, New York

Questions and problems

12.1 Consider the flow of air at standard temperature and pressure in a microchannel whose width is less than 0.1 μm. How do you feel about the continuum assumption in this situation? Why?

12.2 Do you expect most flows in microfluidic devices to be laminar or turbulent? Why?

12.3 In its simplest terms, viscosity is nothing more than a constant of proportionality? Between what two quantities?

12.4 Consider a viscous fluid at rest. Do you expect any viscous forces to show up? Why or why not?

12.5 Show that for a closed system, the macroscopic conservation of momentum equation reduces to $\boldsymbol{F} = m\boldsymbol{a}$.

12.6 Show that $\vec{\nabla} \bullet \left[\vec{\nabla}\phi\right] = \vec{\nabla}^2\phi$.

12.7 Do you think electro-osmotic flow would work if the solid surface were a metal or some other electrical conductor rather than an insulator like glass?

12.8 As we have seen several times, the Navier-Stokes equations for steady, one-dimensional, fully developed flow in rectangular coordinates reduces to a *linear* ordinary differential equation (ODE) of the form

$$d^2u/dy^2 = f(y),$$

where y is the direction normal to the flow.

When a constant pressure gradient is applied in the flow direction with no electric-field, $f(y)$ is a constant equal to $(1/\mu)\cdot dP/dx$, resulting in a parabolic velocity profile. (See section 12.4.) When a constant electric-field is applied in the flow direction with no pressure gradient, $f(y) = -(1/\mu)\cdot(\rho_e E_x)$, resulting in an essentially flat profile. (See Eqs. (12.102) to (12.106).)

Using what you know about the properties of linear ordinary differential equations, sketch what the velocity profile would look like if both a pressure gradient and an electric-field were applied simultaneously. Sketch what this would look like if the applied pressure gradient were positive instead of the usual negative pressure gradient; i.e., if the flow were in the direction of increasing rather than decreasing pressure.[11]

[11] This is what happens in electro-osmotic pumps. The pressure in the fluid is increased across the pump and then sent on its way to flow through some treach-

12.9 Consider the flow of nitrogen (N_2) at 250 kPa and 350 K through a channel with a trapezoidal cross section as shown in the figure.

a. For the case in which $b_1 = 100$ μ and $h = 30$ μ and $\theta = 54.7°$, calculate the hydraulic diameter of the channel.

b. Calculate the approximate speed for which the flow ceases to be laminar.

c. Speculate on how the channel was fabricated and what materials were used. (You may want to review some material from Chapters and 4.)

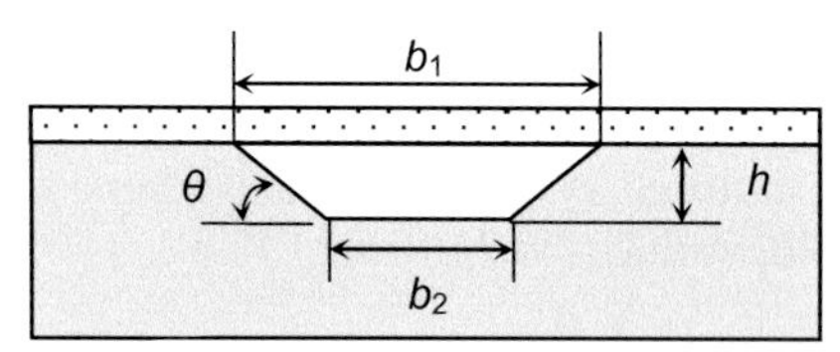

12.10 Pretend you are a firefighter holding onto a firehouse which ejects a constant flowrate of water. The end of the hose is equipped with a nozzle which decreases the cross sectional area. One of your firefighter buddies secretly replaces the nozzle with another nozzle with an even smaller opening. Using the macroscopic conservation equations, determine whether you will have to hold onto the hose with more or less force than with the original nozzle.

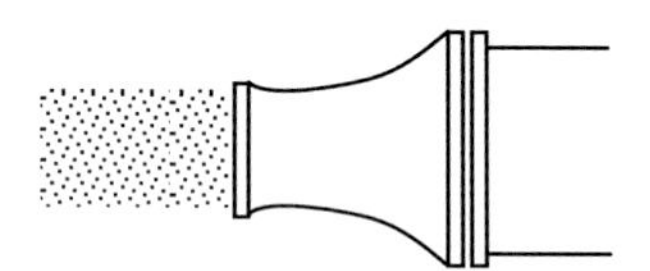

12.11 Consider flow between two parallel plates as shown in the figure below. Both the top and bottom plates move to the right with constant velocity U_0. Consider steady, one-dimensional flow with no applied pressure gradient. Use the continuity and Navier-Stokes equations to find the velocity distribution as a function of the coordinate y.

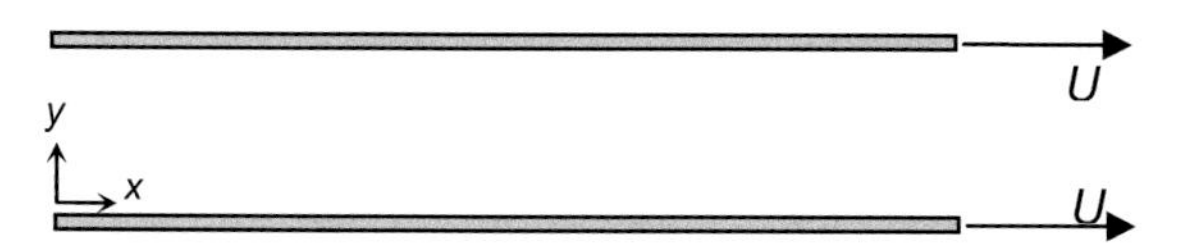

---

erous microfluidic network with all sorts of pressure drops. One nice thing about EO pumps, of course, is that they have no moving parts.

12.12 Find the net force on the point charge in the lower left hand corner in the figure below. Express your answer in terms of the length of a side L and charge q. Be sure to give the direction and magnitude.

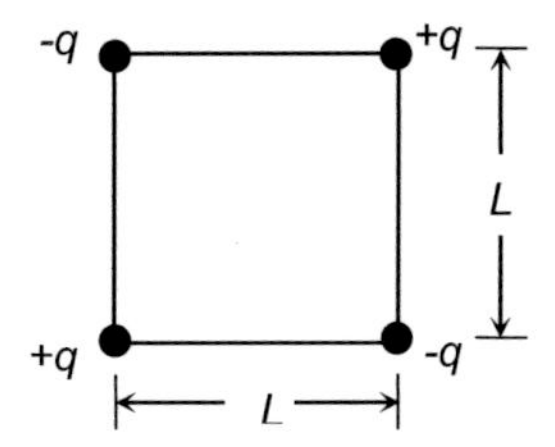

12.13 Find the Debeye length for a 0.5 molar aqueous solution of KCl.

12.14 A microfluidic channel of the type shown in the figure is used to separate type A mitochondria with $\mu_{ep,A}$ and type B mitochondria with $\mu_{ep,B}$ contained in a fluid sample.

a. Due to limitations in the optical detection techniques used in the set-up, a minimum separation distance, d, between the two types of mitochondria is required. Find an expression for the minimum length of the separation column, L, in order to achieve this minimum separation distance.

b. The properties of the electrolyte and the mitochondria are given below.

- Electrolyte: $\varepsilon = 7.09 \times 10^{-10}$ $C^2/N{\cdot}m^2$, μ (viscosity) = 9.0 x 10^{-4} $N{\cdot}s/m^2$,
- Mitochondria: $\mu_{ep,A}$ = -0.80 x 10^{-4} $cm^2/V{\cdot}s$, $\mu_{ep,B}$ = -1.20 x 10^{-4} $cm^2/V{\cdot}s$

Using your results from a., what is the largest (in magnitude) wall potential φ_w (ζ potential) that can be used for an existing channel with $d = 1$ cm and $L = 4$ cm?

c. If the applied voltage in the flow direction is $V_1 - V_2 = 2000$ V, what is the *bulk* fluid velocity in the channel for the wall potential in b)? How long will this separation take, then?

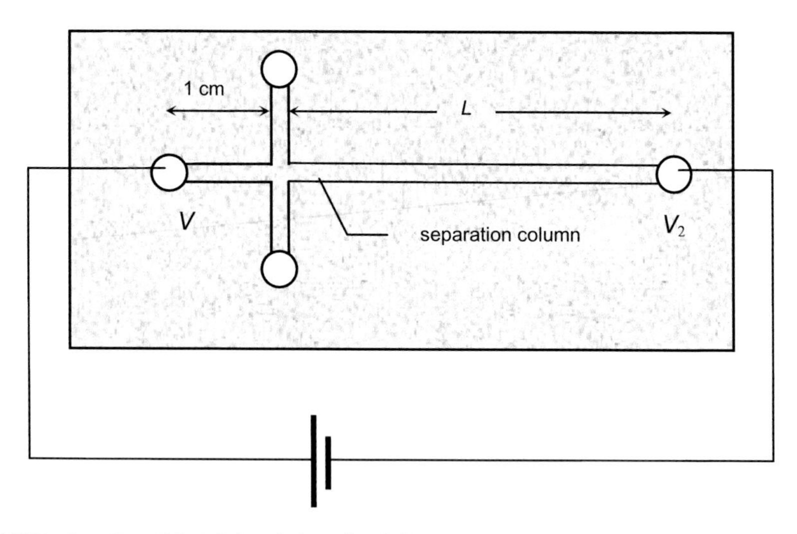

12.15 Derive Eq. (12.54) by doing the following:

a. Write the continuity equation for this case and reduce it to its simplest form.

b. Write the x-direction Navier Stokes equation and reduce it to its simplest form, using the given assumptions and the results of continuity.

c. You should now have a simple, linear ODE in u which can be integrated to find $u = f(y)$. Integrate the equation and apply the appropriate boundary condition(s) to find $u = f(y)$.

Part III—Microfabrication laboratories

Chapter 13. Microfabrication laboratories

13.1 Hot-arm actuators as a hands-on case study

Here we present six field-tested laboratories for the fabrication and characterization of hot-arm actuators. The experiments allow the student to gain some experience with many of the basic microfabrication techniques presented in the text, as well as to observe the operating principles of a functioning MEMS device first hand. The equipment requirements for the laboratories is appropriate for a modest facility, necessitating approximately a class 10,000 clean space for oxidation growth and evaporation, and class 1000 for some of the more sensitive operations, such as photolithography.

A photograph of a hot-arm actuator fabricated using the process flow outlined in these laboratory exercises is given in the figure below. The operating principles of hot-arm actuators themselves were presented in Chapter 7 and again in detail in Chapter 11.

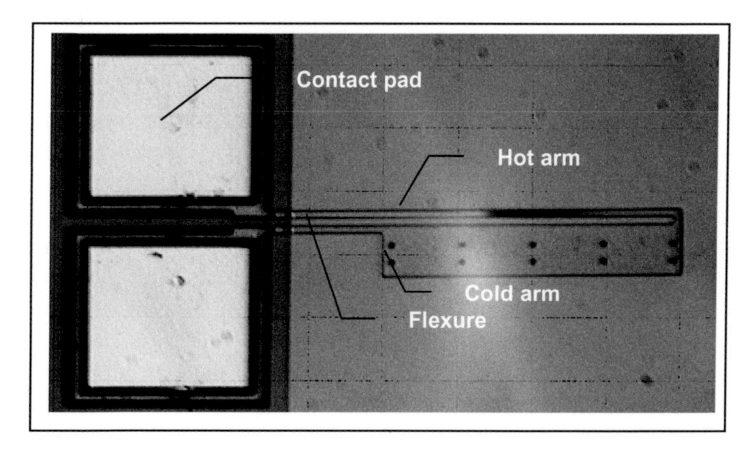

An aluminum hot-arm actuator fabricated at the RHIT MiNDS Facility using the procedures contained in these laboratory exercises.

T.M. Adams, R.A. Layton, *Introductory MEMS: Fabrication and Applications*,
DOI 10.1007/978-0-387-09511-0_13,

13.2 Overview of fabrication of hot-arm actuators

The process flow for creating the hot arm actuators was developed at Rose-Hulman Institute of Technology specifically for teaching the lab portion of an undergraduate MEMS course. The procedure involves surface micromachining with aluminum as the structural layer and photoresist as the sacrificial layer, the sacrificial layer being removed using plasma ashing. A silicon wafer with an oxide insulation layer is used as the substrate. The fabrication steps are the subject of the first five experiments. The sixth experiment involves testing and characterizing the finished hot-arm actuators.

Though the labs were written to be relatively generic and not largely dependent on institutional constraints, specific process steps, equipment and materials are sometimes mentioned. Naturally, these can be changed as needed and the labs modified to reflect the equipment and facilities available. However, such changes are certainly not suggested for anyone without the appropriate level of expertise. Indeed, the laboratories should be carried out under the supervision of someone with at least a moderate level of experience within microfabrication.

The basic process steps for fabrication of the actuators are given by:

1. RCA clean wafer
2. Oxidize wafer to create an isolation layer
3. Photolithography for the sacrificial layer
4. Evaporate a thin film of aluminum for the structural layer
5. Photolithography to pattern the aluminum thin-film to form the actuator
6. Oxygen plasma release to free the actuator

The RCA clean is performed to prepare the wafers for oxidation. The wafers are then oxidized to create an isolating layer, which is needed to prevent electrical current from flowing through the silicon wafer when a voltage is applied across the contact pads of the finished actuator. The oxidation is a wet oxidation process, carried out at 1000°C for one hour. This should yield an oxide layer of approximately 250 nm. After the growth of the oxide, the lab calls for measurement of the oxide film thickness using a tool such as an ellipsometer.

The process flow calls for using Shipley S1813 positive photoresist not only for patterning the structural layer, but also for use as the sacrificial layer. For use as the sacrificial layer, the photoresist is spun on at 55000 rpm for 60 seconds, resulting in thicknesses of 1.25–2 μm after postbaking. The photoresist is then softbaked on a hotplate at 95°C for 30 seconds. After the softbake, the photoresist is exposed using the first mask in a mask aligner. The exposure is typically set to 4 seconds on the aligner with a 350 Watt bulb. A post-exposure bake of two minutes at 95°C follows the UV exposure step. The pattern is developed using an appropriate developer, diluted 1:3 in deionized water. The photoresist is then post-baked in a convection oven for 20-30 minutes at 120-130°C. A thorough post-bake is necessary to ensure the pattern is not damaged in the following physical vapor deposition step, and also to prevent damage to the aluminum layer during the second photolithography step. Furthermore, an insufficient post-bake can cause the photoresist to release gasses, which are trapped under the aluminum layer causing the layer to peel during the second photolithography step.

An electron beam evaporator is used to deposit the aluminum structural layer. The layer has to be thick enough to completely cover the sidewalls of the sacrificial layer; however, excessive thickness can result in extreme undercutting of the pattern. The ideal thickness of aluminum layer is on the order of 1-3 μm.

A second photolithography step is used to pattern the aluminum layer. The positive photoresist is spun on at 5500 rpm, resulting in a thinner mask for etching the aluminum then is used for the sacrificial layer. The process is similar to the first photolithography step; however, care must be taken not to damage the underlying aluminum film. The photoresist must be softbaked in the convection oven. The photoresist is patterned using mask number 2. After a hard-bake at 120°C, the aluminum film is etched. The aluminum layer is chemically etched using a phosphoric-acetic-nitric (PAN) etch at 40°C. Students can observe the wafer using a microscope to ensure excessive undercutting does not occur.

The release step is done using oxygen plasma in a plasma asher. This is a dry release, which is which is used to avoid problems with stiction. A "recipe" for ashing/etching that provides a power/pressure combination for removing all the resist without damaging the heat actuators will be required by the plasma ashing system. Such recipes can be obtained from the manufacturer of the plasma asher, online, or from MEMS information clearing houses. Alternatively the photoresist may be removed in a wet release process with acetone and liquid CO_2

Figure 1 gives a schematic outlining the basic process steps, whereas Fig. 2 gives the mask set used in the fabrication. The first mask opens regions to anchor to the oxide layer. These anchor positions have a reduced footprint in comparison to the second mask. The hot arm should be 5 to 30 times thinner than the cold arm. The flexure may be the same thickness as the hot arm, and may be in the range of one third to one tenth of the overall length. The exact ratio, as well as all final dimensions, will be dependent on whether the use is for maximum deflection, maximum force or some combination thereof, in addition to the resolution capabilities of the fabrication facility at hand.

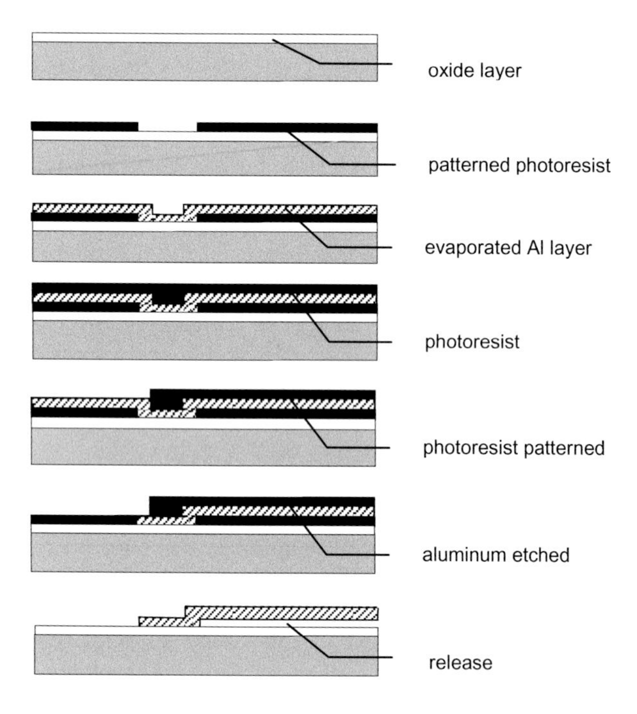

Fig. 1. Process steps for the fabrication of a surface micromachinged hot-arm actuaotor

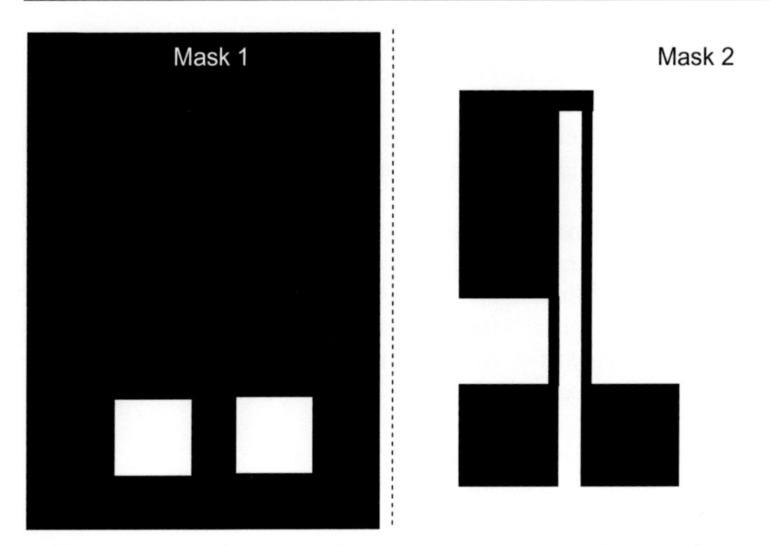

Fig. 2. Schematic of the mask set used in the fabrication of the hot-arm actuators. The first mask generates the anchors for the heat actuator. The second mask is for the heat actuator. The first mask is dark, with clear regions for the anchors. The second mask is clear with dark regions for the actuators.

13.3 Cleanroom safety and etiquette

Students and personnel working in cleanrooms are required to understand all safety rules, regulations, and procedures. It is each individual's responsibility to observe and follow all measures while in the laboratory. Below is a suggested list of requirements to ensure safety while working in a microfabrication facility.

Requirements

1. Always wear approved splash goggles when working with chemicals.
2. Cleanroom garb is required. Long hair (hair that falls in front of the face when leaning forward) should be gathered and restrained.
3. Report all accidents to the lab instructor/technician, regardless of how trivial you might think they are.
4. Always walk slowly and do not lean on equipment.

5. Do not bring anything in with you including PDAs and calculators. Never bring food or beverages into the lab. Do not drink, chew, or smoke in the laboratory.
6. Do no operate equipment without instructor/technician approval. No experiments should be performed in the laboratory without supervision.
7. Know the locations and correct operation of all safety equipment including showers and eye washes.
8. Report physical limitations or physical conditions (sprains, broken limbs, *etc.*) to the instructor/technician prior to the laboratory period. Special arrangements or safety instructions may be required.
9. Do not taste any chemicals in the laboratory. Test odors very carefully.
10. Never use unlabeled bottles of chemicals.
11. Never return unused chemicals to their original containers.
12. Dispose of all waste according to the procedure in the laboratory write-up or the instructions provided by the laboratory instructor/technician.
13. Clean up all spills immediately.
14. Materials Safety Data Sheets (MSDS) should be available upon request for any chemical used in the laboratories. These sheets may be obtained from the appropriate instructor, technician or staff member.
15. Upon completion of an experiment, turn off all equipment, store it properly, and clean your area.
16. Clean your workspace after use and wash your hands at the end of the laboratory period.

13.4 Experiments

Experiment 1: Wet oxidation of a silicon wafer

Purpose:

In this experiment an oxide layer will be grown on a silicon wafer in an oxidation furnace using a wet oxidation process. This experiment begins with an RCA clean, and after oxidation, the thickness of thermally grown oxide layers will be determined using a thin film measurement device.

Materials:

- Silicon wafers
- RCA clean chemicals

Equipment:

- Beakers
- Petri dishes
- Quartz wafer boat and push rod
- Water bubbler
- Oxidation furnace
- Thin film measurement device

Overview and general description:

At high temperatures, silicon reacts with oxygen to form silicon dioxide (SiO_2). In dry oxidation, oxygen gas reacts directly with silicon wafers to produce an oxide layer on the wafer surfaces. In wet oxidation, oxygen gas is first "bubbled" through water before entering the furnace, introducing water vapor to the wafers as well as oxygen. Wet oxidation proceeds faster than dry oxidation but produces a lower quality oxide layer. Oxidation is discussed in detail in Chapter 2.

Essentially an oxidation furnace works by keeping silicon wafers in contact with oxygen at high temperatures. Figure 3 gives a schematic drawing of a typical furnace and shows its major features. A horizontal quartz tube inside the furnace contains a small quartz wafer boat which holds the wafers. The wafer boat is introduced into the tube from the left of the figure by slowly pushing it with a push rod. The required oxygen flows into the tube from the right. Different furnaces will have different arrangement sof the components, but these features should be standard.

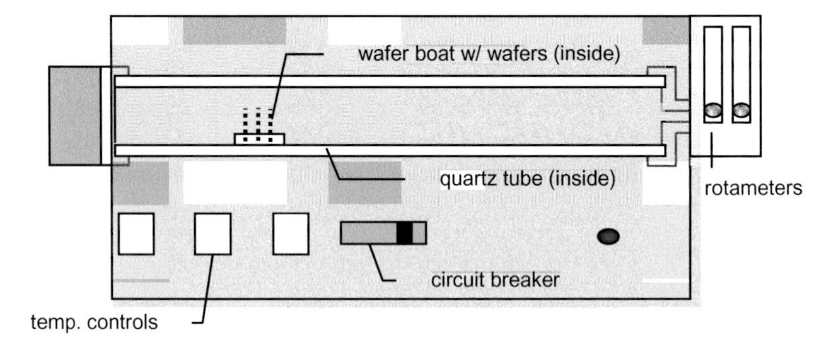

Fig. 3. Schematic of a high-temperature furnace for oxidation

Temperature controller

Different furnaces will have different controls used to achieve and maintain the desired temperature throughout the furnace, as well as to ensure uniform heating. Most furnaces will have a place to input the desired set temperature and a readout for the present actual furnace temperature. When these temperatures are the same, the furnace has reached steady-state and is ready for the introduction of wafers.

Gas flow measurement and control

On the right of the furnace shown in the figure are two meters used to control and monitor gas flow into the furnace. These are typically rotameters, which consist of hollow vertical tubes that contain small floats. Gas enters the meters from the bottom and flows vertically through the rotameter tubes. The level of the floats inside each tube indicates flow rate. For some meters the flow rate is read directly off of the tube in the units indicated on it, while for others only a simple scale will appear on the tube and a conversion to flow rate will need to be performed based on manufacturer specifications. For example, a tube may have a 0-20 scale with 20 corresponding a 1.0 l/min flow of air. Since most rotameters are designed to be used with many different gases and liquids, it is vitally important know how these meters are calibrated for your system.

Typically there is one meter for oxygen and the one for nitrogen. Gas flow is initiated by opening cut-off valves located upstream of the meters. Flow rate can be controlled by adjusting another set of valves at the bottom of each meter.

Water Bubbler

During wet oxidation, oxygen gas is first bubbled through water before entering the furnace. The apparatus used for this process is called the water bubbler. The bubbler will be located in the vicinity of the furnace near the gas flow meters. The bubbler is bypassed completely during dry oxidation.

Figure 4 gives details of the arrangement bubbler. The water is kept at 95-98°C during wet oxidation by means of an electrical heater. This ensures that water vapor rather than liquid enters the furnace tube. A small heater control allows for water temperature adjustment.

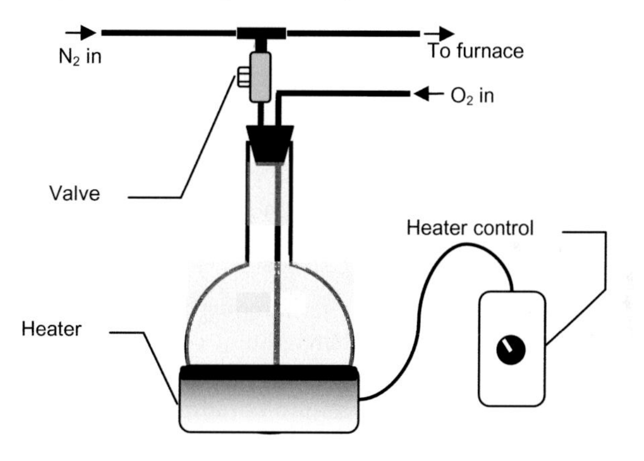

Fig. 4. Water Bubbler

Oxidation procedure:

Wafer cleaning using RCA clean:

Silicon wafers should go through an RCA clean to remove the organic, inorganic, ionic and metallic contamination.

1. RCA1—1:1:5 NH_4OH : H_2O_2: H_2O — This procedure removes organic contaminants.
   a. Add 15 ml NH_4OH (ammonium hydroxide) to 75 ml deionized water in a Petri dish.
   b. Heat to 70-80 °C on a hotplate.
   c. Add 15 ml of H_2O_2 (hydrogen peroxide) .
   d. Immerse the wafer for ten minutes.
2. Rinse wafer in deionized H_2O.
3. RCA2—1:1:6 HCl : H_2O_2 : H_2O – This procedure removes metal ions.
   a. Add 15 ml of HCl to 90 ml deionized water in a Petri dish.
   b. Heat to 70-80 °C on hotplate.
   c. Add 15 ml of H_2O_2.
   d. Immerse the wafer for ten minutes.
4. Rinse wafer in deionized H_2O.
5. Spin wafer dry.

Procedure for wet oxidation:

1. The thickness of the oxide layer depends on many factors including initial oxide thickness, wafer crystal orientation, furnace temperature and time. For example, for a <100> Si wafer with no initial oxide layer, a temperature of 1000°C for 60 minutes should give an oxide layer of approximately 3913 Å. Other oxide thickness requirements can be found using the principles given in Chapter 2, or by using online tools. Thicknesses can be verified by visual inspection using the color chart at the end of this exercise.
2. Turn on the furnace by first ramping the furnace temperature to 200-300°C and then set the temperature controller to the desired set temperature. The suggested set temperature is 1000°C.
3. While the furnace temperature is ramping up to set temperature, start the nitrogen gas flow to the furnace. Nitrogen flow is needed for two things: purging the air inside the furnace and ensuring uniform step oxidation. To start the nitrogen flow, first make sure that the valve on the return from the water bubbler is closed. (Fig. 4) Next, make sure that the pressure regulator on the nitrogen generator reads no more than 15 psi. With the flow adjust on the rotameter turned all the way off; open the cut-off valve for the nitrogen. Open the flow adjust knob on the rotameter until the float in the rotameter is at the correct level for a flow rate of approximately 0.5-0.6 l/min.
4. The water bubbler should be approximately three quarters full of clean deionized (DI) water. The heater should be maintained at the appropri-

ate to keep the water around 95°C. *If the water temperature stabilizes before the furnace set temperature has been reached*, very loosely replace the rubber stop to the flask. Once the furnace set temperature has been reached, tightly secure the rubber stop in the flask. Always check that the bubbler has the ability to release pressure either through the furnace or past the rubber stopper. Without pressure relief, the flask has the potential build pressure and then flash evaporate upon release.

5. Once the desired furnace set temperature has been reached, stop the nitrogen flow. Open the valve on the return from the water bubbler. Start the oxygen flow by first adjusting the pressure regulator valve on the oxygen tank. With the flow adjust on the rotameter turned all the way to off, open the cut-off valve for the oxygen. Next open the flow adjust knob until the float in the rotameter is at the correct level for a flow rate of approximately 0.5-0.6 l/min.
6. Carefully put the clean wafers in quartz wafer boat using tweezers. Make sure the polished side of wafer faces towards furnace tube opening. Push the boat into the furnace tube mouth using the pushrod. Leave the wafers in the mouth for five minutes to allow the wafers to adjust to furnace environment. *Always wear thermally insulated gloves while handling heated items.*
7. Next push the wafers into the center of the furnace tube using the pushrod. This should be done *slowly*, at a rate of 0.5cm/sec. Exceeding this rate may warp the wafers.
8. Place the end cap on the furnace tube.
9. Keep the furnace running at the set temperature for the desired length of time. For the suggested set temperature of 1000°C a required time of 60 minutes is needed to produce an oxide layer of 3913 Å for a <100> silicon wafer. The actual thickness will vary, but most likely will be between 250 and 400 nm.

   If water is seen condensing at the gas flow inlet of the furnace tube, decrease the oxygen flow slightly and/or decrease the bubbler heater temperature. It is highly undesirable to have liquid water enter the furnace tube.

10. Once the wafers have remained in the furnace for the prescribed time, remove the end cap and *slowly* pull out the boat out of the furnace tube. This should also be done at a rate of 0.5cm/sec. Allow the wafers to cool down at the mouth of the tube for five minutes. Carefully remove the oxidized wafers using tweezers and load them into wafer trays.
11. Turn off the oxygen gas flow and turn on the nitrogen gas flow. The nitrogen flow should be kept at ~0.5-0.6l/min. Replace the end cap on

the furnace. Allow the furnace to ramp down to room temperature and then turn off the nitrogen flow. Turn off the furnace by switching off the circuit breaker. *Never* turn off the furnace without performing this step.

13. Repeat RCA Clean of the wafer.

Procecdure for thin film measurement:

The last step in the laboratory is to determine the thickness of the oxide layer. This step, off course, depends on the thin film measurement system that is available. One of the simplest system is a Filmetrics-F20. This instrument measures the spectrum of reflected light from the wafer surface to determine the thickness of the oxide film. To make the measurement with Filmetrics F20, the following steps should be followed.

1. Turn on the light source for the measurement system. This light source needs to be on for several minutes before measurements can be made.
2. A bare silicon reference wafer should be placed below the light from the light source and measured first. This allows the software to obtain a reflectance spectrum as a calibration step.
3. Center the oxidized wafer under the light and make a new measurement. The instrument will now obtain a reflectance spectrum from the oxidized wafer, which should differ significantly from that obtained from the silicon reference wafer. The software then uses this reflectance spectrum to determine the thickness of the oxide layer.
4. Make measurements at several locations across the wafer in order to determine an average thickness of the silicon oxide layer.

The thickness of the oxide layer can also be approximated by simply looking at the color of the wafer from above when illuminated by white light. Observe the color of the wafer by holding the wafer perpendicularly to a white light source and then check the oxide thickness from the color chart in the Appendix.

Suggested content for lab memo:

1. A brief description of the process for obtaining an oxidized silicon wafer.
2. The average measured thickness of the oxide layer and its uncertainty.
3. A comparison of this value to the one obtained via the techniques given in Chapter 2 on an online resource such as the one given the pricedure above.

4. A calculated value of the wafer thickness after oxidation assuming that it has a thickness of 300 μm after the RCA clean.

Experiment 2: Photolithography of sacrificial layer

Purpose:

In this experiment students will learn to spin photoresist onto a silicon wafer using a spin coater. The students will then use a mask aligner to transfer an image of a pattern embossed on a mask (typically chromium film on a glass plate) onto the photoresist film. This step is followed by photoresist development. The photoresist is positive, and so the unexposed regions are removed during development to create the final pattern. The unexposed photoresist will then be hardened using a convection oven.

Materials:

- Clean oxidized wafer (Always handle with clean tweezers.)
- Positive photoresist
- Appropriate photoresist developer
- Mask #1 for the sacrificial layer

Equipment:

- Beakers
- Petri dishes
- Deionized water
- Tweezers
- Resist spinner
- Convection oven
- Mask aligner
- Four point probe microscope/camera

Overview and general description:

The next step in the fabrication of the hot arm actuator is to create a sacrificial layer, in this case using photoresist.

Photoresist is sensitive to temperature as well as UV light, and it is therefore normally kept in a refrigerator to maximize its shelf life. The resist should first be warmed up to room temperature and then a small amount poured into a small beaker. The thickness of the film obtained depends on the properties of the resist and the angular spin speed. Once the film thickness has been chosen, the required spin speed can be determined from material supplied by the manufacturer. Details of the application of photoresist are given in Chapter 3.

In order to obtain a smooth resist coating on the wafer, the chuck of the resist spinner is ramped up to speed in several stages. In the first stage, the chuck is ramped up to 500 rpm within one and a half seconds, and then spun for ten seconds. The first stage is to provide even coverage over the wafer. In the second stage, after a two second dwell, the chuck is ramped up to 3000 rpm within six seconds. That rotation is maintained for sixty seconds, for thicknesses of about 2 microns. In order to generate photoresist thicknesses around 1-1.5 microns, speeds of 5500 rpms and 2 minutes should be used. In the spin down, the chuck is ramped down to 1000 rpm within one and half seconds. After two seconds, the wafer decelerates to zero in ten seconds. This is can sometimes be programmed into the spinner electronics, so that each step happens automatically.

The photolithography procedure has many steps in which the wafer and the photoresist are heated. The first is a *pre-bake*, in which excess moisture and/or solvents are removed from the wafer prior to adding the photoresist, as the excess moisture can interfere with the adhesion of the photoresist. After the resist is applied, a *soft bake* is performed. This step stabilizes the resist film at room temperature, and also removes excess solvent. In addition to improving adhesion, the soft-bake makes the film less tacky, rendering it less susceptible to particulate contamination. The removal of solvent also reduces film thickness and changes post-exposure bake and development properties.

The resist-coated wafer is then exposed. A piece of equipment called a *mask aligner* is used in this step in order to line up the mask and the wafer properly. The outcome of this step will be an oxidized wafer with exposed photoresist wherever the mask was clear. The exposed photoresist is then developed, typically using a developer consisting of an aqueous basic solution. Development is one of the most critical steps in the photoresist process, as the resist-developer interactions to a large extent determine the shape of the photoresist profile and, more importantly, the resolution. After a successful development process, a clear pattern can be observed, as shown in Fig. 5.

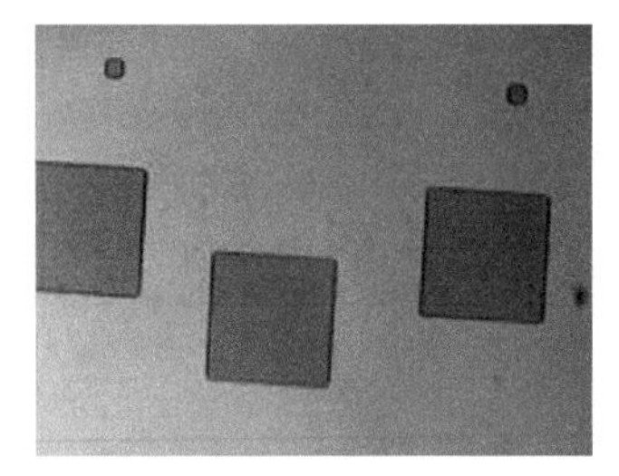

Fig. 5. Patterned photoresist on oxidized silicon wafer after the exposure and development of the first mask. The squares are the locations of the contact pads for the heat actuators.

Finally a *post-bake* is performed to harden the final resist so that it can withstand the harsh environments of etching the underlying oxide layer. The post-bake crosslinks the resin polymer in the photoresist, making the image more thermally stable. The lab procedure here calls for a post bake in a convection oven, but hotplates can be used as well. The temperatures used are in the range of 120-150°C, performed for 20-30 minutes in an oven, but only about a minute on a hotplate.

Photolithography procedure:

1. Perform pre-bake of the wafer on a hot plate for one minute at 95°C.
2. Open the lid on the spinner.
3. Place wafer on center of vacuum chuck.
4. Pour positive photoresist onto center of wafer until it covers about half the wafer.
5. Close the lid.
6. Start the required program on the spinner electronics in order to create a resist layer 1-3 microns thick.
7. (Soft bake) Place wafer in a convection oven at 100° C for half an hour or on hot plate at 95°C for one minute.
8. Place the wafer on the wafer-chuck of the mask aligner and expose Mask 1 to UV light for five seconds.
9. Perform Post Exposure Bake on hotplate for 2 minutes at 95°C.
10. Develop photoresist using appropriate developer diluted with DI water in a ratio of 1:3.
11. Post bake wafer at 130°C for 30 minutes in the oven to help harden resist before the metal deposition.

12. Use the available equipment and/or software to capture some images of the exposed and developed resist layer.

Suggested content for lab memo:

1. A brief description of the process for obtaining a sacrificial layer of photoresist on a silicon wafer.
2. Measured thickness of the resist using the thin-film measurement system.
3. Descriptions of how the wafer looked after each main step in the process.
4. Pictures of the wafer after development.

Experiment 3: Depositing metal contacts with evaporation

Purpose:

In this experiment students will deposit a metal onto a silicon wafer using an electron beam (e-beam) evaporation system. Students will then pattern the metal film using a second photolithography step with a different mask than in Experiment 2 in order to create the heat actuator.

Materials:

- Patterned silicon wafers from Experiment 2
- Aluminum target
- Positive photoresist

Equipment:

- Electron beam evaporation system
- Spinner station
- Convection oven

Overview and general description:

Electron beam evaporation is a technique used to deposit thin films of oxides as well as metals onto substrates. This is done in a vacuum chamber where most of the air has been pumped out, resulting in very low pressures. An e-beam gun emits and accelerates a beam of electrons from a filament and magnetically directs them to impact the center of a crucible containing the evaporant material. The evaporant is vaporized and travels to the substrate where it condenses to form a solid film. Details of the evaporation as well as physical vapor deposition techniques in general are given in Chapter 2.

The e-beam gun typically has a *sweep control*, which sweeps the beam over the evaporant to effect uniform melting rather than boring a hole in one spot of the target. Film thickness is typically controlled with a quartz crystal monitor. As evaporant condenses upon the quartz crystal, the crystal's resonance frequency is measured. The resonance frequency shifts as the mass of the crystal plus solid film increases and the film thickness on the crystal is estimated. Since the crystal is not in the same position as the substrate, a correction factor is determined by measuring the film thickness on the substrate by independent means and comparing it to the film thickness on the crystal monitor display. As long as the crystal is in its linear range, the correction factor remains constant and substrate thickness is controlled. Deposition rates depend upon the material evaporated ranging from a few angstroms per second for dense materials or materials with high melting temperatures to tens of angstroms for lighter materials.

In this lab, the students will prepare a patterned oxidized-silicon wafer to be coated with aluminum, operate the evaporator through its pumpdown cycle, and operate the e-beam gun through deposition. After a cool off period, the chamber will be vented to atmosphere and the sample retrieved.

Metal evaporation procedure

1. Place wafer in the wafer holder in e-beam system. Place the heatsink (1 cm thick disc of aluminum) on top of wafer to prevent damage to the wafer
2. Begin system pump-down of the vacuum chamber.
3. If the system is equipped with ion-gauge controller: Watch the vacuum-gauge controller and when this gauge reads about 5×10^{-6} Torr then turn on the Degas on the ion-gauge controller, let the system degas for about 10 minutes.
4. If the vacuum-gauge reads at least 5×10^{-5} Torr then start the process.
5. Turn on the emission and begin ramping the current up slowly to about 0.01 Amps/30s.
6. Watch the thickness monitor for the deposition rate. At the beginning, the rate will oscillate between positive and negative values. When this begins to read positive, the zero thickness button is pressed and the shutter is opened to allow metal deposition on the wafer.
7. Continue to increase the deposition rate slowly (0.01 Amps/30s) to maintain a rate of 30-40 angstroms/s.
8. The units displayed are typically displayed in kilo-angstroms (10^{-7} m). Deposit until the thickness is 20000Å (2.0 microns).
9. Then the voltage is turned off and the current is back down to zero.
10. Start to vent the chamber. When air is no longer entering the chamber open the door or raise the belljar.
11. Remove the heatsink. *Be careful.* Your wafer will likely lift along with the heatsink.
12. Use tweezers to retrieve the wafer and observe under microscope .(Fig. 6.)

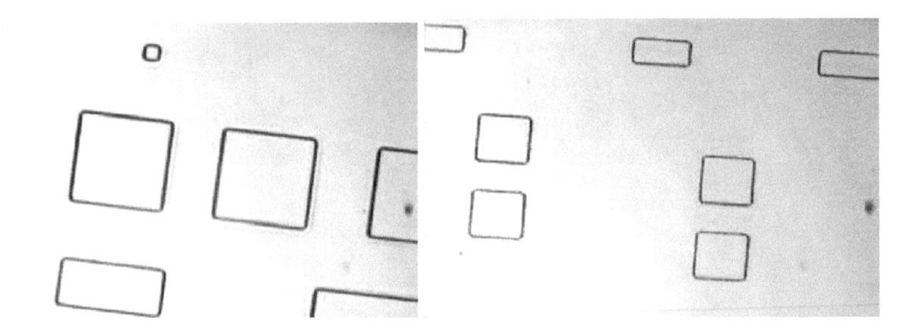

Fig. 6. Pictures of two different regions of the wafer after metal (Al) deposition

Photolithography steps

13. Take wafer and do a pre-bake at 95° C for one minute on the hot plate.
14. Spin photoresist on wafer per instructions in experiment 2 using mask #2.
15. Take wafer and do a soft-bake at 95° C for one minute.
16. Using the wafer that you soft-baked and expose the wafer on the mask aligner using mask 2 again for 5 seconds.
17. Perform a post-exposure bake on hot plate at 95°C for one minute.
18. Develop photoresist using appropriate developer diluted with DI water in a ratio of 1:3
19. After development place wafer in metal cassette to ready it for the hard bake in the oven for 30 minutes at 130°C.

Now the aluminum film on the wafer is patterned and ready to be etched, as shown in Fig. 7.

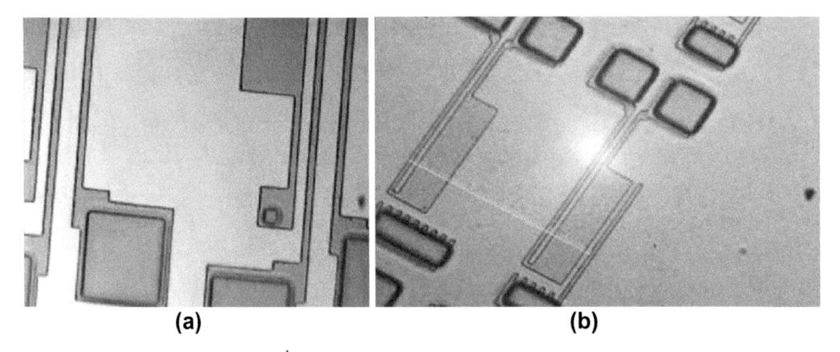

Fig. 7. (a) Wafer after the 2nd exposure for patterning aluminum film and (b) after development and hard bake.

Suggested content for lab memo:

1. Deposition parameters of the physical vapor deposition process: pressure, current, deposition rate, deposition time, final thickness of sluminum film, etc.
2. Answer the question "Why do you have to pump down to low pressures to operate an electron beam gun?"
3. Answer the question "Why is electron beam evaporation limited to lower density, low melting point metals?"

Experiment 4: Wet chemical etching of aluminum

Purpose:

In this experiment students will pattern the aluminum layer using wet chemical etching. The aluminum layer will be etched using a phosphoric-acetic-nitric (PAN) etch solution. The PAN etch will remove the exposed aluminum film yet be unable to etch into the silicon substrate or photoresist mask at any appreciable rate. Students will also observe a small amount of undercutting of the second photoresist layer resulting from the isotropic etching process.

Materials:

- Silicon wafers with an oxide coating, photoresist sacrificial layer, aluminum thin film and a photoresist etch mask from Experiment 3
- Diluted PAN etch (80% H_3PO_4, 5% $HC_2H_3O_2$, 5% HNO_3, 10% DIH_2O plus an equal amount of water to dilute to half strength.)

Equipment:

- Petri dishes
- Dionised water bottles
- Hot plate
- Microscope with CCD digital camera and appropriate imaging software

Overview and general description:

Aluminum etches isotropically in PAN solution. The etch rate for aluminum in a normal PAN etch is typically on the order of 600 nm/min at 40°C. With the diluted PAN etchant used here, the etch rate is roughly half that, 300 nm/min. The etch rates of photoresist and silicon dioxide in PAN are both considerably less than that for aluminum, so that the etch effectively stops as soon as the aluminum layer is removed and the oxide layer is exposed. However, due to the isotropic nature of the etch some of the aluminum forming the contact pads can be undercut beneath the photoresist. It is therefore not advisable to leave the wafer in the etchant for an extended period of time. Rather, the time for which the wafer remains in the etchant solution should be based on the thickness of the deposited aluminum thin film with an additional 10% added to the calculated time in order to account for deviations.

Etching procedure:

1. Use the thickness of the aluminum layer deposited in Experiment 3 to calculate an approximate etch time for the wafer based on a etch rate of 300 nm/min. Add an extra 10% to this time to account for deviations.
2. Immerse the wafer, pattern side up, in a container containing PAN etch.
3. Allow the wafer to remain in the PAN etch for the calculated time from step 1
4. Remove the wafer from the PAN Etch solution carefully using tweezers, and rinse with deionized water using the wash bottle. Catch the rinse solution in a Petri dish.
5. Rinse the wafer with deionized water and spin dry. Dispose the various solutions and byproducts properly according to your lab's protocal.
6. The wafer will now have a patterned aluminum layer, with the photoresist mask still present as shown in Fig. 8. Inspect the surface of the wafer using the available imagining equipment. See Fig. 9 for a typical image of heat-actuators observed under a microscope.

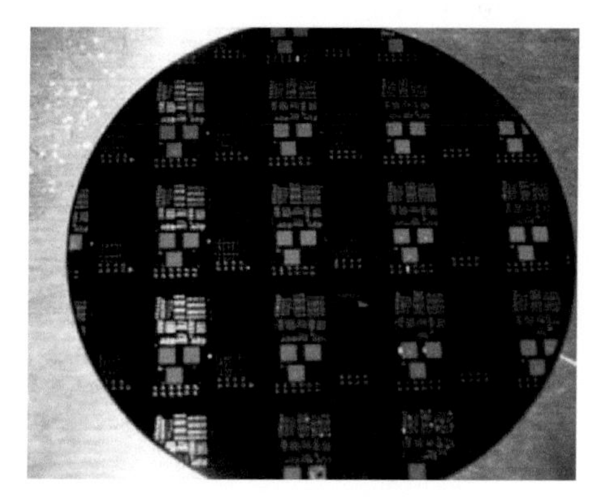

Fig. 8. Wafer after wet metal Etch

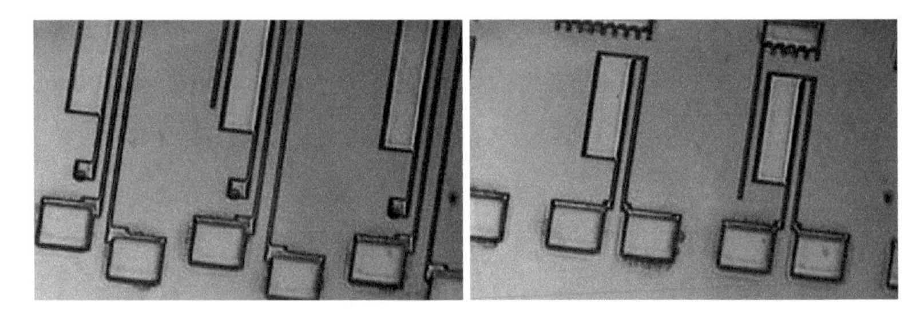

Fig. 9. Heat actuators on the wafer after metal etch step observed under a microscope

Suggested content for lab memo:

The undercutting can be measured using the microscope with the CCD digital camera and the appropriate imaging software. Use the CCD camera and the software to take pictures that show the undercutting, and also to measure the undercut amount. Include images of some heat actuators.

The minimum amount of undercutting will be the same as the amount of the film etched away, as the etch is isotropic. The film thickness is therefore one of the critical dimensions to consider when making a device. The object needs to be at least three times as large as the film thickness in its smallest dimension. The undercutting will be visible as a darker outline. The "teeth" at the top of the above figure on the right show undercutting, as the entire tooth region is dark.

Experiment 5: Plasma ash release

Purpose:

In this experiment a photoresist mask and a sacrificial layer will be removed using an oxygen plasma asher. Students should be able to observe the removal of the photoresist without damage to the aluminum heat actuators.

Materials:

- Silicon wafer with patterned aluminum heat actuators and sacrificial photoresist layer from Experiment 4

Equipment:

- Oxygen plasma asher
- Microscope
- Tweezers

Overview and general description:

Oxygen plasma ashers remove organic materials from a substrate, with little effect on non-organic materials. This makes plasma ashing ideal for removing a sacrificial photoresist layer without damaging the metallic structural layer. If a plasma asher is not available, an alternative release method is to use acetone to remove the photoresist sacrificial layer. However, this method would require a critical point drying step in order to avoid the possibility of stiction.

Oxygen plasma consists primarily of monatomic oxygen created by subjecting O_2 gas to an RF electrical field. This monatomic oxygen is electrically neutral; however it is highly reactive with organic materials. The product of the reaction is an ash, which is removed by the vacuum pump of the asher system. In forming oxygen plasma, oxygen ions with a positive charge are also formed. These ions can recombine to form O_2 with no ill effects, but they can damage the substrate at times.

The asher manufacture can supply you with the appropriate “recipe” for removing a particular type and thickness of photoresist. The recipe will contain the appropriate gas flow levels (typically ≤ 500 sccm), pressure values (typically ≤ 2000 mTorr) and temperature values (typically ≤ 200°C).

Plasma ashing procedure:

1. Ensure the asher's vacuum pump is running.
2. Power on the asher (See Fig. 10 for an example of a commercial unit, a TePla M4L Barrel Plasma Etcher.)
3. Place wafer on either the top or bottom shelf. These are the biased shelves, which result in quicker ashing.
4. Close the door securely
5. Wait for the recipe to complete (Open the door, but allow the shelves and wafer to cool before attempting to remove the wafer

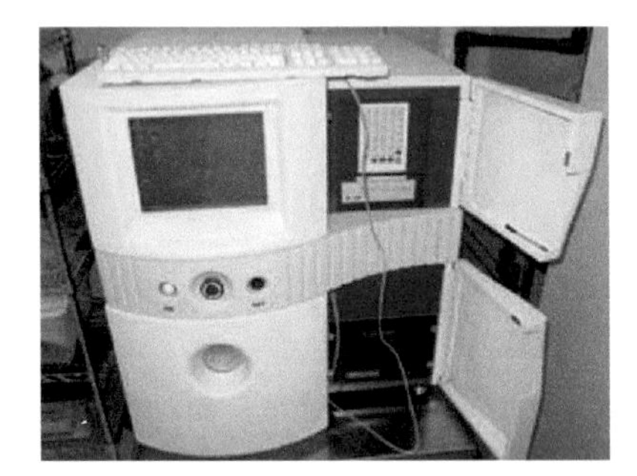

Fig. 10. Photo of TePla M4L Barrel Plasma Etcher

Suggested content for lab memo

Observe the released actuators under the micromanipulator. Capture images with the CCD camera to include in the memo. Discuss the quality of the released actuators. What percentage of the actuators appears to be damaged? Are there regions on the wafer where the yield appears to be better or worse than others?

Experiment 6. Characterization of hot-arm actuators

Purpose:

In this lab students will test the hot-arm actuators fabricated in Experiments 1-5. The voltage and current response during deflection of the thermal actuators will also be experimentally determined.

Materials:

- Hot-arm actuators fabricated in Experiments 1-5.

Equipment:

- Micromanipulator station
- Micromanipulator probes
- Microscope with CCD digital camera and appropriate imaging software
- Vacuum pump to secure wafer to microscope stage and counteract vibrations
- Arbitrary waveform generator (AWG)
- Two digtal multimeters (DMMs)

Overview and general description:

This experiment will be performed using a setup similar to that shown in Fig.11. In addition to a micromanipulator station, the setup includes two micromanipulator probes, an arbitrary waveform generator (AWG), a digital camera connected to a computer via a frame-grabber, a vacuum pump, two digital multimeters and a wafer with the released aluminum hot-arm actuators. A vacuum pump is critical to the setup, as the provided suction holds the MEMS chip containing the actuators to the microscope's stage plate and counteracts any vibrations present.

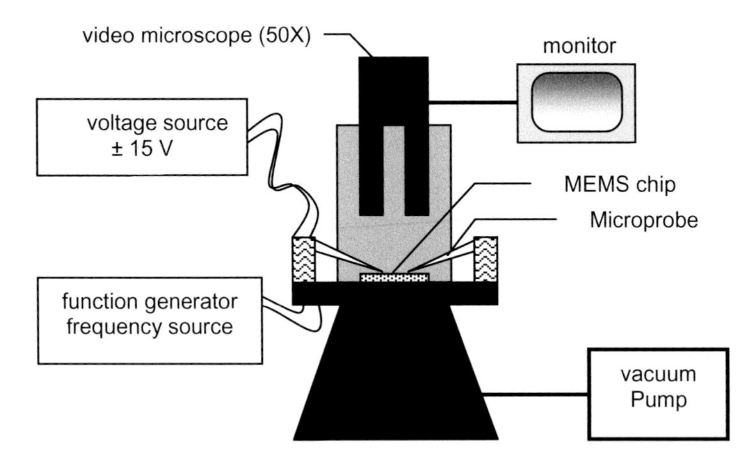

Fig. 11. Schematic of a basic experimental setup used in a Microfabrication laboratory. All electrical voltages are connected to the MEMS chip through microprobes which contact at metallic pads.

The probe tips are connected to the AWG in order to provide a DC signal to the actuators and cause actuation. The AWG will also be used to provide a square wave signal to the actuators in order to assess frequency response. The probe tips will be placed on the contact pads of the hot-arm actuators to supply the required voltage.

Procedure:

Part 1: Software setup

Set up the software that will be used to apply signals to the thermal actuators. (This part is optional since all testing can be done without the use of a computer).

Part 2:

This part is purely informative, and it deals with a few of the features used in the image-processing software. The software illustrated here is called EPIX® PIXCI®, and is used to capture images during the experiment.

1. When the frame capture is set to "Live" the current frame buffer is being constantly updated, while all others keep the last frame that was displayed.
2. The imaging software provides a tool that can assist to take deflection measurements. It can paint a ruler on the screen as shown in the Fig. 12.
3. The Ruler Tick Interval must be set such that at ×50 total zoom, the scale is 1 micron/increment. On the microscope, the three objectives are typically at ×8, ×25, and ×50. There may also be an adjustable zoom knob near the eyepiece that goes between ×1 and ×2. The table below gives the ruler scale corresponding to each zoom level.

Ruler scales for typical zoom levels

Objective	Scale with ×1 zoom (μm/increment)	Scale with ×2 zoom (μm/increment)
×8	6.25	3.125
×25	2	1
×50	1	0.5

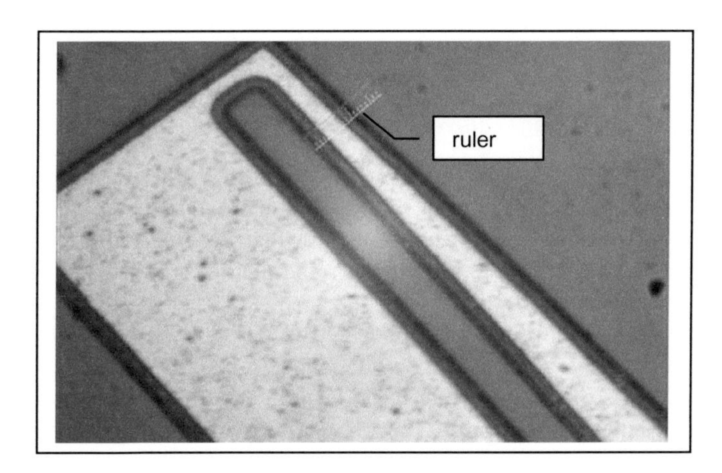

Fig. 12. Microscope view of a heat-actuator with the ruler superimposed

Part 3: Putting probe tips on the contact pads of thermal actuators

1. After focusing the microscope one should be able to see two probe tips at one focal plane and the heat actuators on the chip at another focal plane.

2. The probe-tip wires are connected to a multi-meter that is set for measuring resistance. It reads infinite resistance at this point.
3. The probe tips are *slowly* lowered by using the dial on the front of the manipulator station until the probe tips and their shadows are only a small distance apart. Note that the probe tips and heat actuators can be damaged by excessive force.
4. One probe tip is carefully positioned over one of the pads of the heat actuator. This should be done by using the XYZ controls on each manipulator.
5. After the first probe tip is placed on the contact pad. The actuator is vertically (perpendicular to the substrate) moved at this point. When the probe makes contact, the very tip will be in sharper focus. *If one tries to lower it more, the probe tip will slide forward on the contact pad.* Excessive motion of the tips on the surface creates buildup on the probes, and damages the actuators.
6. Steps 4 and 5 are repeated for the other probe tip and contact pad. When both probe tips are in contact with the pads, the resistance will no longer read infinite.
7. If after placing the second probe down, there is still infinite resistance, the first probe should be lowered by a very small amount, and then check your resistance again.

Part 4: Taking, voltage, current and deflection measurements.

For this step voltages from 0V-10V in steps of one volt are applied to the hot-arm actuator and the current and deflection are measured.

1. The wires need to be reconfigured so that the first multi-meter is in series with the waveform generator and the hot-arm actuator to measure current.
2. The second multimeter is connected in parallel with the hot-arm actuator to observe the applied voltage.
3. An image of the tip of an un-deflected actuator is obtained with the ruler superimposed on it. (Fig. 13.) There are varying geometries of actuators on the wafer but all are 25-30 μm wide from the outside edge of the hot arm to the outside point of the nub. Magnification (objective and zoom) value and the ruler tick interval as well as the sub ticks must be recorded.
4. Using the AWG, apply 1V and record the current displayed on the multi-meter and save the deflected image with ruler. Then turn the output off. Save the deflected image. Repeat steps 4 for voltages from 1V to 10V in 1V steps.

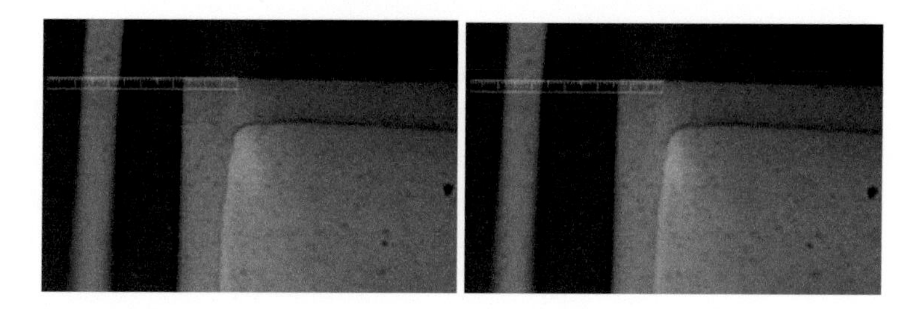

Fig. 13. Images taken at ×50 magnification and 9V showing a deflection of about 4.5 μm

Part 5: Frequency response

The hot-arm actuator's response should be different for different amplitude and frequency of the square wave.

1. The square wave generated by the AWG switches between 0V and +A V where A is the Amplitude parameter.
2. Start with a square wave with Amplitude 4 V, and frequency 1 Hz. Estimate the maximum deflection by the hot-arm actuator.
3. Repeat step 2 for frequencies of 10 Hz, 25 Hz, 50 Hz, 500 Hz, 1 kHz, 5 kHz, and 10 kHz. What do you notice about the maximum deflection of the hot-arm actuator as you change the frequency?
4. Repeat steps 2 and 3 for an Amplitude of 6 V.
5. Repeat steps 2 and 3 for an Amplitude of 8 V. What do you notice about the maximum deflection of the hot-arm actuator as you change the amplitude?

Suggested content for lab memo

You will need to use the images you saved to determine the deflection produced by each applied voltage.

1. Include in your report the following graphs and comment on them. Include trendlines for each curve, as well as an R^2 value.

    a. Current vs. Voltage. (*i-e* curve) An *i-e* curve for a resistor is linear with a positive slope.

    b. Deflection vs. Current. What is the maximum recorded deflection and for what current value?

    c. Deflection vs. Voltage. What is the maximum recorded deflection and for what voltage value?

   d. Deflection vs. power. What is the maximum recorded deflection and for what power value?

2. In your memo, describe how you determined the deflection for each data point.
3. Comment on how the frequency response was different for different amplitudes of applied voltage. What does this tell you about the hot-arm actuator as a system? Did you notice a difference between the response curves?

References and suggested reading

Beey S, Ensell G, Kraft M, White N (2004) MEMS Mechanical Sensors, Artech House, Inc.

Bryzek J, Roundy S, Bircumshaw B, Chung C, Castellino K, Stetter JR, Vestel M (2007) Marvelous MEMS, IEEE Circuits and Devices Magazine, Mar/Apr, pp 8-28.

Busch-Vishniac I (1998) Electromechanical Sensors and Actuators, Springer.

Doebelin EO (2004) Measurement Systems: Application and Design 5th edn. McGraw Hill.

Fraden J (2004) Handbook of Modern Sensors 3rd edn. Springer.

Madou MJ (2002) Fundamentals of Microfabrication, 2nd edn. CRC Press.

Roth Z, Kirkpatrick SR, Siahmakoun A (2004) Side-by-side Comparison of Thermal Actuators. In Proc. SEM 2004 MEMS and Nanotechnology Symposium, Costa Mesa, CA

Appendix A. Notation

a	acceleration
a	constant in resist profile equation
A	area
A	arrival rate
A	constant in Deal-Grove model of oxidation
b	damping coefficient
b	body force per unit volume
b_{min}	minimum feature size
B	constant in Deal-Grove model of oxidation
B	magnetic field
Bi	Biot number
C	capacitance
C	concentration
C_P	peak dopant concentration
C_s	surface concentration
C_{bg}	background concentration
C_{ij}	stiffness coefficients
d	distance, gap distance
d_{ij}	piezoelectric constant
D	diameter
D	dose
D	diffusion constant
D	flexural rigidity
D_h	hydraulic diameter
D_o	frequency factor
D_p	dose to clear and lithographic sensitivity for positive photoresist
D_p^0	reference dose for positive photoresist
D_g^0	dose to clear for negative photoresist
D_g^i	critical dose for negative photoresist
D_g^x	lithographic sensitivity for negative photoresist
e	effort variable (forces, voltages, and pressures)
e	voltage (electricity)
e	piezoelectric constant
E	Young's modulus

T.M. Adams, R.A. Layton, *Introductory MEMS: Fabrication and Applications*,
DOI 10.1007/978-0-387-09511-0,

E electric field
E energy
E_a activation energy
f flow variable (velocities, currents, and flow rates)
f force
f focal length
f frequency
F force
F flux
F effective F number (ratio of a lens's focal length to its diameter)
F gage factor
G shear modulus
h height
h_{conv} convective transfer coefficient
H height
i current
I light intensity
I mass moment of inertia
I second moment of area
j mass flux
J current density
k Boltzmann's constant
k constant in resist profile equation
k_{eq} equivalent spring constant
k_{sp} spring constant
k_1 experimentally determined parameter projection printing resolution
k_2 process dependent contrast in photolithography
K experimental parameter in photoresist spinning
K bulk modulus
Kn Knudsen number
$\mathscr{L}$ angular momentum
L inductance
L length
m mass
$\dot{m}$ mass flow rate
M moment or torque
M molecular weight
M magnetomotance
n index of refraction
n number of moles

n_i	concentration of species *i* in an electrolyte
N	number of molecules
NA	numerical aperture
N_0	temperature dependant parameter in evaporant flux
P	pressure
P	force
P	permeance
$\mathscr{P}_w$	wetted perimeter
P_v	vapor pressure
P	applied the load per unit area
q	charge
Q	amount of dopant per area
$Q, \dot{Q}$	heat, and heat transfer rate
Q_i	total implanted ion dose
r	position vector
r	radius
R	radius of curvature
R	resistance
R	resolution
R_P	projected range
R_u	universal gas constant
Re	Reynolds number
s	object distance from the lens
s	gap distance
s'	image distance
S_{ij}	compliance coefficient
t	time
t	film thickness
w	width
T	temperature
T	photoresist thickness
T	substrate thickness
T_{amb}	ambient temperature
u	displacement
u, v, w	fluid velocities in coordinate *x*, *y*, and *z* directions, respectively
U	velocity
$\mathscr{U}$	electrical potential energy
v	velocity
V	velocity
V	voltage
V	volume

V shear force
$\dot{V}$ volumetric flow rate
W, $\dot{W}$ work/power
x_j junction depth
x_{add} added oxide thickness
x_{diff} diffusion length
x_{ox} oxide thickness
z resist thickness

α constant in resist profile equation
α_T thermal expansion coefficient
β angle
δ depth of focus
γ shear strain
γ elastoresistance coefficient
γ_p resist contrast
ε strain
ε electrical permittivity
ε emissivity
ζ damping ratio
η dynamic viscosity
η sensitivity
θ angular displacement
κ thermal conductivity
λ mean free path
λ wavelength
λ_D Debeye length
μ magnetic permeability
μ electric mobility
μ_{ep} electrophoretic mobility
ν Poisson's ratio
π piezoresistance coefficient
ρ density
ρ electrical resistivity
ρ_e charge density
σ interaction cross section
σ normal stress
σ Stefan-Boltzmann constant
σ electrical conductance
τ time parameter in Deal-Grove model of oxidation
τ shear stress

τ	time constant
ϕ	magnetic flux
φ	electric potential
Φ	photochemical quantum efficiency
Φ_e	activation energy
ω	angular velocity, angular frequency
ω	deflection

Appendix B. Periodic table of the elements

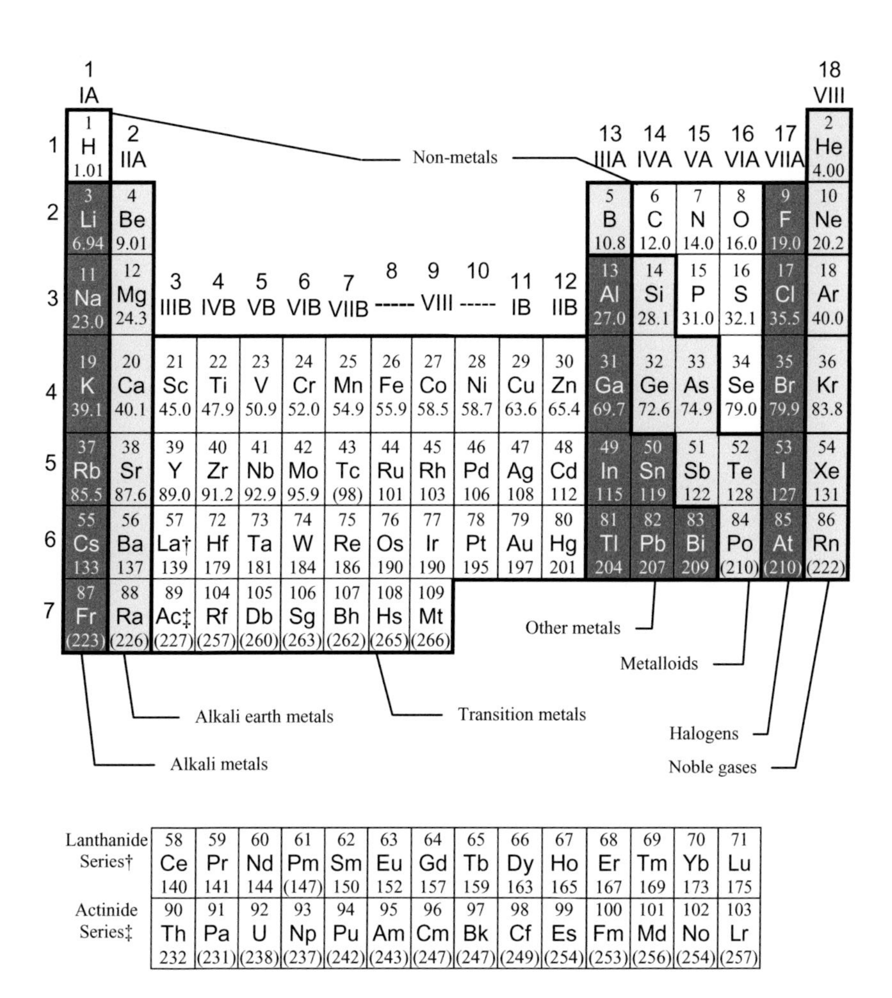

	1 IA	2 IIA	3 IIIB	4 IVB	5 VB	6 VIB	7 VIIB	8 -----	9 VIII	10 -----	11 IB	12 IIB	13 IIIA	14 IVA	15 VA	16 VIA	17 VIIA	18 VIII
1	1 H 1.01																	2 He 4.00
2	3 Li 6.94	4 Be 9.01											5 B 10.8	6 C 12.0	7 N 14.0	8 O 16.0	9 F 19.0	10 Ne 20.2
3	11 Na 23.0	12 Mg 24.3											13 Al 27.0	14 Si 28.1	15 P 31.0	16 S 32.1	17 Cl 35.5	18 Ar 40.0
4	19 K 39.1	20 Ca 40.1	21 Sc 45.0	22 Ti 47.9	23 V 50.9	24 Cr 52.0	25 Mn 54.9	26 Fe 55.9	27 Co 58.5	28 Ni 58.7	29 Cu 63.6	30 Zn 65.4	31 Ga 69.7	32 Ge 72.6	33 As 74.9	34 Se 79.0	35 Br 79.9	36 Kr 83.8
5	37 Rb 85.5	38 Sr 87.6	39 Y 89.0	40 Zr 91.2	41 Nb 92.9	42 Mo 95.9	43 Tc (98)	44 Ru 101	45 Rh 103	46 Pd 106	47 Ag 108	48 Cd 112	49 In 115	50 Sn 119	51 Sb 122	52 Te 128	53 I 127	54 Xe 131
6	55 Cs 133	56 Ba 137	57 La† 139	72 Hf 179	73 Ta 181	74 W 184	75 Re 186	76 Os 190	77 Ir 190	78 Pt 195	79 Au 197	80 Hg 201	81 Tl 204	82 Pb 207	83 Bi 209	84 Po (210)	85 At (210)	86 Rn (222)
7	87 Fr (223)	88 Ra (226)	89 Ac‡ (227)	104 Rf (257)	105 Db (260)	106 Sg (263)	107 Bh (262)	108 Hs (265)	109 Mt (266)									

Lanthanide Series†	58 Ce 140	59 Pr 141	60 Nd 144	61 Pm (147)	62 Sm 150	63 Eu 152	64 Gd 157	65 Tb 159	66 Dy 163	67 Ho 165	68 Er 167	69 Tm 169	70 Yb 173	71 Lu 175
Actinide Series‡	90 Th 232	91 Pa (231)	92 U (238)	93 Np (237)	94 Pu (242)	95 Am (243)	96 Cm (247)	97 Bk (247)	98 Cf (249)	99 Es (254)	100 Fm (253)	101 Md (256)	102 No (254)	103 Lr (257)

Appendix C. The complimentary error function

Table C.1. The complimentary error function

λ	erfc(λ)	λ	erfc(λ)	λ	erfc(λ)	λ	erfc(λ)	λ	erfc(λ)	λ	erfc(λ)
0.02	0.9774	0.4	0.5716	0.78	0.2700	1.16	0.10090	1.54	0.02941	1.92	0.00662
0.04	0.9549	0.42	0.5525	0.8	0.2579	1.18	0.09516	1.56	0.02737	1.94	0.00608
0.06	0.9324	0.44	0.5338	0.82	0.2462	1.2	0.08969	1.58	0.02545	1.96	0.00557
0.08	0.9099	0.46	0.5153	0.84	0.2349	1.22	0.08447	1.6	0.02365	1.98	0.00511
0.1	0.8875	0.48	0.4973	0.86	0.2239	1.24	0.07949	1.62	0.02196	2.00	0.00468
0.12	0.8652	0.5	0.4795	0.88	0.2133	1.26	0.07476	1.64	0.02038	2.10	0.00298
0.14	0.8431	0.52	0.4621	0.9	0.2031	1.28	0.07027	1.66	0.01890	2.20	0.00186
0.16	0.8210	0.54	0.4451	0.92	0.1932	1.3	0.06599	1.68	0.01751	2.30	0.00114
0.18	0.7991	0.56	0.4284	0.94	0.1837	1.32	0.06193	1.7	0.01621	2.40	0.00069
0.2	0.7773	0.58	0.4121	0.96	0.1746	1.34	0.05809	1.72	0.01500	2.50	0.00041
0.22	0.7557	0.6	0.3961	0.98	0.1658	1.36	0.05444	1.74	0.01387	2.60	0.00024
0.24	0.7343	0.62	0.3806	1.00	0.1573	1.38	0.05098	1.76	0.01281	2.70	0.00013
0.26	0.7131	0.64	0.3654	1.02	0.1492	1.4	0.04771	1.78	0.01183	2.80	0.00008
0.28	0.6921	0.66	0.3506	1.04	0.1414	1.42	0.04462	1.8	0.01091	2.90	0.00004
0.3	0.6714	0.68	0.3362	1.06	0.1339	1.44	0.04170	1.82	0.01006	3.00	0.00002
0.32	0.6509	0.7	0.3222	1.08	0.1267	1.46	0.03895	1.84	0.00926	3.20	0.00001
0.34	0.6306	0.72	0.3086	1.1	0.1198	1.48	0.03635	1.86	0.00853	3.40	0.00000
0.36	0.6107	0.74	0.2953	1.12	0.1132	1.5	0.03389	1.88	0.00784	3.60	0.00000

Appendix D. Color chart for thermally grown silicon dioxide

Table D.1. Color chart for thermally grown silicon dioxide

Film Thickness (μm)	Color and Comments
0.05	Tan
0.07	Brown
0.10	Dark violet to red violet
0.12	Royal Blue
0.15	Light blue to metallic blue
0.17	Metallic to very light yellow green
0.20	Light gold to yellow; slightly metallic
0.22	Gold with slight yellow orange
0.25	Orange to melon
0.27	Red violet
0.30	Blue to violet blue
0.31	Blue
0.32	Blue to blue green
0.34	Light green
0.35	Green to yellow green
0.36	Yellow green
0.37	Green yellow
0.39	Yellow
0.41	Light orange
0.42	Carnation pink
0.44	Violet red
0.46	Red violet
0.47	Violet
0.48	Blue violet
0.49	Blue
0.50	Blue green
0.52	Green (broad)
0.54	Yellow green
0.56	Green yellow

Table D.1. Color chart for thermally grown silicon (cont'd).

0.57		Yellow to "yellowish" (not yellow but is in the position where yellow is to be expected; at times appears to be light creamy gray or metallic)
0.58		Light orange or yellow to pink borderline
0.60		Carnation pink
0.63		Violet red
0.68		Bluish" (not blue but borderline between violet and blue green; appears more like a mixture between violet red and blue green and looks grayish)
0.72		Blue green to green (quite broad)
0.77		"Yellowish"
0.80		Orange (rather broad for orange)
0.82		Salmon
0.85		Dull, light red violet
0.86		Violet
0.87		Blue violet
0.89		Blue
0.92		Blue green
0.95		Dull yellow green
0.97		Yellow to "yellowish"
0.99		Orange
1.00		Carnation pink
1.02		Violet red
1.05		Red violet
1.06		Violet
1.07		Blue violet
1.10		Green
1.11		Yellow green
1.12		Green
1.18		Violet
1.19		Red Violet
1.21		Violet red
1.24		Carnation pink to salmon
1.25		Orange
1.28		"Yellowish"
1.32		Sky blue to green blue
1.40		Orange
1.45		Violet
1.46		Blue violet
1.50		Blue
1.54		Dull yellow green

Color Chart Table for thermally grown silicon dioxide films observed perpendicularly under daylight fluorescent lighting.

Glossary

Acceptor A type of dopant that accepts electrons from a semiconductor into its valence band. Resulting semiconductors are p-type.

Accuracy Degree to which a transducer output conforms to the true value of the measurand (sensor) or the desired output effect (actuator)

Activation energy In chemistry the energy that must be overcome in order for a chemical reaction to occur. Also called threshold energy.

Actuator A transducer with a mechanical output. Used to move or control a system, such as controlling the position of a movable microstructure

ADC Analog-to-digital converter.

Aerial image An image of an unmodified mask pattern projected onto the surface of a photoresist layer by an optical system.

Amplifier Device used to increase (or decrease) the amplitude of a signal.

Amplitude ratio The ratio of the output amplitude to the input amplitude of a transducer. Usually reported in units of decibels.

Analog signal A signal that is continuous in amplitude and over time.

Analog-to-digital converter Device used to convert continuous signals to discrete digital numbers.

Anisotropic etching An etching process used with crystalline materials for which etching occurs at different rates depending upon the crystal face exposed. Whether or not anisotropic etching occurs and to what degree is strongly dependent on the etchant-crystal combination. In single-crystal materials (e.g. silicon wafers), this effect can allow very high anisotropy.

Anodic bonding A method of hermetically and permanently joining glass to silicon without the use of adhesives. The silicon and glass wafers are typically heated to a temperature of 300-500°C, depending on the glass type, while a voltage is applied across the silicon/glass interface.

Aspect Ratio The ratio of the length of a structure in the plane of a wafer to its width in that plane (lateral) or the ratio of the height of a structure perpendicular to a wafer's surface to its depth in the wafer's plane (vertical).

Backside processing A method of processing that uses both sides of the substrate.

Batch fabrication Any of the various fabrication techniques in which a quantity of devices are produced at one time using parallel operations.

Beam deflection equation Solid mechanics relationship that relates deflection to the internal reactions of a loaded beam.

Biaxial modulus Modulus of elasticity relating biaxial stress and strain. Given by $E/(1-\nu)$.

Biaxial stress/strain Two dimensional stress/strain in which only equal normal stresses are present.

Binary mask Mask made up of opaque and transparent regions such that light is either completely blocked or transmitted at certain locations.

BioMEMS MEMS systems with biological and/or chemical applications

Biot number A dimensionless quantity that represents the relative resistance to internal conduction within a body to the external resistance to convection.

Black body A perfect emitter of thermal radiation.

Body force A force on a body that arises due to the attraction or repulsion of some other body. The bodies need not make physical contact. Contrast **Surface force**.

BOE Buffered oxide etch

Bonding Pads A pad placed around the perimeter of the chip used to attach the chip to package leads.

Brittle Describes a material that fractures before it can be plastically deformed.

Bulk micromachining A process used to produce MEMS in which structures are defined by selectively etching inside a substrate. The resulting structures are produced inside the substrate itself. This is distinguished from surface micromachining, which uses a succession of thin film deposition and selective etching techniques to create structures on top of a substrate.

Bus A group of conductors that interconnect individual circuitry in a computer or digital device with which data is transferred between various components of the device.

Calibration Process by which the output of a transducer is related to a known input, or standard.

Capacitance Measure of the amount of stored electric charge in a device subject to a given electric potential.

Capacitive half-bridge An arrangement of two variable capacitors such that as one capacitance decreases, the other increases. Used in MEMS sensing applications.

Charge carriers The quantities responsible for the movement of charge in an electrical conductor. Typically electrons in metals, electrons or “holes” in semiconductors, and ions in electrolytes.

Charge density The net electric charge per unit volume at a point in a substance.

Chemical etching The process of using acids, bases or other chemicals to dissolve away unwanted materials such as metals, semiconductor materials or glass.

Chemical Vapor Deposition (CVD) A chemical process used to deposit high-purity thin films onto surfaces. Typical CVD processes expose a wafer (substrate) to one or more volatile precursors, which react and/or decompose on the substrate surface to produce the desired deposit. Frequently, volatile byproducts are also produced, which are removed by gas flow through a reaction chamber.

Clean room A space intentionally kept at a certain level of cleanliness in which most microfabrication processes take place.

Closed system A system consisting of a fixed mass. Also called a control mass.

CMOS Complimentary Metal Oxide Semiconductor: A type of integrated circuit/fabrication process based on insulated gate field effect transistors, similar to MOS, but uses complimentary MOS transistors to provide the circuits basic logic functions. The process layers from top to bottom are: metal, insulating oxide, semiconductor. CMOS is finding many applications in surface micro machined MEMS fabrication.

Comb-drive A MEMS device consisting of electrically charged interdigitated fingers similar to a comb. Primary applications include inertial sensors and RF resonators.

Compliance coefficient The elements of the matrix that relates the stress tensor to the strain tensor for an anisotropic elastic material.

Conduction Transport of electrical charge through a medium.

Conduction (Heat transfer) Mode of heat transfer in which thermal energy is transported through a medium without being accompanied by relative motion of the conducting molecules. Predominant mode of heat transfer in solids and stationary fluids.

Conduction band Range of electron energy, higher than that of the valence band, sufficient to make electrons free to accelerate under the influence of an applied electric field, thereby constituting an electric current.

Conservation of energy Principle that states that energy can neither be created nor destroyed, although it may change forms.

Contact printing Lithographic printing process in which the mask is in physical contact with the photoresist layer. Also called hard-contact printing.

Continuity equation An equation representing the concept of conservation of mass expressed at an infinitesimally small point in a continuum.

Continuum A model of a physical material in which the discrete molecular structure of the material is unimportant and in which intensive properties are thought of as having values at a point.

Convection Mode of heat transfer in which thermal energy is transported by conduction accompanied by bulk motion of the heat transfer medium. Predominant mode of heat transfer within flowing liquids and gases.

Convective heat transfer coefficient Quantity relating the rate of heat transfer between a surface and a moving fluid to surface area and temperature difference. The convective heat transfer coefficient is a function of the fluid properties, fluid velocity, surface-to-fluid geometry, and other parameters.

Coulomb's Law States that the force that one electrically charged particle of charge q_1 exerts on another electrically charged particle of charge q_2 is directly proportional to the charges and inversely proportional to the square of the distance between them.

Current density Time rate of transport of electric charge per unit area normal to flow direction.

Crystal plane In crystalline solids, a plane defined by the alignment of certain atoms.

CVD Chemical Vapor Deposition.

Czochralski method Standard process for growing ultra pure crystalline silicon.

Deal-Grove model Most widely used model of oxidation kinetics. Relates oxide thickness to reaction time.

Debeye length Exponential decay distance for wall potential within an electric double layer.

Depth of focus Tolerance within a layer of photoresist in which a latent image is in focus to an acceptable degree.

Developer A chemical that when brought in contact with an exposed photoresist chemically attacks exposed and unexposed regions at much different rates.

Die The square or rectangular section of a wafer onto which an single integrated circuit or device is fabricated.

Die separation Cutting a wafer apart in order to separate its individual devices.

Dielectric A material typically used as an insulator that contains few free electrons, has low electrical conductivity and supports electrostatic stresses.

Diffusion constant A temperature dependent constant that characterizes the mass diffusion of one material in another.

Diffusion length A characteristic length in a diffusion process given by $\sqrt{(4Dt)}$. The diffusion length gives a rough estimate of how far a dopant has diffused into a substrate material at a given time.

Diffusivity Diffusion constant.

Digital signal Signal with discrete levels of amplitude that change only in discrete increments of time.

Disk method Method for measuring stress in thin films in which substrate deflection is measured and related to the film stress.

Distributed load Load on a mechanical component that is distributed over length or area.

Distributed parameter modeling Modeling scheme in which spatial variations of physical quantities are sought.

Divergence The operation $\vec{\nabla} \bullet (\vec{x})$, where $\vec{x}$ is a vector quantity. Can be interpreted as the net out-flux of the $\vec{x}$ per unit volume.

Divergence theorem States that the integral of a vector dotted with an outward normal area of an enclosing surface is equal to the volume integral of the divergence of the vector within the volume of the surface.

Donor A dopant that donates electron(s) to a semiconductor's valence band. The resulting semiconductor is n-type.

Dopants The elements introduced in a doping process.

Doping Process of introducing impurity atoms into a semiconductor in order to modify its electrical properties.

Dose In photolithography the amount of energy per unit surface area incident on a surface during exposure.

Dose to clear In photolithography the amount of optical energy per unit surface area required to completely expose a layer of photoresist.

DRIE Deep Reactive Ion Etching: Highly anisotropic etch process used to create deep, steep-sided holes and trenches in wafers. Can result in aspect ratios of 20:1 or more.

Drift Motion of particles in a medium relative to the medium.

Drift velocity Velocity of a particle in a medium relative to medium.

Drive-in The second step in a two step implantation process. The first step consists of the bombardment of a wafer surface with dopant ions and is called pre-deposition. In drive-in the dopant is further diffused into the wafer by thermal diffusion.

Dry etching Process in which the reactive components of an etchant are contained in a gas or plasma.

Dry oxidation Oxidation process in which only oxygen diluted with nitrogen makes contact with the wafer surface. Slower than wet oxidation, but results in higher quality oxide layers.

Ductile Describes a material that is capable of being plastically deformed in at least some range of strain.

Einstein relation Equation relating mass diffusivity of charge carriers to electric mobility.

Einstein summation convention Summing values for a repeated index.

Elasticity The property of a some substances that enables them to change their length, volume, or shape in direct response to a force and to recover their original form upon the removal of the force. Elastic solids obey **Hooke's Law**.

Elastoresistance coefficient One of several coefficients for a piesoresistive material that relates relative change in electrical resistivity to strain.

Electric field A region characterized by a distribution of electric charge. The numerical value of an electric field at a point gives the value of the net electric force exerted on particle of unit charge per unit charge.

Electric flux The lines of force that make up an electric field.

Electric mobility Measure of how likely a charge carrier is to move within a medium.

Electric potential Electric potential energy per unit charge a point in an electric field. Also the driving force for electric current. Measured in volts.

Electrical conductivity The measure of a material's ability to support an electric current density when subjected to an electric field.

Electrical resistivity The inverse of electrical conductivity

Electrodeposition Electrochemical deposition of a thin film material usually used with metals.

Electrolyte Fluid capable of conducting electricity.

Electromigration Movement of charge due to a potential difference.

Electron beam (e-beam) evaporation Physical vapor deposition in which the electron beam evaporative the source material.

Electron beam (e-beam) lithography A method of fabricating submicron and nanoscale features by exposing electrically sensitive surfaces to an electron beam. Similar to photolithography, but using electrons rather than photons in order to take advantage or the former's far smaller wavelengths.

Electro-osmotic flow Fluid motion caused by the application of an electric field. Motion is caused by the electric body force on charged particles within the fluid, which in turn viscously drag the surrounding fluid along with them.

Electrophoresis Drift of particles within a fluid due to the influence of an electric field.

Electrophoretic mobility Coefficient of proportionality between drift velocity of a particle and electric field strength.

Electrophoretic separation Technique used to separate chemical species in a fluid sample due to the application of an electric field.

Emissivity A dimensionless number measuring the ability of a surface to emit thermal radiation as compared to a blackbody. Ranges from zero to one.

Entrance length Region near the entrance of a flow channel characterized by flow parameters changing in the flow direction.

Etchant Chemical used to etch a material.

Etching See chemical etching.

Etch rate The amount of material removed per unit time during etching.

Etch stop One of several techniques used to cause the chemical reaction in a etching process to stop.

Evaporation Thermal evaporation.

Extrensic stress Stress resulting from some externally imposed factor.

FEA Finite Element Analysis. Also FEM, Finite Element Method.

FEM Finite Element Method. Also FEA, Finite Element Analysis.

Field mask Type of mask used to protect a region of the substrate during an additive process.

Flats Straight edges appearing on otherwise circular wafers used to identify wafer crystalline orientation and the presence/type of doping.

Flux A physical quantity expressed per unit area in the direction normal to that quantity. In some cases (most notably the flux associated with an electric field) flux refers to a physical quantity multiplied by the area in the direction normal to the quantity.

Foreline The line connecting the exhaust of a high vacuum pump to a rough pump.

FSO Full-Scale Output.

Full-scale output Range of output values of a transducer corresponding to the span.

Fully developed flow Fluid flow in which the velocity distribution does not change in the flow direction.

Frequency factor A parameter related to vibrations of the atoms in a lattice. Often used to calculate mass diffusivity.

Gage factor The ratio of the relative change in resistance in a piezoresistive sensor to applied strain.

Gain Amplitude ratio.

Gas entrapment The trapping of a gas within the thin film.

Growth stress See intrinsic stress

Hall effect Production of a voltage difference across an electrical conductor, transverse to an electric current in the conductor and a magnetic field perpendicular to the current.

Hard bake The baking of wafers after resist patterning in order to remove solvents and moisture, and to provide for better adhesion during subsequent etch or implant processes.

Hard mask Mask through which etching of a substrate can occur.

HARM High Aspect Ratio Micromachining.

Heat transfer Energy transport due to a spatially occurring temperature difference.

High vacuum Region of vacuum pressure from 10^{-5} to 10^{-8} torr.

High vacuum pump One of a specialized class of pumps required to achieve high vacuum pressures.

Hole Void in a lattice structure caused by a missing valence band electron. Can be a charge carrier in some semiconductor materials.

Hooke's Law In solid mechanics, an approximation that states that the amount a material body is deformed is lineally related to the stress causing the deformation.

Hydraulic diameter Calculated value used in place of diameter in fluid channels with non-circular cross sections. Value is four times the cross sectional area divided by perimeter.

I-line For ultraviolet light emitted by an Hg or Hg-Xe lamp, local peak of ultraviolet light with a wavelength of 365 nm.

Implantation A method of doping by which a beam of energetic ions implants charged dopants atoms into the substrate.

Intrinsic stress Stresses resulting directly from the nucleation process of creating a thin film. Also called growth stresses.

Ion dose The amount of dopant per unit area delivered to a substrate.

Ionic double layer The thin layer next to a solid surface in contact with an electrolyte in which the electric charge density has a nonzero value.

Isotropic etch Wet or dry etching that occurs in all directions at the same rate.

Junction A p-n junction.

Junction depth Location in a wafer in which the background concentration of dopant matches the concentration of a newly implanted or diffused dopant.

Knudsen number Ratio of the mean free path of a molecule to some characteristic dimension of a system.

Lab-on-chip MEMS technology applied to analytical instrumentation. Instruments are reduced in size such that they can fit onto a single chip.

Laminar flow Streamlined flow with no turbulence occurring at low Reynolds numbers.

Laplacian operator The differential operator $\vec{\nabla}^2$.

Latent image Reproduction of the aerial image in the resist layer as a spatial variation of chemical species.

Lattice constant The distance characterizing the side length of the unit cell of a crystal.

LIGA A type of High Aspect Ratio Micromachining (HARM); it is a German acronym for Lithographie, Galvanofomung, Abformung, meaning Lithography, Electroforming, Molding.

Lift-off Method of selectively depositing a functional layer through a temporary mask.

Light intensity Amount of power per unit surface area delivered to a surface via a light source in photolithography.

Linear rate constant A constant in the Deal-Grove model of oxidation kinetics.

Lithographic sensitivity The dose needed to completely clear the resist layer.

Load An externally applied force to a mechanical component.

Low vacuum Region of vacuum pressure up to 10^{-3} torr.

Lumped element modeling Modeling scheme in which spatial distributions of physical parameters are ignored and individual components are treated as simple devices such as springs and dashpots.

Lumped parameter Simple parameter of a device (e.g., a mass, spring constant, damping factor) that can be used as an analytical representation of the real device.

Macroscopic conservation equation Conservation equation (such as conservation of mass or linear momentum) applied to a finite region of space and from which distributions of physical quantities within the region cannot be determined.

Magnetic flux Product of magnetic field and an element of area perpendicular to it.

Magnetic permeability Property characterizing the degree to which a magnetic field can permeate a material.

Magnetomotance A physical force that produces a magnetic field. Also called magnetomotive force (mmf)

Magnetoresistive material Material whose electrical resistance changes when subjected to a magnetic field.

Magnetostriction Property of a ferromagnetic material to change its shape when subjected to a magnetic field.

Mask An opaque border or pattern placed between a source of light and a photosensitive surface to prevent exposure of specified portions of the surface.

Mask alignment Procedure of aligning the pattern on a mask with patterns and/or features already on a wafer. Also referred to as registration.

Mass diffusion Transfer of mass due to a spatially occurring differences in species concentration.

Mass diffusivity Diffusion constant.

Mass flow rate The time rate at which mass is crosses a system boundary.

Mass flux Mass flow per unit area normal to the flow direction.

Material derivative The time derivative of a fixed mass of fluid as it is followed in a flow field.

Mean free path In physics and kinetic theory, the average distance a molecule or atom travels between collisions with other molecules or atoms.

Measurand A physical parameter being quantified by measurement.

MEMS Micro-Electro-Mechanical-Systems.

Mesh A network of finite points distributed in space used in finite element analysis (FEA).

Microfluidics General term for MEMS fluidic devices.

Micromachining One or more of the processes of thin film deposition, pattern transfer and etching processes used to create the structures that make up a MEMS device.

Micron A micrometer; i.e., 10^{-6} m. Standard unit of length for MEMS devices.

Microstrain (μ-strain) A dimensionless pseudo-unit of deformation having magnitude 10^{-6}.

Microvoids Gaps in thin films left by escaping gas byproducts.

Miller indices A notation used to describe lattice planes and directions in a crystal.

Modulation Process of varying one waveform in relation to another waveform

MOEMS Micro-Optical-Electro-Mechanical-Systems

MOS Metal Oxide Semiconductor: A type of integrated circuit fabrication process based on insulated gate field effect transistors originating in the very large scale integration (VLSI) industry. Process layers from top to bottom are: metal, insulating oxide, semiconductor.

MOSFET Metal Oxide Semiconductor Field Effect Transistor.

MST Micro-System-Technology: A more general term used for MEMS including optical and fluidic systems. MST is a more common term than MEMS in Europe and Japan.

MUMPS Multi-User MEMS Process: Foundry process and standard developed by Cronos Integrated Microsystems, Inc. The process allows several devices and/or system designs to be fabricated simultaneously on one wafer.

Nano Prefix meaning ten to the minus ninth power, or one one-billionth.

Nanometer (nm) A unit of measurement equal to one billionth of a meter.

Nanotechnology Refers to the science and technology of devices ranging in size from a nanometer to a micron.

Navier-Stokes equations Partial differential equations representing conservation of linear momentum for a Newtonian fluid.

Negative photoresist Type of photoresist in which the portion of the photoresist exposed to light becomes relatively insoluble to the photoresist developer. The unexposed portion of the photoresist is subsequently dissolved by the photoresist developer. Creates the negative of the mask image on the surface after development.

Newtonian fluid A fluid exhibiting a linear relationship between shear stress and the rate of deformation. The constant(s) of proportionality is viscosity.

Neutral axis Special location in a beam that has experienced no deformation after bending has occurred.

Normal strain In its simplest terms, change in differential length divided by original differential length of an element undergoing deformation.

Normal stress In its simplest terms, force per unit area oriented normal to that area.

No slip Most common boundary condition used in fluid mechanics. S states that there is no relative velocity between a fluid and a surface with which makes contact.

N-type Indicates a doped semiconductor in which the donated charge carriers are negatively charged.

Open system A system in which mass can cross the system boundaries. Also called a control volume.

Oxidation Controlled formation of silicon dioxide layers on a silicon wafer as a result of the chemical reaction between the silicon and oxygen. Typically carried out at high temperatures within an oxidation furnace.

Oxide A silicon dioxide layer.

Package Protective enclosure for a MEMS device.

Peel force Net force usually occurring near the edges of thin films that tends to peel thin films away from the underlying layer.

Permanent resist A patterned resist that has become a permanent part of the MEMS device.

Permeability Magnetic permeability.

Permeance Measure of a magnetic circuit's ability to conduct magnetic flux. Reciprocal of reluctance.

Permittivity Property characterizing a material's ability to transmit or an electric field associated with the separation of charge.

Phase-shift mask Mask containing spatial variations in intensity and phase transmittance of light.

Photolithography Typically, process for creating patterns in which a photosensitive surface (a photoresist) is selectively exposed to light.

Photomask See mask.

Photoresist A light-sensitive material used in photolithography and photoengraving to form patterned coatings on a surface.

Physical vapor deposition Any of the various techniques used to deposit thin films of various materials onto surfaces by physical means, as compared to chemical vapor deposition (CVD).

Piezoactuator An actuator typically producing a force or displacement in response to an electrical input signal.

Piezoelectric charge coefficient One of several constants that relate various mechanical quantities to various electrical quantities in a piezoelectric material.

Piezoelectricity Property of certain materials to develop an internal electric field when deformed.

Piezoresistive coefficient One of several coefficients for a piesoresistive material that relates relative change in electrical resistivity to stress.

Piezoresistance Property of certain materials to exhibit changes in electrical resistance when subject deformation.

Planarization A process of evening out the topography of a MEMS device after certain fabrication steps.

Plasma ashing A dry stripping process of photoresist removal in which a plasma source-generated monatomic reactive species combines with the photoresist to form ash, which is then removed via a vacuum pump.

Plasma etching Etching process in which the chemically reactive etchant is contained in an ionized gas, or plasma.

Polarization The total electric dipole moment per unit volume of dielectric materials.

P-n junction The area where n-type and p-type materials in a semiconductor come together.

Point load Load applied at a finite point to a mechanical component.

Poisson's ratio The measure of the tendency of a material as it is stretched in one direction to get thinner in the other two directions. More precisely, the ratio of the relative contraction strain, or transverse strain (normal to an applied load), divided by the relative extension strain, or axial strain (in the direction of the applied load). For a perfectly incompressible material, Poisson's ratio is 0.5.

Poly etch An wet etchant consisting of hydrofluoric, nitric and acetic acids. Typically used to etch silicon isotropically.

Polyimide A polymer of imide monomers commonly used as a permanent resist and a sacrificial layer.

Polymer shrinkage Shrinking of polymers due to curing.

Polysilicon Polycrystalline silicon. Used as a structural material for many MEMS devices.

Positive photoresist Type of photoresist in which the portion of the photoresist exposed to light becomes soluble to the photoresist developer. The portion of the photoresist that is unexposed remains insoluble to the photoresist developer. Creates an image identical to the mask image on the surface after development.

Precursor In chemical vapor deposition, a vapor containing the to-be-deposited material in a different chemical form.

Pre-deposition The first step in a two step implantation process consisting of bombardment of a wafer surface with dopant ions. The second step is drive-in in which the dopant is further diffused into the wafer by thermal diffusion.

Principal axes Unique set of orthogonal axes in a material for which only normal stresses are present.

Process flow All the necessary process steps in the correct order needed to produce a MEMS device.

Projected range The depth of peak concentration of dopant in wafer due to implantation.

Projection printing Photolithographic printing method in which the photo mask is imaged by a high-resolution lens system onto a resist-coated wafer.

Proximity printing Photolithographic printing process in which the mask is not pressed against the wafer. Typically, a gap of 10 to 50 μm is left between the mask pattern and the wafer.

P-type Indicates a doped semiconductor in which the donated charge carriers are positively charged.

Pull-in An undesired phenomenon encountered in electrostatic actuators in which the actuator plates venture too close to one another, resulting in an overwhelming electrostatic force that slams the plates together.

PVD Physical Vapor Deposition.

Quantum dot Grouping of atoms so small that the addition or removal of an electron will change its properties in a significant way.

Radiation In heat transfer, thermal radiation.

RCA clean A standard set of wafer cleaning steps performed before high temperature processing steps (e.g., oxidation, diffusion, CVD) of wafers in semiconductor manufacturing. RCA cleans involve the following steps:

1. removal of organic contaminants using a 1:1:5 solution of NH4OH + H_2O_2 + H_2O at 80°C
2. removal of thin oxide layers using a 1:10 solution of HF
3. removal of ionic contamination using a 1:1:5 solution of HCl + H_2O_2 + H_2O at 80°C

Reactions In solid mechanics, forces and/or moments that supports supply to a mechanical structure.

Release See sacrificial layer.

Resist contrast Describes the sharpness of the pattern in the resist formed during the lithographic process.

Resistive heating In thermal evaporation, the process of creating a vapor out of the source material by passing a large electric current through a highly refractory metal structure containing the source.

Resolution (Photolithography) The smallest distinguishable feature size of a transferred pattern.

Resolution (Transducers) The smallest change in input that produces a detectable change in the output of a transducer.

Resonator Structure such as a thin beam that vibrates with small displacements at high frequencies.

Resonance Tendency of some systems' outputs to exhibit large magnifications of inputs at certain frequencies.

Resonant frequency Frequency corresponding to resonance.

Response time Time the output of a transducer takes to reach its new steady state value when subject to a step input.

Reynolds number Dimensionless number that measures the relative importance of inertia to viscous forces in a flowing fluid. Can be used to predict whether flow fields are laminar or turbulent.

RIE Reactive Ion Etch.

Rough pumps A positive displacement pump used to initially lower the pressure of a vacuum chamber and to back (connected to the outlet of) other pumps.

Sacrificial layer A temporary layer used in surface micromachining. Usually a structural layer is deposited on top of a sacrificial layer, which is then selectively etched at certain locations, creating the surface structure. This process is called release.

Scaling Process of relating physical quantities of interest to powers of length scale. Used to assess the relative importance of different physical phenomena.

Scanning electron microscope (SEM) An electron microscope that forms a three-dimensional image on a cathode-ray tube by moving a beam of focused electrons across an object and reading both the electrons scattered by the object and the secondary electrons produced by it. Electron microscopes use electrons rather than visible light to produce magnified images, especially of objects having dimensions smaller than the wavelengths of visible light, with linear magnification approaching or exceeding a million (106).

Selectivity Relative etch rate of an etchant solution with one material compared to another. Also in anisotropic etching, relative etch rate of an etchant solution in one crystalline direction compared to another.

SEM Scanning Electron Microscope

Semiconductor Any of various solid crystalline substances, such as germanium or silicon, having electrical conductivity greater than insulators but less than good conductors. Used especially as a base material for computer chips and other electronic and MEMS devices.

Sensitivity Constant of proportionality between output and input of a transducer.

Sensor A transducer used as a measuring device. The output is usually an electrical quantity.

Shadowing In line-of-sight deposition methods, the prevention an evaporant from being deposited on a surface due to the blocking effect of a structure in the line-of-sight of the deposition.

Shadow printing Photolithographic printing method that relies on the opaque and transparent regions of the mask to directly transfer (without aid of a lens) a mask pattern to a resist. Includes both contact and proximity printing methods.

Shape memory alloy (SMA) Alloy in which crystalline phase transformations result in an easily deformable material at low temperatures and a rigid, hard-to-deform material at high temperatures. When an SMA in its rigid state is cooled to its deformable, low temperature state it may be plastically deformed by as much as 10%. Upon heating back to the high temperature phase, an SMA will return to its original rigid shape. This is known as the shape memory effect.

Shear strain Strain occurring with no change in volume.

Shear stress In its simplest terms, force per unit area tangent to that surface area.

Shear modulus The constant of proportionality of shear stress to the shear strain. Also referred to as the modulus of rigidity.

Signal conditioning The methods used to help convert the output of a transducer to a useful form.

Silicon The chemical element with atomic number 14 and given the symbol Si. Elemental silicon is the principal component of most semiconductor devices.

SMA Shape Memory Alloy.

SOI Silicon on Insulator: A type of fabrication process in which silicon is deposited on an insulating substrate.

Solid mechanics The branch of physics dealing with quantities such as force, deformation, stress, and strain in bodies capable of deformation.

Source material The material to be deposited in physical vapor deposition techniques.

Span The range of input values over which a transducer produces output values with acceptable accuracy. Also called full scale input.

Spinning Method commonly used to apply thin layers of photoresist or other materials on a substrate. In spinning a small amount of liquid material is placed in the center of a substrate which is then rotated at high angular velocities in order to propel the material across the surface.

Sputtering A type of physical vapor deposition in which as ion beam in a vacuum bombards a target in order to eject material for deposition.

Statics Field of mechanics in which systems so not accelerate.

Step coverage Ability of a deposition technique to create films with uniform surface features over various topographies.

Step input An instantaneously applied input to a transducer from one constant value to another constant value.

Step response The output of a transducer corresponding to a step input.

Stiction The usually unwanted adhesion of micromachined structures to a substrate. The word stiction comes from the combination of stick and friction.

Stiffness coefficients The elements of the matrix that relates the strain tensor to the stress tensor for an anisotropic elastic material.

Strain Dimensionless formulation of deformation of a material under the action of stress.

Stress The internal distribution of force per unit area that balances and reacts to external loads applied to a body. Stress is a second-order tensor with nine components, but can be fully described with six components due to symmetry in the absence of body moments.

Stiffness coefficients Elements of the tensor relating strain to stress for a non-isotropic material.

Stringers Pieces of unetched material unintentionally left behind after an etching step.

Stripping Removal of patterned photoresist from a wafer.

Structural layers Layers which form structural elements in a MEMS device.

Substrate A wafer.

Surface force Force resulting solely from a body's being in physical contact with another body. Contrast **body force**.

Surface micromachining A process used to produce MEMS in which structures are created on top of a substrate using a succession of thin film deposition and selective etching techniques. This is distinguished from bulk micromachining in which structures are produced inside the substrate itself.

System partitioning Issue of whether or not to integrate a MEMS device and any associated electronics on the same chip, as well as how to accomplish it.

Target A solid or molten source in an evaporation or sputtering process.

Tensor Formally, a mathematical entity with components that change in a particular way during a transformation from one coordinate system to another. Most tensors can be represented as matrices and can be thought of as something resembling a higher order vector. Common examples of tensor quantities include stress and strain.

Thermal actuator An electro-thermally actuated MEMS device.

Thermal conductivity Material property that measures the ease with which thermal energy is conducted through a material.

Thermal diffusion

Thermal evaporation A form of physical vapor deposition in which a target material is first changed to the gaseous phase under high vacuum before being deposited onto a surface. Evaporation is accomplished by using resistance heating or an electron beam.

Thermal expansion coefficient Material property that represents the amount of strain produced per unit temperature change upon heating or cooling.

Thermal mismatch stress A residual stress introduced within a body resulting from a change in temperature.

Thermal radiation That part of the electromagnetic spectrum emitted by a body due to its temperature.

Thermal resistance Thermal analogue to electrical resistance in which heat transfer rate is the analogue of electrical current and its driving potential is temperature, the counterpart of voltage.

Thermocouple Device consisting of two wires of dissimilar thermoelectric materials. Usually used as a temperature sensor.

Thermoelectric material Material that exhibits a voltage distribution in response to a flow of thermal energy through it.

Torr Unit of pressure commonly used for vacuum systems. 1 atm = 760 torr.

Transducer A device, usually electrical, electronic, or electro-mechanical, that converts one type of energy to another for various purposes including measurement or information transfer.

Transport phenomena The collective studies of the fluid mechanics of a Newtonian fluid, heat transfer, and mass transfer in which the constitutive relations for the transport of momentum, thermal energy and mass, respectively, all obey a form of Fick's Law of diffusion.

Turbulent flow Fluid flow in which the velocity at any point varies erratically.

Ultra high vacuum Region of vacuum pressure from 10^{-9} to 10^{-12} torr.

Undercutting The phenomenon of an etching process proceeding in the horizontal direction under a mask as well as the direction perpendicular to it.

Uniaxial stress/strain Special case of the stress/strain relation in which only stress and strain in one linear direction is important..

Unit cell A set of atoms arranged in a particular way and which is periodically repeated in three dimensions to form a crystal lattice.

Valence band The electron energy band that contains the valence electrons in solid materials.

Vias Through-holes in MEMS devices and layers used for various purposes.

Viscosity The property of a fluid that resists the force tending to cause the fluid to flow. In a Newtonian fluid, the constant of proportionality between shear stress and fluid deformation.

VLSI Very-Large-Scale-Integration: process of creating integrated circuits by combining thousands of transistor-based circuits into a single chip.

Wafer A thin slice of semiconducting material, such as a silicon crystal, upon and/or in which microcircuits and/or microstructures are constructed.

Wet oxidation Oxidation process occurring in the presence of water vapor. Faster than dry oxidation, but results in lower quality oxide layers.

Wheatstone bridge A certain type of electrical circuit designed for detecting small changes in resistance.

Yield Percentage of the total number of die available on a wafer that function to specification.

Young's modulus In solid mechanics, a measure of the stiffness of a given material. For small strains it is defined as the ratio stress to strain. Also called the modulus of elasticity, elastic modulus or tensile modulus.

Zeta potential The wall potential $\varphi_{w.}$ in electro-osmotic flow.

Subject Index

A

Accelerometer, 13, 120, 127, 160–162, 220, 231, 247–248, 250–251, 262
Actuators
  capacitive, 193–194, 231, 236–239
  definition, 168, 169
  electric and magnetic, 194–196, 202–204
  electric, *see* Electric actuators
  electrochemical, 58
  hot-arm, 287, 294–310, 371–375, 397–401
  magnetic, 202–204
    See also Magnetic actuators
  piezoelectric, 171, 194–196, 255, 259, 263, 276–280
    See also Piezoelectric actuators
  thermal, 196, 197, 200, 283, 284, 294, 397, 398, 399–400
    See also Thermal actuators
  thermo-electric cooling, 201–202
  thermo-mechanical, 167, 196–201
Additive process(es)
  chemical vapor deposition, 57–58
    See also Chemical vapor deposition (CVD)
  physical vapor deposition, 40–56
    See also Physical vapor deposition (PVD)
  silicon doping in, 26–27
  silicon oxidation as, 35–40
  sol-gel deposition, 58
    See also Sol-gel-deposition
Alignment, 19, 88–89, 119, 120, 121, 123, 175
Alloys, 55, 58, 198–200
Amplitude modulation, 245
Anisotropic wet etching
  in (100) oriented silicon, 101–104
  in (110) orientated wafers, 102, 103–104
  etch rate (s), 101–102
Anodic bonding, 59, 127
Ashing, 79, 106, 107, 108, 372, 373, 395, 396
Assay(s), 359

B

Biaxial modulus, 139, 148, 276
Bimetallic thermal actuators, 197
Biot number, 293, 294
Block diagram, 160, 161, 163, 211, 212, 250, 256, 283
Bonding
  eutectic, 127
  silicon fusion, 127
  thermal, 59
Boron etch stop, 105, 106
Bulk micromachining, 10, 11, 34, 95, 96–108

C

Calibration, 172, 173, 382
Cantilevers, 4, 11, 12, 98, 109, 110, 146–147, 207, 220, 221, 264–268, 272–274, 276–280, 299, 364
Capacitance, definition, 232–233
Capacitors, 55, 161, 181, 182, 183, 195, 205, 207, 231, 232–239, 243, 245, 249, 250, 251
  parallel plate, 231, 232
Chemical vapor deposition (CVD)
  low pressure, 58, 122
  plasma enhanced, 58, 122
Clean room classifications, 68
CMOS (complementary metal oxide silicon), 106, 109
CNC (computer numerically controlled) machining, 8–9
Compliance coefficient, 260
Conductivity
  electrical, 24–25, 348–349
  thermal, 286–288, 302, 305, 310, 311
Contact printing, 74–75, 81, 82, 91
Continuum theory, 319–320
  breakdown, 320
Contrast, 24, 51, 52, 76, 84, 85, 86, 88, 106, 309
Convective heat transfer, 288, 289, 291, 297, 309
CVD, *see* Chemical vapor deposition (CVD)
Czochralski method, 18

D

Dangling bonds, 102
Deal-Grove model, 37, 38
Debeye length, 347–355, 356, 357
Deep UV lithography, 87
Demodulation, 245, 250
Deposition methods, *see* Additive process(es)
Depth of focus (DOF), 83, 84, 87
Diffusion length, 30
Disk method, of stress measurement, 147–148
Dopant(s), 18, 26–35, 59, 99, 120, 122, 143
  See also Silicon
Dry etching process(es)
  deep reactive ion etching, 107–108
  ion beam etching, 107
  ion beam milling, 96
  plasma etchants, 106–107
Dynamic response, 171

E

EDP (ethylenediamine pyrocatechol), 101
Einstein relation, 352
Einstein summation convention, 140, 260
Electric actuators
  magnetic, 202–204
    See also Magnetic actuators
  piezoelectric, 194–196
    See also Piezoelectric actuators
  thermoelectricity, 200–202
Electrochemical etch stop, 106
Electrolyte, 346, 347, 349, 353, 358, 359, 360
Electron beam lithography, 77, 80
Electro-osmosis, 318, 339–357, 358
Electrophoresis, 358, 359
Electroplating etch rate, 58
Etchant(s), 96, 97, 98, 99, 100, 101, 102, 104, 105, 106, 108, 109, 110, 111, 115, 118, 119, 124, 392–393
Etching
  chemical, 96–106
  dry, 106–108
    See also Dry etching process(es)
  wet, 39, 96–106
    See also Wet etching
Etching profiles, 96–108, 392–394
Etch stop process(es)
  with boron, 105–106
  electrochemical, 106
Eutectic bonding, 127
Evaporation, 40, 46–51, 52, 371, 388–391
  thermal, 40, 46–51, 53
    See also Thermal evaporation

F

FEA (finite element analysis), 138, 306–311
Fick's law, 27, 350
Filter, low-pass, 67, 205–206, 207, 245, 246, 251
Fluidics
  flow continuum breakdown, 320
  Poiseuille law, 339
  Stokes-Navier equations, 329–336
Fourier's law, 306, 310
Frequency modulation, 244–245
Frequency response, 173, 175, 177, 184, 242, 249, 398, 401, 402
Full scale output, 172, 173, 222, 225

G

Glass, 6, 25, 43, 59, 76, 79, 100, 111, 124, 346, 347, 359, 362, 384
Gyroscopes, 4, 13

H

Heat transfer
  conduction, 286–288
  convection, 288–289
  radiation, 289–291
Hooke's law, 136, 138

I

Implantation, silicon doping by, 26, 27, 31–33, 125
Ink jet cartridges, 255
Ink jet printing, 4
Integrated circuit (IC), 9, 71, 90, 91, 99, 106, 120, 126, 161, 188, 251
Isolation layer, 372
Isotropic wet etching, 100

K

Kirchhoff's current law, 241
Knudsen number, 47

L

Lab-on-a chip (LOC), 318, 362, 363
Laser(s), 40, 73, 90, 91, 92, 96, 311, 363
Lattice, 13, 19, 20, 21, 26, 28, 311
Lattice constant, 19, 20, 21
Lattice planes, silicon
(100) orientation, 101, 102, 103
(110) orientation, 102, 103, 104
Lead zirconate titanate (PZT), 182, 195, 256, 258
Lift-off, 87, 112–113
Linearity, 172
Lithography
history of, 90–92
masks in, *see* Masks
resists in, *see* Resist(s)
sensitivity of, 85, 86, 87
x-ray, 91
Loading effects, 169, 183, 205
Low pass filter, 245, 246, 251
Low pressure chemical vapor deposition (LPCVD), 58, 122

M

Machining
chemical vapor deposition (CVD) methods, 111, 118, 122
micro, 95–128
oxidation, 35–40
Magnetic actuators, 202–204
Magnetostriction, 204
Masks
alignment of, 88–89
design of, 79, 80–81
phase-shifting, 79, 88
Mass spectrometry (MS), 363–364
Mean free path, 44, 45, 46, 47, 311, 319
MEMS (microlectromechanical systems)
machining, 95–128
packaging in, 126–128
partitioning, 120, 121
scaling in, 5–7, 112, 293, 294
Micromachining, 10, 11, 34, 95–129, 203, 372
Micro-motor(s), 238
Modeling, 20, 34, 86–87, 114, 122, 141, 157–164, 167, 196, 212–221, 238, 239–249, 256–276, 277, 295, 306, 309, 347
Modulation, 244, 245, 250
MST (microsystems technology), 4

N

Natural frequency, 243, 248, 249, 251
Negative resists, 9, 65, 70, 78, 84, 85, 86, 88
Newton's law of cooling, 288
Nickel, 19, 46, 53, 197, 199
Notation, 20, 140, 162–163, 242, 277, 278, 290, 324, 331, 335

O

Ohm's law, 178, 212, 214, 287, 296–297, 305, 348, 349
op-amp rules, 243
Oscillator, 162, 250, 251
Overshoot, 174

P

Packaging, 114, 115, 123, 124, 126–128, 232
Parallel plate capacitor, 231, 232
Partitioning, 120, 121
Patterning, 108, 113, 123–124, 372, 390
PECVD (plasma enhanced chemical vapor deposition), 58, 122
Periodic table, 24, 26
Photolithography
clean room in, 66–69
contrast, 85, 86
development process in, 77–79
exposure process in, 65, 69, 70, 72–77, 81, 84, 85, 86, 87, 88, 90, 91, 123, 373, 385, 386, 390
intrinsic sensitivity of resist in, 86
lithographic sensitivity in, 85, 86
mask making in, 79–81, 88–89
post baking in, 71, 72, 74, 78, 87
resist profiles in, 69, 84–87
resists in process, 86, 89–92
resist stripping in, 78–79
resolution in, 81–89

soft baking in, 372–373, 385
spinning resist in, 9, 58, 72
wafer cleaning in, 67
Photoresists, 9, 10, 11, 27, 58, 65, 66, 69–70, 71–72, 73, 74, 78, 79, 80, 81, 82, 84, 90, 91, 96, 97, 99, 101, 106–107, 108, 111, 112, 113, 115, 116, 117, 118, 122, 123, 372, 373, 374, 384, 385, 386, 387, 388, 390, 392, 393, 395
Physical vapor deposition (PVD)
sputtering in, 40, 51–56, 106, 111, 143
thermal evaporation, 40, 46–51, 53
Piezoelectric actuators, 171, 195, 255, 259, 261, 263, 276–280, 298
Piezoelectric coefficient, 258, 260
Piezoelectric constant, 183, 195, 260, 261, 277
Piezoelectric crystals, 182–183
See also Piezoelectric materials
Piezoelectricity, 178, 193, 256, 257–258, 260
Piezoelectric materials, 182, 183, 194, 195, 204, 255, 256–261, 277, 278, 279
Piezoresistance, 178, 179, 212, 218
Piezoresistivity coefficients, 218
Piezoresistors, 179, 180, 211, 212, 213, 214, 215, 216, 218, 219, 220, 221, 222, 223
Planarization, 123
Plasma etching, 106–107
Plasma(s)
DC, 54
in dry etching processes, 96, 106, 111
ionization, 107
RF (radio frequency), 106
Poisson's ratio, 137, 217, 218
Polyimide(s), 11, 12, 89, 90, 111
Polymers, 11, 58, 69, 71, 78, 89, 107, 123, 127, 143, 256, 362, 386
Polysilicon, 52, 100, 106, 108, 110, 111, 118, 119, 125, 126
Positive resists, 9, 65, 69, 70, 74, 78, 81, 84, 85, 86, 87, 88, 117
Postbaking, 372
Post-exposure treatment, 69, 74, 87, 373, 385, 386, 390
Power, 7, 32, 43, 54, 74, 91, 108, 111, 127, 128, 158, 159, 167, 168, 169, 170, 171, 183, 189, 194, 205, 233, 235, 236, 238, 287, 289, 297, 306, 309, 317, 363, 373, 396, 402
Pressure sensor, 9, 11, 96, 115, 121, 127, 172, 173, 185, 221–226, 263
Projected range, 32, 33
Projections printing
depth of focus and resolution, 83, 84
mask alignment, 83, 88–89
mathematical expression of resolution, 82
resist profiles, 86–87
Proof mass, 253
Pull-in, 236
PVD, *see* Physical vapor deposition (PVD)

R

Radiation, heat transfer, 283, 284, 285, 289–291
Radio frequency plasma (RF), 4, 40, 53, 55, 106, 107, 108, 111, 207, 395
Range, 9, 13, 19, 26, 32, 33, 36, 40, 42, 46, 52, 71, 72, 73, 80, 82, 92, 106, 172, 173, 180, 198, 206, 207, 224, 242, 251, 285, 373, 386, 389
RCA1, 380
Reactive ion etching (RIE), 107–108, 111
Residual stress, 122, 123, 142, 279
Resistance change, 34, 179, 180
Resistive coefficients, 218, 219, 224
Resistive heating, 48, 203
Resistivity, 25, 26, 89, 178, 179, 215, 216, 218, 296, 301, 304, 307, 309, 348
Resistors, 178, 179–180, 212, 213, 214, 215, 216, 219, 220, 287, 349, 401
Resist(s)
contrast and resolution of, 85, 88
intrinsic sensitivity of, 86
mathematical expressions, 82–83
negative, 70, 78
permanent, 89–90
positive, 69
profiles for, 84–87
spinning, 58, 71, 72
Resist stripping, 78–79
Resolution, 73, 75, 79, 80, 81–89, 90, 91, 92, 105, 111, 172, 173, 206, 208, 374, 385
Resonant frequency, 184–186, 189
Resonator, 184–186, 207, 238, 239

S

Sacrificial layer, 11, 12, 17, 35, 36, 106, 108, 109, 110, 116, 118, 123, 146, 147, 372, 373, 384–387, 392, 395
Scaling, 5–7, 8, 112, 293, 294
Scanning electron microscopy (SEM), 98, 147, 194, 238, 239
Seebeck coefficient, 187
Siedel, model of anisotropic chemical etching, 101–105
Semiconductors, conductivity, 26, 34, 38
Sensitivity, 73, 84–86, 87, 172, 173, 174, 180, 185, 218, 221, 225, 226, 237, 243, 250, 362
Sensor(s)
  capacitive, 181–182, 239–249
  definition, 168
  magnetic, 12–14, 48, 58, 179, 186, 189–193, 202–204
  piezoelectric, 182–183
  resistive, 178–180
  resonant, 184–186
  thermoelectric, 167, 186–189
Shadowing, 40, 49, 50, 51, 112
Shadow printing, 75, 76
  resolution in, 76
Shape memory alloy (SMA), 198–200
  in thermal actuators, 200
Shear modulus, 137
Shear strain, 134, 135, 137
Shear stress, 132, 133, 137, 138, 139, 319, 321, 322, 323, 324
Signal conditioning, 161, 169, 171, 179, 180, 185, 188, 204–206, 222, 226, 240, 241, 243, 245
Silicon dioxide, 9, 17, 27, 35, 37, 96, 97, 98, 99, 100, 101, 105, 106, 109, 118, 123, 310, 377, 392
Silicon fusion bonding, 127
Silicon-on-insulator (SOI) surface micromachining, 434
Silicon
  crystalline
    lattice planes in, 20
    Miller indices of, 20–23
    orientation of, 22, 23, 38, 98
  doping
    conductivity of semiconductors, 26–27
    by diffusion, 27–31
  elasticity constants, 136
  etch rate, 96, 99, 101
  lattice plane orientation of
    (100) orientation, 22, 23, 102
    (110) orientation, 22, 23, 102, 103–104
  oxides of, color table and applications, 40, 415, 416
  piezoresistivity, 179
  residual stress, 122, 123, 142
  in sensors, 179, 197, 219, 228, 286, 310, 362, 372
  single crystal, 18, 52, 59, 60, 179
  stiffness coefficient and compliance coefficient, 140, 260
  thermal properties, 310–311
SMA (shape memory alloy), 198–200
  in thermal actuators, 198–200
Soft baking, 372–373, 385
Software, 30, 34, 80, 122, 138, 295, 306, 309, 382, 387, 392, 394, 397, 398, 399
Sol-gel-deposition, 58
Spinning resist, 58, 72
Sputtering, 40, 51–56, 106, 111, 143
Static response, 172–173
Stiction, 95, 111–112, 119, 124, 373, 395
Stiffness coefficient, 140, 260
Surface micromachining
  examples of, 11, 109, 110, 115–119
  history of, 108–109
  stiction in, 111–112
Surface tension, 6, 7, 112

T

Tensor, 139, 140, 260, 323
Thermal actuators
  bimetallic, 197
  Newton's law of cooling, 288
  shape memory alloy, 198–200
  thermopneumatic, 197–198
Thermal evaporation
  heat sources for, 46–47, 48
  thin film deposition by, 47, 108–109
Thermocouple, 187, 188, 201
Thermoelectricity, 167, 186–189, 201, 202
Thin film(s)
  adhesion of, 142
  deposition of, 95, 108, 142, 143–144

stress measurement in, 142–148
Transducer, 141, 163, 167–208, 211–226, 231–251, 255–280, 283–311
Transistor gates, 34, 91, 124, 125, 126

U

Undercutting, 74, 97, 98, 101, 102, 103, 104, 111, 119, 123, 373, 392, 394
Units, 3, 19, 20, 21, 25, 27, 30, 41, 44, 46–47, 49, 74, 90, 96, 126, 127, 132, 133, 136, 158–159, 163, 172, 173, 175, 185, 225, 248, 257, 258, 260, 268, 276, 286, 287, 288, 289, 290, 309, 319, 322, 329, 331, 333, 336, 340, 341, 342, 343, 345, 348, 350, 356, 378, 396

V

Vacuum, 41–46, 47, 52, 54, 55, 59, 60, 71, 72, 81, 91, 106, 107, 124, 386, 388, 389, 395, 396, 397, 398

Via, 118
Viscosity, 28, 71, 247, 320–324, 334, 350

W

Wafers, silicon
(100) orientation, 22
(110) orientation, 22
anisotropic, 101–105
in (100) oriented silicon wafers, 101–102
in (110) orientated wafers, 102–103
chemical models for, 101, 102
convex corners in, 102, 103
etch stop techniques, 105–106
isotropic, 100–101
etchants for, 100
Wet etching, 39, 96–106, 111
Wheatstone bridge, 179, 205, 211, 212, 213

X

X-ray lithography, 91

Y

Young's modulus, 136–137, 139, 163, 219, 300

Z

Zeta potential, 354, 357, 361